Springer Japan KK

S. Nishisato, Y. Baba
H. Bozdogan, K. Kanefuji (Eds.)

Measurement and Multivariate Analysis

With 90 Figures and 51 Tables

 Springer

Shizuhiko Nishisato
Professor Emeritus
Measurement and Evaluation Program
CTL, OISE/UT
The University of Toronto
252 Bloor Street West, Toronto, Ontario, Canada M5S 1V6

Yasumasa Baba
Professor
Center for Information on Statistical Sciences
The Institute of Statistical Mathematics
4-6-7 Minami Azabu, Minato-ku, Tokyo 106-8569, Japan

Hamparsum Bozdogan
McKenzie Professor in Business
Information Complexity & Model Selection
Department of Statistics, 336 SMC
The University of Tennessee
Knoxville, TN 37996-0532, USA

Koji Kanefuji
Associate Professor
Center for Information on Statistical Sciences
The Institute of Statistical Mathematics
4-6-7 Minami Azabu, Minato-ku, Tokyo 106-8569, Japan

ISBN 978-4-431-65957-0   ISBN 978-4-431-65955-6 (eBook)
DOI 10.1007/978-4-431-65955-6

Library of Congress Cataloging-in-Publication Data applied for.

© Springer Japan 2002
Originally published by Springer-Verlag Tokyo Berlin Heidelberg New York

Typesetting: Editors and authors

SPIN: 10869171

Preface

The International Conference on Measurement and Multivariate Analysis was held on May 12–14, 2000, in Banff, Alberta, Canada. It was a memorable beginning with a beautiful snowy scene on the day of a preconference workshop on May 11, which pleasantly surprised many of the participants from 15 countries.

The idea of the conference in Banff was conceived when Shizuhiko Nishisato was a Visiting Professor at the Institute of Statistical Mathematics (ISM) in Japan in 1998. With strong support from Yasumasa Baba, Takemi Yanagimoto, Yoshiyuki Sakamoto, Koji Kanefuji, Shinto Eguchi, Masakatsu Murakami, Ryozo Yoshino, Takahiro Tsuchiya, and Tadahiko Maeda of the ISM, among others, a proposal was prepared by Nishisato and Baba and was presented to Dr. Ryoichi Shimizu, the Director General of the ISM, for sponsorship. It was enthusiastically accepted by the ISM. Later, the Centre for the Advancement of Measurement, Evaluation, Research and Assessment (CAMERA), University of Toronto, joined the ISM in sponsorship.

When the name of the conference was chosen by Nishisato, Baba, and Yanagimoto, the paper by David Hand (1996) on the roles of "measurement" in statistics was much in our minds, in the hope that our conference would provide an interdisciplinary international forum for those involved in a wide area of research to discuss a balanced "interplay between measurement research and statistics", which roughly translates as cooperative work between the social sciences and the statistical sciences. In retrospect, the title was appropriate for reflecting the interests of the participants from all sides of this multidisciplinary field. The editorial committee believes that the conference was successful in stimulating cooperative work among international participants from different disciplines, and that it met our expectations.

There were three key-note speeches, by John Gower, David Hand, and Chikio Hayashi; 22 invited sessions; and numerous individual contributed papers. The topics of the papers covered a wide range of theoretical work in statistics and related disciplines and a number of applications.

Hamparsum Bozdogan initially spent countless hours in search of a publisher. In the meantime, Ejaz Ahmed and Bruno Zumbo also offered help in looking for a possible publisher. Almost 8 months after the conference, publication eventually became possible when Yasumasa Baba and Koji Kanefuji negotiated with Springer-Verlag in Tokyo and, at the same time, secured the financial assistance of the ISM. Kanefuji acted as liaison among the participants, collecting all revised papers, and also acting as a copy editor, while Baba did all the negotiations with the publisher and the ISM. Although peer reviews were conducted, the editorial committee is ultimately responsible for any inadvertent errors or omissions. In particular, responsibility rests with Nishisato as the Chief Editor who reviewed most of the revised papers for final editorial corrections.

The papers for the proceedings were grouped into the following categories: Greetings from Dr. Ryoichi Shimizu (Director General, ISM) and Dr. Heather Munroe-Blum (Vice-President, University of Toronto), key-note speeches (two papers), introduction (one paper), scaling (eight papers), structural analysis (eight papers), statistical inference (seven papers), algorithms (five papers), and data analysis (four papers). As we believe our readers will agree, the editorial committee was blessed with high-quality papers. In addition, we included lists of the conference committees with their members, and those of the reviewers of the papers.

The organizers of the conference are indebted to three corporations that supported the conference:

> The Institute of the Japanese Union of Scientists and Engineers
> S-PLUS Group, Mathematical Systems, Inc.
> Pasco Corporation

We would like to acknowledge with our sincere thanks a countless number of offers of help from the participants, the sponsors, and the Banff Centre for Conferences. In particular, we would like to thank the ISM for the funding for the publication of the proceedings, a number of participants for volunteering to purchase more than a copy of the proceedings to off-set expenditures, and the assistance of the staff of Springer-Verlag, Tokyo.

Editorial Committee:

Shizuhiko Nishisato, Chief Editor (University of Toronto)
Yasumasa Baba (Institute of Statistical Mathematics)
Hamparsum Bozdogan (University of Tennessee)
Koji Kanefuji, Copy Editor (Institute of Statistical Mathematics)

December 2001

Ladies and gentlemen,

It is my great pleasure and honor to join Dr. Heather Munroe-Blum, the Vice-President of the University of Toronto, in welcoming each and everyone of you to the International Conference on Measurement and Multivariate Analysis, held in this beautiful city of Banff, right in the middle of magnificent Canadian Rockies.

I am delighted to learn that this is truly an international and interdisciplinary gathering represented by world's top-class scholars and many young researchers of great promise. When Professor Shizuhiko Nishisato of the University of Toronto served as Visiting Professor at the Institute of Statistical Mathematics three years ago, he and Professor Yasumasa Baba of my Institute approached me with a proposal for this conference. I wholeheartedly endorsed it since I believed in its positive implications for further international communication, cooperation and potential intellectual benefits for researchers and students throughout the world.

Now that all the preparations for the conference are completed, we can see such a wide representation of the delegates from many countries in both statistical and social sciences that makes this conference a rare and unique international forum. As such, I have no doubt that this will prove to be one of the most memorable and fruitful events for all of you in this millennium year. Canada is the first foreign country I have ever visited in my life, and the only personal regret I have now is that I would not be able to revisit the beautiful city of Banff to greet each one of you in person and engage in pleasant and stimulating dialogue.

I would like to thank all the delegates for your invaluable contributions to the international scientific community. I am certain that you will find this conference very stimulating and personally of great success. Have a wonderful conference. Finally, I would like to congratulate all those who have made the initial proposal for this conference a realilty.

With very best wishes,

R. Shimizu

Ryoichi Shimizu, Professor
Director-General
The Institute of Statistical Mathematics

University of Toronto

May 11, 2000

Dear Delegates,

It is with great pleasure that I welcome you to the International Conference on Measurement and Multivariate Analysis, and the Dual Scaling Workshop. I am honoured to have been chosen, along with Dr. Ryoichi Shimizu, Director General of the Institute of Statistical Mathematics (ISM) in Tokyo, to offer opening remarks for this unique and internationally significant event. My only regret is that I could not join you at the conference, which promises to offer valuable insights into this fascinating field.

I was especially excited to learn of Professor Shizuhiko Nishisato's plans to organize this event, both because of his internationally important work and his dedication to such a lofty task in the year of his retirement, and because we share a common bond – as Ph.D. alumni of the University of North Carolina. This connection is particularly relevant at this conference, because it is often at the doctoral stage in our academic careers that our analytic frameworks become more mature and crucial to the work we do as researchers.

The challenge facing researchers in all disciplines is to conduct research that is fresh, innovative and of tangible benefit to society. This conference is in keeping with that goal, both in terms of the research that is being presented and the cross-disciplinary nature of these proceedings. The organizers have brought together an imaginative mix of researchers from a variety of disciplines to address overarching issues in data analysis. Moreover, these researchers have come from around the world, adding a valuable international depth in perspective to the issues before you this week.

Each of you has been drawn to this conference by a common goal: to have access to the most sophisticated and relevant analytic research tools to apply in your own research – whether your field is psychology, mathematics, sociology, statistics, computer science, music or agriculture. The techniques that you will learn about throughout the week will no doubt enrich your understanding of how to approach analysis in new and innovative ways.

I know that the week will be fruitful for presenters and delegates alike. This conference will live on through greater collaboration among the disciplines represented here, among research and academic institutions, and among cities and countries. It is this kind of unique forum that enriches our own roles as researchers and contributes significantly to the body of knowledge we offer to society.

Sincerely,

Heather Munroe-Blum, Ph.D.
Vice-President, Research & International Relations

Conference Committee

Sponsoring Organizations

The International Conference on Measurement and Multivariate Analysis was held under the joint auspices of the Institute of Statistical Mathematics and the Centre for the Advancement of Measurement, Evaluation, Research and Assessment (CAMERA), University of Toronto.

Organizers

Shizuhiko Nishisato (University of Toronto)
Yasumasa Baba (Institute of Statistical Mathematics)

Scientific Program Committee

Shizuhiko Nishisato, Chair (University of Toronto)
Yasumasa Baba (Institute of Statistical Mathematics)
Charles Mayenga (Government of Manitoba)
Takemi Yanagimoto (Institute of Statistical Mathematics)

Local Arrangement Committee

Charles Mayenga (Government of Manitoba)
Lorraine A. M. Nishisato (University of Toronto)
Shizuhiko Nishisato (University of Toronto)

List of Reviewers

Peter Allerup
Naohito Chino
Jose Garcia Clavel
Alain De Beuckelaer
Mark De Rooij
Hironori Fujisawa
Tadashi Imaizumi
Makio Ishiguro
Koji Kurihara
Tony Lam
Daniel Lawrence
Soonmook Lee
Tadahiko Maeda
Masahiro Mizuta
Yuichi Mori

Hans-Joachim Mucha
Fionn Murtagh
Akinori Okada
Tatsuo Otsu
Robert Pruzek
James O. Ramsay
Kazuo Shigemasu
Yoshio Takane
Yutaka Tanaka
Takahiro Tsuchiya
Mark Wilson
Richard G. Wolfe
Takemi Yanagimoto
Haruo Yanai

Contents

Keynote Papers

Introduction

Scaling

Structural Analysis

Statistical Inference

Algorithms

Data Analysis

Categories and Quantities

John C. Gower

Department of Statistics
The Open University
Walton Hall
Milton Keynes
U. K. MK7 6AA

Summary: The rank of a matrix of categorical variables is defined. This may be used as a basis of multivariate methods for approximating categorical data in a similar way that the rank of a quantitative data matrix may be used to define standard methods such as multidimensional scaling, multiple correspondence analysis, principal components analysis and non-linear principal components analysis. Several problems are outlined that depend on the notion of categorical rank. The algorithmic tools for handling categorical rank need to be developed.

1. Introduction

Many well-known multivariate methods work in terms of r-dimensional approximations with associated visual representations. These often depend on the best rank-r least-squares approximation to a matrix. For quantitative variables, the Eckart-Young theorem answers the question: What rank r matrix $\mathbf{Y}$ best approximates a given data matrix $\mathbf{X}$? What is the equivalent for categorical variables? For example, if:

$$\mathbf{X} = \begin{pmatrix} \text{red} & \text{male} & \text{Canadian} \\ \text{brown} & \text{male} & \text{British} \\ \text{black} & \text{male} & \text{Canadian} \\ \text{red} & \text{female} & \text{French} \end{pmatrix}$$

what meaning, if any, can we put to an approximation:

$$\mathbf{Y} = \begin{pmatrix} \text{red} & \text{male} & \text{Canadian} \\ \text{red} & \text{male} & \text{French} \\ \text{black} & \text{male} & \text{Canadian} \\ \text{red} & \text{male} & \text{French} \end{pmatrix}.$$

An obvious thing to do is to count the number, or proportion, of agreements. This is too simplistic because we can always get exact agreement merely by setting $Y = X$. Therefore, counting agreements is not enough and we also need some notion similar to that of rank. We would like to make a statement like Y has less rank than X but nevertheless attains 75% agreement. As with numerical data, we are interested in visual representations. Rank is defined algebraically in terms of linear combinations or, equivalently, by a geometrical interpretation in terms of linear subspaces. The notion of linear combination depends fundamentally on quantitative information so is not available for categorical variables. Can we exploit the notion of dimensionality to assign a meaning to *rank*(Y) and, if so, can we say to what extent X departs from, say, a rank 2 matrix? This paper is concerned with establishing a framework for answering such questions.

First, we shall need some notation. X refers to n = 4 cases and p = 3 categorical variables. We may code X in Indicator Matrix form:

$$
\mathbf{G} = \begin{pmatrix}
100 & 10 & 100 \\
010 & 10 & 010 \\
001 & 10 & 100 \\
100 & 01 & 001
\end{pmatrix}.
$$

$$
\mathbf{L} = \quad 211 \qquad\quad 31 \qquad\quad 211
$$

The column-sums of $\mathbf{G}$ are shown. They give the number of cases for each category-level of each categorical variable. These may be arranged into a diagonal matrix $\mathbf{L}$, which itself may be partitioned to give the frequency of each category-level for each of the three variables:

$$
\mathbf{L}_1 = diag(2,1,1) \quad \mathbf{L}_2 = diag(3,1) \quad \mathbf{L}_3 = diag(2,1,1)
$$

with the number categories for each variable given by the scalars: $L_1 = 3 \quad L_2 = 2$

$L_3 = 3$ and the total umber of levels $L = L_1 + L_2 + L_3 = 8$. The individual columns of $\mathbf{G}$ may be partitioned as: $\mathbf{G} = \begin{pmatrix} \mathbf{G}_1 & \mathbf{G}_2 & \mathbf{G}_3 \end{pmatrix}$.

2. Category-Level Points

To overcome the difficulty that the notion of rank is not immediately available with categorical data, it is common to seek quantifications of category-levels and then the quantitative approaches become available again. Quantifications may be assigned in some fixed way or may be determined by optimising some criterion. Correspondence Analysis (especially Multiple Correspondence Analysis/Homogeneity Analysis), Optimal Scaling and Non-linear Principal Components Analysis are examples of quantification methods which have features of both approaches, as will be explained.

The notion of a co-ordinate axis, which is fundamental to all quantitative methods. The coordinates to be associated with any point are obtained by the nearest positions on the axes; that is by orthogonal projection. Then, the optimal approximation is naturally defined as a subspace containing a set of n points with coordinates that best match the coordinates in the full space. One measure of best match is a function of the difference between the distances between the points in their true positions and as approximated in the subspace. This naturally leads to consideration of least-squares estimation techniques and their generalisations. Another measure, used in multidimensional scaling, is a function of the differences between distances in the true space and their approximations. The notion of distance is important in both approaches. When categorical variables are quantified, they may be associated with co-ordinate axes in precisely the same way as for natural quantitative variables. The main difference is that whereas for a quantitative variable the whole axis (or at least a continuous segment of such an axis) is available, a quantified categorical variable only has as many associated values L_k as it has category levels. In these circumstances, techniques that involve projection are questionable, for they usually predict non-existent values. However, the concept of the nearest valid value remains an attractive predictor; indeed, it is hard to see what alternatives there may be. Further, for categorical variables, there is no reason in general why the L_k points associated with a categorical variable should be on any linear axis so, usually, they will be represented by a simplex of L_k points – one point for each category-level. The simplex of L_k points associated with all the category-levels of a categorical variable will be referred to as *category-level-points* (CLPs). Ordered categorical variables are a special case where the CLPs *are* arranged on a linear axis, their positions determining the ordering only and not numerical values. It is true that sometimes a continuous scale may be considered to underlie an ordinal scale, in which case values intermediate to the ordinal scores may have credible interpretations, but this is not always so. In general, there is not a continuum and the regions between the linearly arranged CLPs have ill-defined meaning; again the use of projection concepts is questionable.

The rows of a matrix $\mathbf{C}$ give the positions of the CLPs and $\mathbf{GC}$ gives coordinates of the n cases. It follows that the squared distance between every pair of the n cases may be obtained from the inner-product matrix $\mathbf{GCC'G'}$. The following are the CLPs for some well-known multivariate methods.

(a) Extended Matching Coefficient (p = 6)

$$
{}_L\mathbf{C}_L =
\begin{pmatrix}
\mathbf{I}_{L_1} & & & & & \\
& \mathbf{I}_{L_2} & & & & \\
& & \mathbf{I}_{L_3} & & & \\
& & & \mathbf{I}_{L_4} & & \\
& & & & \mathbf{I}_{L_5} & \\
& & & & & \mathbf{I}_{L_6}
\end{pmatrix}
$$

Thus the CLPs for each variable form a regular simplex. The squared distance between two cases is the compliment of the number of matches of the categorical variables – hence the term *Extended Matching Coefficient*. The distance matrix of all these coefficients may be analysed by any form of multidimensional scaling. If classical scaling/principal coordinates analysis is used, then the cases are approximated in a subspace of a full exact *n-1* dimensional representation.

(b) Chi-squared Distance (MCA) (p = 6)

$$_LC_L = \begin{pmatrix} \mathbf{L}_1^{-\frac{1}{2}} & & & & & \\ & \mathbf{L}_2^{-\frac{1}{2}} & & & & \\ & & \mathbf{L}_3^{-\frac{1}{2}} & & & \\ & & & \mathbf{L}_4^{-\frac{1}{2}} & & \\ & & & & \mathbf{L}_5^{-\frac{1}{2}} & \\ & & & & & \mathbf{L}_6^{-\frac{1}{2}} \end{pmatrix}$$

Now the CLPs for each variable form an irregular simplex related to the inverses of the square-roots of the category frequencies. The squared distance between two cases is *Chi-squared Distance*. It is usual to adopt a special form of analysis of **G** weighted by **L** but the same result can be reached by a PCA of **GC**; a different analysis may be obtained by using any form of multidimensional scaling of the chi-squared distance matrix. In MCA approximations the cases are approximated in a subspace of a full exact *n-1* dimensional representation but this is not so with most MDS representations. There is a close relationship with the optimal scores of Homogeneity Analysis. Essentially, the optimal scores are given by the principal axes of **GL**; note that the optimal score differ from the CLPs.

(c) Category scores of Non-linear PCA

$$_LC_p = \begin{pmatrix} \mathbf{z}_1 & & & & \\ & \mathbf{z}_2 & & & \\ & & \mathbf{z}_3 & & \\ & & & & \\ & & & & \\ & & & & \mathbf{z}_p \end{pmatrix}$$

In non-linear PCA the vectors $\mathbf{z}_k$ give optimal score for the levels occurring in the *k*th variable. Thus, with this method, the CLPs coincide with optimal scores. Optimality is defined as giving the best *r*-dimensional approximation (i.e. largest proportion of *trace*($\mathbf{C'G'GC}$)) where *r* is *specified in advance*. Once the scores are obtained, analysis is as for ordinary PCA. Variants allow for ordinal constraints on the scores and for s_k sets of scores for each variable when $\mathbf{z}_k$ is replaced by $\mathbf{Z}_k$ with s_k columns. See Gifi (1990) for further details. The CLPs for $\mathbf{z}_k$ are one-dimensional; those for $\mathbf{Z}_k$ will be s_k-dimensional.

We have seen that if projection methods and the implicit use of linear co-ordinate axes are to be avoided for categorical variables, then we need to try to ascribe a meaning to the "rank" of a matrix of categorical variables. If we can define categorical rank, we may seek directly an optimal approximating matrix of low categorical rank. Ordered categorical variables should be handled as special case. An approach to such problems will be described which is based on the use of CLPs rather than co-ordinate axes. The basic idea is that every set of CLPs defines a set of Neighbour Regions, one region being associated with each category level. The region consists of all points nearer to a particular CLP than to any other CLP. All points in the same Neighbour Region are assigned the same category level.

In the full space, all *n* cases lie in their correct neighbour regions. In *r*-dimensional approximations, we may define *prediction-regions* that are often, but not necessarily, the intersections of the neighbour regions with the *r*-dimensional subspace L_r, say. Supposing that the cases are represented by *n* points in L_r, we may count how many lie in their correct prediction-regions, i.e. predict the true nominal values. When all predictions are correct we may say that $\mathbf{G}$ has rank *r*, or less.

Thus we define: *the rank of a categorical matrix* $\mathbf{G}$ *is the dimensionality of the smallest subspace in which admissible prediction-regions can be constructed that correctly predict all nominal values of* $\mathbf{G}$.

Three different geometrical situations concerning L_r may be identified. Firstly L_r may be a subspace of the space holding the exact representation of the *n* cases; this holds with standard multivariate methods such as PCA, MCA and NLPCA. Secondly, L_r may contain *n* points giving an approximation to the distances between the *n* cases arising, say, from some form of multidimensional scaling. Then L_r may not be a natural subspace, but it may be embedded in the full space and then treated as a subspace (see Gower, Meulman and Arnold, 1999, for two methods of embedding). Thirdly, L_r with its *n* points may exist in isolation and we are faced with constructing prediction regions using only the information in L_r itself. The construction and allowable forms for prediction-regions will be discussed further, below.

3. Quantitative and Qualitative Reference Systems

The L_k dimensions of the CLPs of a categorical variable are the counterpart of a one-dimensional coordinate axis for quantitative variables. In this section the close relationship between the two reference systems is explained by exhibiting side-by-side the notation for numerical and categorical information.

The data

NUMERICAL	CATEGORICAL
$\mathbf{X} = (\mathbf{x}_1, \ldots, \mathbf{x}_p)$	$\mathbf{G} = (\mathbf{G}_1, \ldots, \mathbf{G}_p)$
$\mathbf{x}_i = (x_{i1}, \ldots, x_{ip})$	$\mathbf{g}_i = (g_{i1}, \ldots, g_{ip})$
$1 \times P$	$1 \times L$ $1 \times L_1 \ldots 1 \times L_p$

We start with the form of the data. Numerical data is given in an $n \times p$ matrix $\mathbf{X}$, the ith row $\mathbf{x}_i$ of which. refers to the ith case, with x_{ik} the numerical value of the kth variable for the ith case. The notation for coding categorical information in an indicator matrix $\mathbf{G}$ has already been described. Parallel with the notation for $\mathbf{X}$, we may write the ith case $\mathbf{g}_i$ with kth categorical value g_{ik}. Now, however, g_{ik} is not a scalar but is a row-vector with L_k columns. This difference is very superficial and arises solely from the coding in the indicator matrix. Essentially, there is only the single categorical value corresponding to the sole indicator value of "1". Thus, the parallel of $\mathbf{G}$ with $\mathbf{X}$ is very close as is clear from the table.

The reference systems

NUMERICAL	CATEGORICAL
$y_i = \sum_{k=1}^{p} x_{ik}\mathbf{e}_k \ (\text{i.e. } y_i = \mathbf{x}_i)$.	$y_i = \sum_{k=1}^{p} g_{ik}\mathbf{C}_k$ orthogonal spaces for h and k.
$\mathbf{e}_k$ is unit vector along continuum of axis ξ_k.	$\mathbf{C}_k$ is finite set of L_k CLPs ξ_k.
$\mathbf{e}_h$ and $\mathbf{e}_k$ are orthogonal.	$\mathbf{C}_h$ and $\mathbf{C}_k$ are in orthogonal spaces.
The axes are calibrated and labelled with a numerical scale.	The CLPs are labelled by the category level names.
The point y_i is nearest (by projection) the marker x_{ik} on ξ_k.	The point y_i is nearest (shortest distance) to CLP $g_{ik}\mathbf{C}_k$ of ξ_k bearing the label of the categorical variable indicated in g_{ik}.

Next the above table shows how the values contained in $\mathbf{X}$ and $\mathbf{G}$ are represented relative to the two reference systems. For quantitative variables, we use orthogonal Cartesian axes whose directions are given by unit vectors $\mathbf{e}_k$ ($k = 1, \ldots p$). Each $\mathbf{e}_k$ is a row-vector with p columns, the kth of which is unity and the remainder zero. The position of the ith case is given by the vector-sum shown in the table. For categorical variables, the unit vector $\mathbf{e}_k$ is replaced by the CLPs given as the rows of $\mathbf{C}_k$ and the position of the ith case is again given by a similar form of vector-sum. Thus, $\mathbf{e}_k$ and $\mathbf{C}_k$ are different kinds of reference systems but with similar roles; we shall use a common name ξ_k to refer to the kth reference system. Orthogonal Cartesian axes could be replaced by oblique axes and, similarly, although it is convenient to choose the CLPs of two categorical variables to be in orthogonal subspaces; this is not essential. As we have already seen with NLPCA, neither need the CLPs occupy as many as L_k dimensions. Markers calibrating a numerical axis are numbers; markers labelling CLPs are names. The crucial thing about Cartesian axes is that all points with a particular value x_k lie in the normal plane to ξ_k at the marker labelled x_k. This implies that the value of the kth variable to be associated with any point z is obtained by reading off the *nearest* label to z on ξ_k, that is it is obtained by orthogonal projection. For categorical variables projection is inappropriate but every set of CLPs define their labelled neighbour regions. z must lie in one of these regions and the associated label gives the required category. That is the nearest label to z is chosen for each of the sets of CLPs, ξ_k.

Approximation

NUMERICAL	CATEGORICAL
The predicted value for the kth variable is given by reading off the nearest (by projection) marker z_k on. ξ_k. Z is a rank r approximation to X.	Prediction is from the nearest (not by projection) CLPs to z and reading off the associated label h_k, say, among the CLPs of ξ_k. This predicts G by $H = (H_1, H_2, ...,H_p)$, a rank r approximation to G.

Suppose z is a given point in some subspace L_r. Then, we can predict the values of all associated p variables by using the shortest distances to the ξ_k (whether they refer to Cartesian axes or to CLPs). If now there are n points in L_r then predicted numerical values may be arranged in a matrix Z $(n{\times}p)$ which may be regarded as a rank r approximation to X. Similarly, predicted nominal values may be arranged in a nominal matrix H, say, which may be regarded as a rank r approximation to G.

Optimal approximation

FUNDAMENTAL PROPERTY (NUMERICAL VARIABLES)

Given any subspace L_r then: $\text{Min} \sum_{i=1}^{n} \| x_i - z_i \|$ for $z_i \in L_r$ is given by $z_i = \text{proj}(x_i)$

onto L_r. This is a global minimum when L_r is spanned by the r leading eigenvectors of $X'X$ (or from the Singular Value Decomposition of X - Eckart-Young (1936) theorem).

FUNDAMENTAL PROPERTY (CATEGORICAL VARIABLES)

Define *mismatch*(g_i, h_i) to be the number of disagreements between the L observed and predicted category levels for the ith case.

Given any subspace L_r then: $\text{Min} \sum_{i=1}^{n} mismatch(g_i, h_i)$ for $z_i \in L_r$ defines the

best rank r
approximation to G, the minimisation being made over all H.
An exact rank r fit occurs when the criterion is zero. Then G has categorical rank r.

Before closing this section, some simple illustrations of some of the ideas involved might be helpful.

Figure 1 shows table **X** of section 1. The representations are both exact. For example, we see that case number 1 lies in the regions Canadian, red, male as required. Of course, **X** is a very small table so it should be no surprise that it is so easy to represent in few dimensions. Indeed, the points representing the cases can be moved around within their regions without prejudicing correct prediction. With more variables and more cases low dimensional representations would be more tightly constrained and exact fits in, say, two dimensions would be unavailable.

We have not represented **Y**, but it too has a rank 1 representation. We have seen that **Y** gives only 75% correct prediction, so clearly it is a very poor representation because we have already shown that an exact rank 1 representation exists.

Exact Representations of the n = 4, p = 3 data

Rank 2 representation

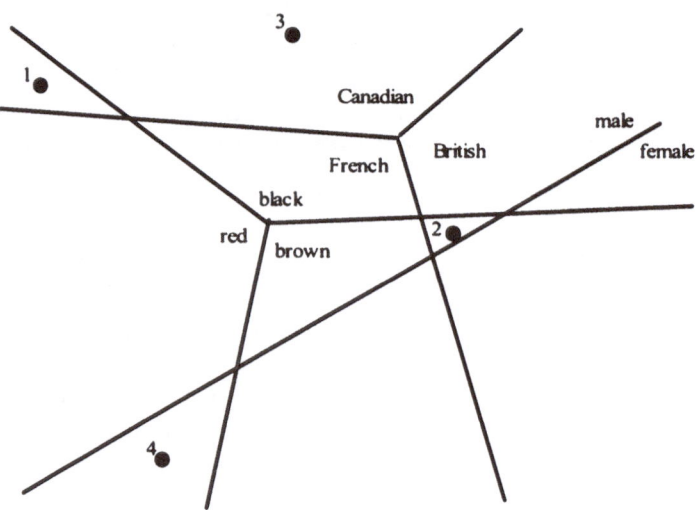

Rank 1 representation

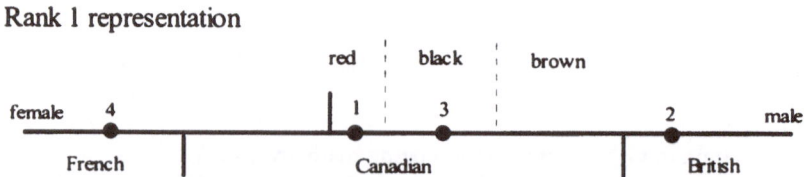

Figure 1: Representations of the matrix **X** of section 1.The top figure is an exact rank 2 representation and the lower figure is an exact rank 1 representation.

4. Some Problems

Associated with this concept of categorical rank are many multivariate data analysis problems. We shall take these problems in order of increasing complexity.

(i) At its most simple, we may take the CLPs implicitly or explicitly associated with any classical method of multidimensional categorical data-analysis together with the prediction-regions of the optimal subspace, or embedded subspace, solution given by the method. Then we may count the number of correct predictions of category-levels. A solution to this problem allows us to compare the relative performances of existing methods with their ability to minimise categorical rank. Even at this level, we have interesting new visualisations of categorical variables in terms of their prediction regions and interesting algorithmic problems in constructing prediction regions. Ordered categorical variables may be displayed in what looks superficially like conventional displays of numerical variables but the interpretation is different.

Gower (1993) and Gower and Hand (1996) outline an algorithm for constructing prediction regions formed by the intersection of a specified subspace with the neighbour regions generated by given CLPs. Such prediction-regions are necessarily convex. A single but effective computer algorithm to give two-dimensional prediction regions is to evaluate for every pixel the nearest CLP and to colour the pixel accordingly. The computer screen will then be divided into coloured prediction regions – See Gower and Harding (1998) for an example.

Figure 2 illustrates the situation for three variables. In this figure it is assumed that the points representing the cases have been found by some standard method and that what is shown are the prediction-regions in Lr. There are eight misclassifications arising from points that lie in either one or two incorrect regions. As just stated, the regions could be coloured but, to avoid confusion, then it is usually better to show the regions for each variable on separate diagrams

(ii) Secondly, we may take the CLPs of classical methods and try to find the subspace L_r that maximises the correct predictions for these CLPs. We need to know how the cases lie in L_r. In classical methods these are usually given by projection, and we may use this here too. A formal solution to this problem would provide a categorical variable analogue of the Eckart-Young theorem, now minimising prediction error rather than its numerical analogue of a residual sum-of-squares. The difference from (i) is that there L_r is given by the chosen multivariate method whereas in (ii) L_r is allowed to take up the position giving optimal predictions.

(iii) The most challenging problem is to try to find prediction-regions that maximise the number of correct predictions for any r-dimensional space L_r. The cases may be placed freely in L_r. Edwards (1990) has shown that it is always possible to construct a two dimensional Venn diagram showing all possible 2^n regions associated with the intersections of n sets. The regions may be written as the binary numbers $0, 1, 10, 11, 100, 101, ...,11...1$. The n cases may then be assigned to the n regions corresponding to the rows of **G** interpreted as binary numbers. To eliminate this trivial solution, some constraint must be put on the form allowed for the prediction regions; convexity of the prediction regions is a natural requirement. Note that with a Venn diagram, it is easy to make all the 2^n

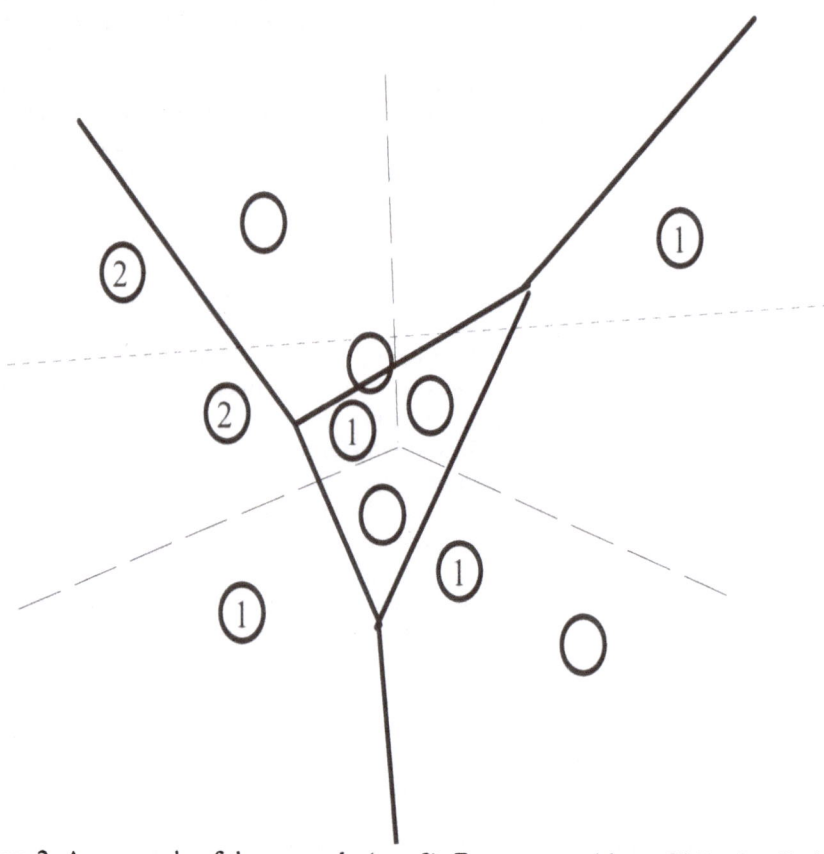

Figure 2: An example of the space *Lr* (*r* = *2*). Error rates without CLPs. Prediction-regions for three variables are shown with 2, 3 and 4 categories, respectively. The circles represent cases and the figures within represent the number of errors, totalling 8.

individual regions convex but then boundaries cannot be made to define convex regions. An easy way to see this is with a conventional Venn diagram where a set is denoted by an oval shape. Although the oval may be convex, the region outside the oval is not. Even if one confines oneself to the interior of the oval, the intersection of two ovals divides space into five regions, rather than the three representing A&notB, B&notA, A&B – in fact, the first two sets are split into two disjoint regions and hence are not convex. Thus, the convexity constraint eliminates trivial solutions but it will no longer be possible to represent all 2^n regions. The hope is that it will be possible to exhibit all the regions associated with the n cases, or at least most of them.

Figure 2 continues to illustrates the geometry. Now, however, no CLPs are assumed and there is complete freedom in arranging the convex prediction-regions and the points representing the cases. Points representing those cases that belong to a set that is exhibited are not uniquely defined and may be placed anywhere in the proper region; nevertheless they may be tightly constrained. One could assign a unique point in a region, say its centroid, but the topology of the prediction regions is invariant to stretching transformations. Can this freedom be exploited to determine a transformation where all the prediction regions are generated as neighbour regions of some set of CLPs within the approximation space? If so, these CLPs might be used to locate the cases uniquely. It might be best to abandon point representation and use the whole of a prediction region to represent a case. Cases that do not belong to an exhibited region are a problem. They have to be put into a region with fewest errors but this might not give a unique region. The resulting diagrams could become very confusing.

(iv) A more constrained version of (iii) is to require the prediction-regions to be generated by sets of CLPs either required to lie in L_r or allowed to lie outside L_r. It is known that only in special circumstances is a set of convex regions generated as the neighbour regions of some set of points, so the constraint is quit a severe one. Thus, the prediction-regions are constrained to be, or to be part of, the neighbour-regions generated by the CLPs. Note that neighbour-regions play no part in (iii).

(v) An algorithm for constructing CLPs generating convex neighbour regions giving prediction regions in L_r with minimal prediction-error would provide a useful addition to the multidimensional scaling armoury.

(vi) A variant of (v) would not specify the dimensionality of the prediction regions but would require no prediction errors. This algorithm would be of mathematical interest, allowing the categorical rank of a matrix to be ascertained.

These are difficult problems and I provide few solutions but hope that others will be interested to develop these ideas further. Some additional information and preliminary numerical results are given in the references cited below.

References:

Eckart, C. and Young, G. (1936) the approximation of one matrix by another of lower rank. *Psychometrika*. **1**, 211-218.

Gifi, A.A.(1990). *Nonlinear Multivariate Analysis*. Chichester: John Wiley and Sons, 579 + xx pp.

Gower J. C. (1993). The construction of neighbour-regions in two dimensions for prediction with multi-level categorical variables. In: *Information and Classication: Concepts - Methods - Applications Proceedings 16th Annual Conference of the Gesellschaft fur Klassifikation, Dortmund, April 1992* Eds. O. Opitz, B. Lausen and R. Klar. Springer Verlag: Heidelberg - Berlin, 174-189.

Gower, J.C. and Hand, D. J. (1996). *Biplots*. London: Chapman and Hall, 277 + xvi pp.

Gower, J.C. and Harding, S.A. (1998). Prediction regions for categorical variables. In: *Vizualisation of Categorical Variables*. Eds J. Blasius and M. J. Greenacre. Academic Press, London. 405-419.

Gower, J.C., Meulman, J. J., and Arnold, G. M. (1999). Non-metric linear biplots. *Journal of Classification*, **16**, 181-196.

Questionnaire Construction, Data Collection and Data Analysis: An Approach by the Idea of Data Science

Chikio Hayashi

The Institute of Statistical Mathematics
Sakuragaoka Birijian 304, 15-8 Sakuragaoka,
Shibuya-ku, Tokyo 150-0031, Japan
(E-mail;kazue@med.Teikyo-u.ac.jp)

Summary: Data design, data collection and data quality evaluation are crucial to data analysis if we are to draw out useful relevant information. Analysis of low-information data never bears fruit; however, data analytic methods can be refined. In spite of the importance of this issue in actual data mining and data analysis, I am forced to ask why these problems cannot be discussed at its most essential level. Perhaps it is a matter of the laborious practical work involved or the otherwise plodding pace of research. Indeed, these problems are rarely addressed because in academic circles it is regarded as unsophisticated. In the present paper, I dare to take up these problems, regarding it as one very important to data science.

1. Fundamental Idea of Data Science

First, the out-line of data science [Hayashi, C. 2001a] is explained as an introduction. Data Science is not only a synthetic concept to unify statistics, data analysis and their related methods but also comprises its results.   It includes three phases, design for data, collection of data and analysis on data. It is essentially important to treat the actual problem and to theorize the methods.   This process leads to the dynamic unification of application and theory into an entity as data science. Data Science (DS) intends to analyze and understand actual phenomena with `data`.   In other words, the aim of data science is to reveal the features or the hidden structure of complicated natural, human and social phenomena with data from the established or traditional theory and method.   This point of view implies multidimensional, dynamic and flexible ways of thinking. We would like to explain the position of DS in the realm of science or scientific methods in Figure 1. So, we take up the views of the World by science, and we would like to explain a classification of `Understanding of phenomena or view of the World` and the position of `data science` in science or scientific methods.   The details will be read in Figure 1. In my opinion, people, who insist on SS, consider mainly the area as SS, while people, who insist on DS, consider mainly this area as DS. I dare to repeat, in DS, Data themselves are very important.

The process of data science in practice is explained in Figure 2. In the first stage of specialization, we work to solve practical problems (PP). PP means   to treat phenomena and fact (what it is). Thus, on the base of general theory, specialization begins and new useful ideas and methods are born through the

14

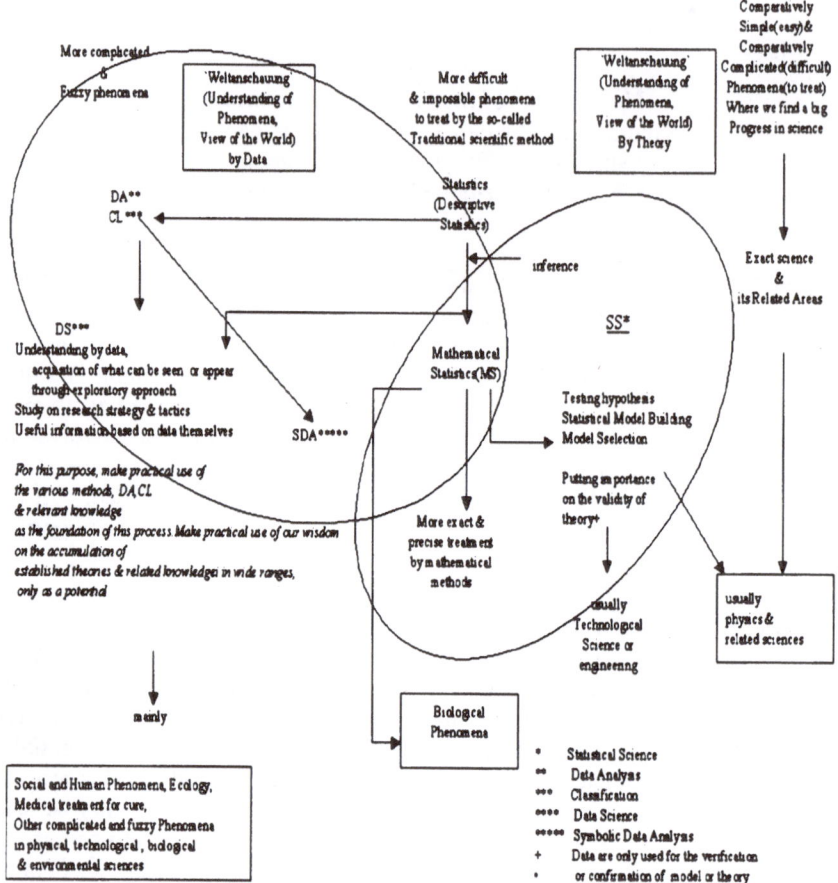

Figure 1:Data Science in General Science

solution of actual problems. On the basis of these ideas and methods, generalization begins and the development of them are expected. These two process occur in turn in DS. Remark that wisdom is personal and not easily communicable, being a problem of scientist, whereas knowledge is communicable and science itself. Knowledge can not be well-developed by scientists without their wisdom. This process is summarized in Figure 2. It is essentially important to treat the actual problems and to develop the methodology and method (theory). This process leads to the dynamic unification of application and theory into an entity DS.

'Data Science' is *an ever-expanding closed set*, for unifying beyond each other in turn.

As previously mentioned, data science consists of three phases:

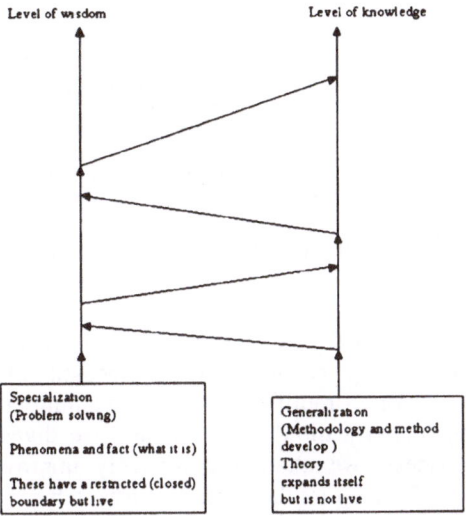

Figure 2:The Process of Data Science in Practice

design for data, collection of data, and analysis on data.   The three phases must be treated as a unified whole, based on a fundamental philosophy of science as explained below.   In each of these phases, the accompanying methods, which must fit and be valid for each objective, must be studied in this unified whole perspective.   Figure 2 summarizes the strategy for research in data science through these three phases.

The former two phases, the discussions will be given in practical problems in the sections 2 and 3.   In this stage, an idea in the third phase is described in Figure 3.

Generally speaking, phenomena are multifarious.   First, the phenomena are formulated, then the survey or experiment is devised, based on data science (the data design phase).   Thus phenomena are expressed as multidimensional and, often, as time-series data.   The data collection phase reveals the characteristics and properties of the data.   But, the obtained data are too complicated to draw concise conclusions.

Next, then, classification and multidimensional data analysis methods, as well as other mathematical/statistical methods, reveal the data structure.   We can characterize this process as simplification and conceptualization.   Nonetheless, the resulting information generally turns out to be incomplete and unsatisfactory, even though it does reveal the data structure.

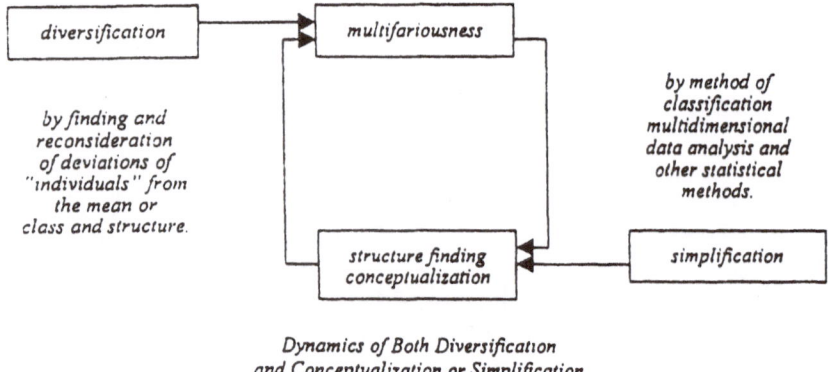

| diversification | → | multifariousness |

by finding and reconsideration of deviations of "individuals" from the mean or class and structure.

by method of classification multidimensional data analysis and other statistical methods.

| structure finding conceptualization | ← | simplification |

Dynamics of Both Diversification and Conceptualization or Simplification

Figure 3:The leading idea of data treatment in the research process is:

At this stage, then, by finding and reconsidering the deviations of 'individuals' (which gives us a vivid account of the roughness of our conceptualization or simplification) by examining the mean values or class-belonging (classification) and structure, we diversify the data. Based on this multifariousness, we attain progressively improved conceptualizations or simplifications. Such a circular pattern of research continues. The dynamics of both simplification or conceptualization and diversification begin in turn. Thus, having been able to solve one problem, we fully expect to find a new problem. This development process: design-collection-analysis, design-collection-analysis, design-collection-analysis, continues in phase as a dynamic process, all with progress and regress, which are indispensable aspects of the data science process. It can be said that the methodology of data science develops, as it were, in an ascending spiral; research proceeding as a climb up a spiral staircase. See Figure 2 and Figure 3 for a schematic depiction.

Data science, then, comprises not only the results of theory and method but also the methodological results related to the various processes necessary to derive the results. The former are called 'hard results' and the latter 'soft results'. Data science includes, simultaneously, both hard and soft results.

The three phases must be consistently and synthetically treated in order to understand phenomena through data. This is the fundamental concept of data science. Of course, each subject is studied separately. However, each subject must be studied in the context of data science. This idea will lead the development of statistics and data analytic procedures in new directions. Thus their foundations will be enhanced, a new horizon will appear and innovative methods and theory will be created in three phases.

2. Questionnaire Construction in Cross-Cultural Social Surveys

In comparative study, the following items are indispensable:

1. Securing comparability in a scientific sense

Design;
       Sample;
       Question selection and questionnaire construction
   Translation
   Data collection
2. Data analysis using familiar logic and scientific methods
3. Clarifying particularity and commonality

In the present section, the topics are only discussed.   First, we present the idea of Cultural Link Analysis (CLA) as a new tool of comparative survey research.
(i) A spatial link inherent in the selection of the subject culture or society.   The connections seen in such selection may be considered along the dimensions of social environment, culture and ethnic characteristics.
(a) Unidimensional Linkage and, (b) Multidimensional Linkage give two examples.
(ii) An item-structure link inherent in the commonalities and differences in item response patterns within and across cultures.
(iii) A temporary link inherent in longitudinal analysis.   This primarily corresponds to analysis by comparison of time-series data.   Further, the idea is expanded to comparison among many peoples in different cultures with time-series data.
CLA is, in detail, explained in [Hayashi 1996a, 1996b, Hayashi, Suzuki and Sasaki 1992]
The tools of measurements in the study are (i) making of questions and (ii) construction of a questionnaire.   The questions cover the area shown as below.
1)   Fundamental Attribute
2)   Religion
3)   Family
4)   Social Life
5)   Interpersonal Relations
6)   Politics
7)   Individual Attitude toward Other Unclassified Social Issues.

In our comparative study, we use the questionnaire from consisting of questions from several sources, based on the idea of Cultural Link Analysis (CLA).   These sources include questions from continuing surveys of the Japanese National Character Study, GSS, ISR, CREDOC, EC and ALLBUS*.   Pilot surveys were conducted by first translating each question into each nations' language.   The questions were then translated back to the original language for comparison and contrast of the results.   The discussion of translation is in details described in [Hayashi 2000a]

*The General Social Surveys of the National Opinion Research Center.   The Institute for Social Research at the University Michigan.   Centre de Recherche pour L'Etude et L'Observation des Conditions de Vie, The 'Eurobarometre' of the Commission of European Communities.   Allegemeine Bevoelkerungsumfrage der Sozialwissenschaften Mannheim; Zentrum fur Umfragen, Methoden und analysen e.V.

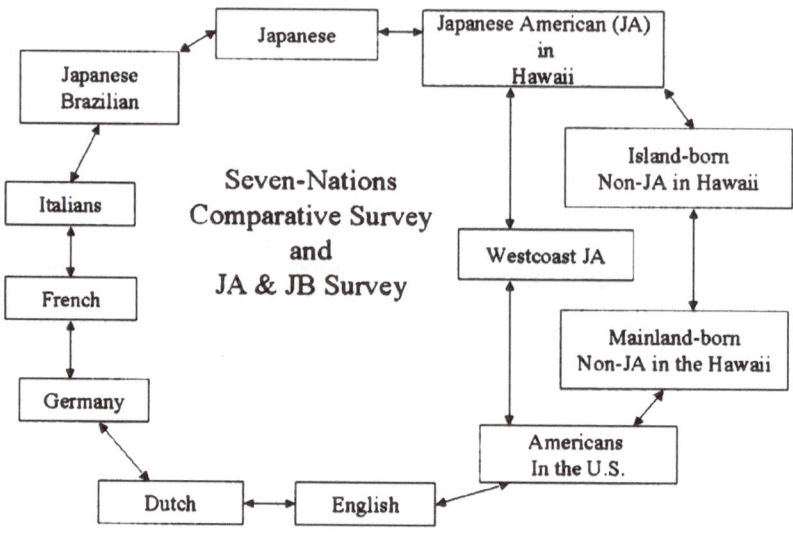

Figure 4: Conjecture of Culture Link

In our surveys, the following nations or races are taken up and their linkage who conjectured in to primary step of survey design   (Figure.4).

The existence of this linkage has been vividly verified in the results of the practical data analysis [Hayashi, 1998, 2000a, 2001b].   The outline is depicted in the following figures.

Applying the quantification method III to the data of simple tabulations of almost all questions for all groups, Figure 5 and Figure 6 is obtained.

In the case including JA in West Coast, there are a few common questions. Applying quantification method III to these data, we find the similar relative positions although the configuration is somewhat distorted.

3. Questionnaire Construction in the Survey of Non-Profit Corporations and Evaluation of Data Quality [Hayashi et al. 2000b]

The preliminary survey had been carried out.   The survey was incomplete.

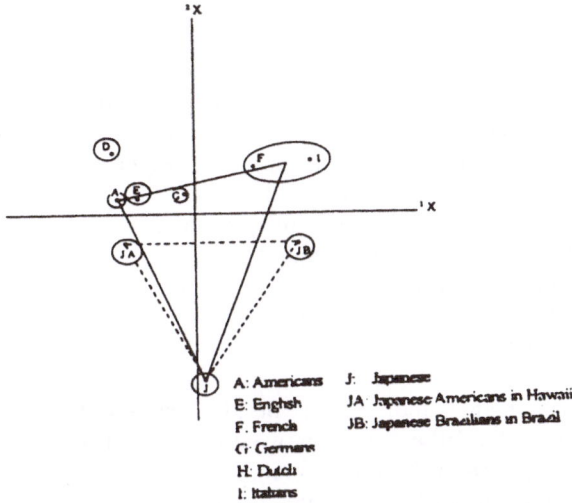

A: Americans   J: Japanese
E: English   JA: Japanese Americans in Hawaii
F: French   JB: Japanese Brazilians in Brazil
G: Germans
H: Dutch
I: Italians

Figure 5 :Configuration of Nations

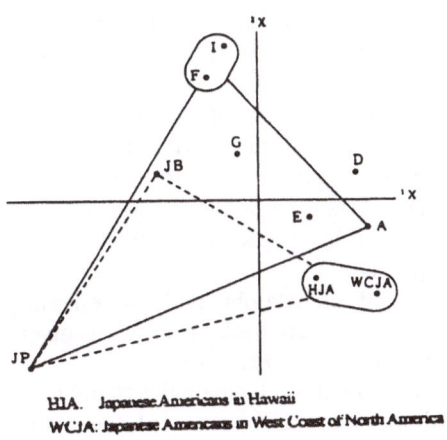

HJA. Japanese Americans in Hawaii
WCJA: Japanese Americans in West Coast of North America

Figure 6: Configuration of Nations including Japanese American in West coast

However I imagined that some important information might be hidden, although the data were incomplete and biased.
I am a computer. My experience and knowledge are software. I investigated every questionnaire filled by sample one page by one. It took about one minute

in average to read one questionnaire filled. I paid attention to the characteristics and classification of various response patterns. Through this process, an idea for the questionnaire construction of a new survey flashed across my mind.

This computer is very slow. However a valuable solution emerges through the multifarious and dynamic intuitive data analysis even though it is not quantitative. The importance of careful reading of complicated response patterns existing in the responses, by full use of brains, can be understood.

Through such a deep analysis of pre-surveys, the essential points of questionnaire construction have been found.

    (a)   contents themselves

    (b)   devices to be responded easily

    (c)   lay-out

As a measurement tool of (a), besides the questions usually used in enterprise surveys, the following items were considered to be necessary for the penetrating study of non-profit organization.

    1 The importance of ideal, vision, purpose and planning for future

    2 Activities

    3 Management

        (i) Income

        (ii) Budget control

        (iii) Organization

    4 Incorporated foundation vs Incorporated association

    5 Private section initiated (B) vs Quasi-governmental (A)

Particularly the penetrating feature of this item has been clearly verified by the data analysis by quantification method III [Hayashi et.al. 2000b] to be the classification of non-profit organization into Private section initiated (B) vs Quasi-governmental (A). This is shown as in Table 1.

Sampling procedure is very simple and so is abridged. Here we treat only the evaluation of data quality.

In order to test the quality of the data obtained, the reliability of the questionnaire responses was confirmed by cross-checking answers given to several questions (money amounts, numbers of employees, etc).

A procedure was also carried out to elucidate the characteristics between response and non-response corporations. The names of the response corporations had been compiled into a register at the time the questionnaires were returned. First, the names of corporations, which did respond to the survey, were identified and, by inference from their names, classified into groups of seemingly similar organizations, within each of which the distribution of organizations by types (see Table 1), which had been determined by the response pattern, was obtained. The non-response corporations which did not respond were also classified into groups only by the inference from their names and then, in those groups, their distributions by type were estimated using the information mentioned above of response group. The distributions by type in response and non-response corporations were compared. This revealed that, within the respective

Table 1:Type by Coding for Activities of Public-interest Corporations

Code		Nature of activities		Notes
110		1. Purely independent basis		
121	I Private-sector initiated*	2. Diversity, discretionary autonomy. freedom	(1) Non-impartial	Activities requiring violation of the principles of impartiality and equality (grants. awards, research, etc.)
122			(2) Free of legal and other constraints	Support associations, academic societies. etc.
140		3. Non-practical, socially important activities		Welfare services, environmental protection, scholarship, promoting cultural activity, international exchange, invigorating rural communities, instruction/training
150		4. Commercially unviable but socially necessary		Instruction/training classrooms, vocational training centers, special medical facilities
160		5. Networking		Mediating to facilitate various services
170		6. Activities virtually indistinguishable from profit-making enterprise		
180		7. Other activities		
231	II Quasi-governmental	1. Activities on behalf of government	(1) Theoretically impossible for government to conduct	Protective support, subsidies for scholarship, science and the arts, promotion of Japanese culture, exchange of ethnic Japanese overseas, Japanese language education, overseas technical training, promoting Japan, promoting historical awareness
232			(2-1) Vicarious operation for greater ease of operation and economic efficiency	
233			(2-2) Channeling funds for greater ease of operation and economic efficiency	
234			(3) By-passing legal impediments (Public Accounts Act)	e.g. sales agents selling medical products which cannot legally be sold by the government or by national hospitals
240		2. Non-practical, socially oriented activities		Welfare services, environmental protection, scholarship, promoting cultural activity, international exchange, invigorating rural communities
250		3. Commercially unviable but socially necessary activities		Bathing facilities and hairdressing services in remote areas, stores in special hospitals/medical facilities for the handicapped and the aged
260		4. Networking		Mediating to facilitate various services

* The distinction between "private-sector initiated" and "quasi-governmental" public-interest
corporations is based on whether or not the corporation in question
  (a) received funds for its establishment from,
  (b) receives funds for operating expenses from. or
  (c) has representatives who currently or formerly held posts in public organizations.

corporations, returned questionnaires (response group) and the questionnaires of (non-response group) had more or less the same characteristics. (Table 2)   This practical procedure is concretely explained as below.

We classify the respondents, by their names of corporations, into I groups,

$$L = \sum_{i}^{L} Li,$$

where   Li is the size of corporations belonging to the i-th group and L is the size of respondents, and every   i   means the group which is similar in names.

We define

$$Kij = Lij / Li \quad i = 1, 2, \cdots I, \quad j = 1, 2, \cdots J$$

$$\sum_{j}^{J} Kij = 1 \quad \text{for all} \quad i.$$

where Lij is the size of groups which belong to the j-th category of coding in group i, J being the number of categories of coding, and Kij means the distribution of categories of coding in group i. Mi is the size of corporations belonging to the i·th group in the non·response group.

Even though they are non-response, they have names of corporation and so the classification by names is possible as in the response groups.

$$\sum_{i}^{I} Mi = M,$$

where Mi / Kij = Mij   is define for all i and j .

This means the size of the j-th category of the coding in Mi .

$$Lij + Mij = Nij$$

$$\sum_{i}^{I} Nij = Nj \quad , \quad \sum_{j}^{J} Nj = N.$$

Thus we can estimate the size of j-th category in the total.

Here, we compare the distributions of categories of coding in response and non-response groups.

That is to say,

$$\sum_{i}^{I} Lij / L \quad \text{and} \quad \sum_{i}^{I} Mij / M$$

are compared in all j.

For this discussion, it was felt generally acceptable to apply the findings of the response group to the non-response group, and the accuracy of the survey is almost confirmed.

Table 2:Characteristics of non-response by Type

code total J	response group	non-response group	
distribution (%)	distribution (%) $\sum L_{ij} / L$	distribution (%) $\sum M_{ij} / M$	$N_j / N \cdot 100$
110	0.1	0.1	0.1
121	14.7	13.2	14.0
122	1.1	1.1	1.1
140	36.1	38.1	37.1
150	1.8	1.8	1.8
160	12.2	*16.6	14.3
170	3.3	3.7	3.5
180	0.1	0.1	0.1
231	1.4	1.3	1.4
232	15.3	*11.1	13.3
233	0.9	0.6	0.7
234	1.1	1.1	1.1
240	8.8	8.6	8.7
250	1.4	1.5	1.4
260	1.7	1.2	1.4
Total	1616	1529	3145

*significant under the confidence level   0.95

References:

Hayashi, C. et al. eds.   (1995). Data Science and Its Applications,   Academic Press, Tokyo.

Hayashi, C. (1996). Cultural Link analysis (CLA) for Comparative Quantitative Social Research and its Applications, *Quantitative Social Research in Germany and Japan*, Hayashi, C. and Erwin K. Scheuch eds., Leske + Budrich, Opladen, 209-229.

Hayashi, C. (1997). What is Data Science?, *Student* Vol. 2, No.1, March, Presses Académiques Neuchâtel, Switzerland. 47-51.

Hayashi, C. (1998). What is Data Science? Fundamental Concepts and a Heuristic Example, *Data Science, Classification, and Related Methods*, Hayashi, C, et al. eds., Springer, Tokyo, 40-51.

Hayashi, C. (1998). The Quantitative Study of National Character, Interchronological and International Perspectives, *Values and Attitudes Across Nations and Time*, Sasaki, M. ed., Brill, Leiden, 91-114.

Hayashi, C. (2000a). Evaluation of Data Quality and Data Analysis, *Data*

24

Analysis, Classification, and Related Methods, Henk A. L. Kiers, *et al.* eds. Springer, 335- 340.

Hayashi, C. *et al.*, (2000b). Public-Interest Corporations in Japan Today: Data-Scientific Approach, *Behaviormetrika*, Vol. 27, No. 1, 67-88.

Hayashi, C. (2001a). DETA NO KAGAKU (Data Science), Asakura Shoten, Tokyo.

Hayashi, C. (2001b). NIHON JIN NO KOKUMINSEI KENKYU (Data – Scientific Study on Japanese National character), Nanso sha, Tokyo.

Hayashi, C., Suzuki T., and Sasaki, M. (1992). *Data Analysis for Comparative social Research: International Perspectives.* Elsevier, North-Holland. The original Japanese version: Hayashi, C. and Suzuki, T., 1986, *Data Aanalysis in Social Survey and Quantification Method*, Iwanami Shoten.

Measurement and Multivariate Analysis

Shizuhiko Nishisato

University of Toronto
252 Bloor Street West, Toronto, Ontario M5S 1V6, Canada
snishisato@oise.utoronto.ca

Summary: This paper is prepared to serve the purpose of introducing the readers to the main theme of this conference. The emphasis is placed on the introduction of desirable properties of measurements so as to make the outcomes of multivariate analysis meaningful. Starting with such fundamental requirements for measurements to be valid as metric axioms and the Young-Householder theorem, the paper is then extended to the discussion of two topics where we see important interplays between measurement and multivariate analysis. The paper is concluded with a note on the importance of a well-balanced interplay between measurement and multivariate analysis, to which end this international and interdisciplinary conference was planned and successfully organized.

1. Introduction

Hand (1996) presented a stimulating paper on the roles of "measurement" in statistics, pointing out that this topic has been largely ignored by statisticians. Because statistics deals with numbers, one would immediately agree with Hand that the numerical nature of data should be an important topic in statistics. In contrast to the scene described by Hand, measurement has been a very popular topic in the social sciences. Thus, the content of this paper may lack novelty to the eyes of social science researchers, but considering the nature of this conference, that is, interdisciplinary and international, I would like to present an introduction to measurement as related to multivariate analysis.

2. Univariate Case

Measurement is defined as the assignment of numbers to objects according to certain mathematical rules. One of the most popular classification schemes of measurement as seen in the social sciences is the one proposed by Stevens (1951), who identified four types of measurement:

1. Nominal measurement, defined by (A) one-to-one correspondence.
2. Ordinal measurement, defined by (A) and (B) order relations.
3. Interval measurement, defined by (A), (B) and (C)the unit of measurement.
4. Ratio measurement, defined by (A), (B), (C) and (D) the origin.

Nominal and ordinal measurements are called *categorical, non-metric or qualitative* data, and interval and ratio measurements *continuous, metric or quantitative.*

There is a branch of measurement research, called *Scaling.* It represents a family of techniques to *upgrade* the level of measurement. If the data at hand are ordinal measurement such as paired comparison responses, the role of scaling is to transform ordinal measurement to the level of interval measurement or ratio measurement.

Thus, the main realm of scaling is often restricted to categorical data analysis, with the object being to convert them to interval or ratio measurement. Considering the ubiquitous existence of categorical, qualitative or non-metric data in the social sciences, it is no wonder why the topic of measurement is a familiar one among psychologists, sociologists and related disciplines. In this process, what properties are desirable for scaled measurement?

2.1 Metric Axioms

The first property that comes to our mind is what the metric axioms state. A scaled measurement of distance d_{jk} between point j and point k is said to be metric if:

(1) $d_{jj} = 0$, that is, the distance from point j to itself is zero.
(2) $d_{jk} \geq 0$
(3) $d_{jk} = d_{kj}$, that is, the distance is symmetric.
(4) $d_{ik} \leq d_{ij} + d_{jk}$: triangular inequality.
(5) That $d_{jk} = 0$ means $j = k$.

There are measurements that satisfy only a subset of the axioms. For instance, d_{jk} is called *pseudometric* if it satisfies (1) to (4), *semi-metric* if (1), (2), (3) and (5) hold. and *semi-pseudometric* if only (1), (2) and (3) are true.

From the scaling point of view, we would like to have metric measurement that satisfies all five conditions. What will happen if measurement does not satisfy the metric axioms? If the triangular inequality is not satisfied, a negative value of the sample variance, for example, may be one of the consequences, for then the norm of the projected vector may become larger than that of the original vector. This results in a negative value of the variance. given by

$$s^2 = \frac{\sum_{i=1}^{n}(x_i - \bar{x})^2}{n-1} = \frac{x'x - x'Px}{n-1},\tag{1}$$

where P is the mean projection operator, $\mathbf{11}'/n$.

Distance defined in the Euclidean space satisfies all the axioms. but this is only one of many examples. More generally, there is a family of measurements that satisfy all the metric axioms. It is called the Minkowski power metric (Minkowski. 1896).

2.2 The Minkowski Power Metric

The Minkowski power metric $d_{jk}^{(p)}$ is given by

$$d_{jk}^{(p)} = (\sum_{i=1}^{K} |d_{ji} - d_{ki}|^p)^{1/p} \qquad (2)$$

When $p \geq 1$, it satisfies all the metric axioms. Note that the value of p can be a real number such as 3.65. There are, however, some preferred choices. For example, when $p = 1$ and the variables are binary, it is called the Hamming distance (Hamming, 1950); when $p = 1$ and the data are continuous, it is called the city-block metric or the Manhattan metric (Torgerson, 1958), and; when $p = 2$, it yields the Euclidean distance. In most cases, scaling aims to produce the Euclidean distance for at least two reasons, one because it is well understood by most of us and one because the distance between any two points remains invariant over the orthogonal rotation of axes. This latter point is important for multivariate analysis because in many instances we look for principal axes as the most economical Cartesian coordinates in describing data.

3. Multivariate Case

In order to remain in Euclidean space, it is good to know the so-called Young-Householder theorem. Torgerson (1952) described how one can examine if we are dealing with the Euclidean space.

3.1 Young-Householder Theorem

Young and Householder (1938) described the following, which is referred to as the Young-Householder theorem in the social sciences:

(1) If an $n \times n$ inner-product matrix $\mathbf{A}$ calculated from distance measurements (see Torgerson, 1952) is positive semi-definite, the distance between any two points may be considered as distance between points lying in a real Euclidean space.
(2) The rank of any positive semidefinite matrix $\mathbf{A}$ is equal to the dimensionality of the space required to represent n points.
(3) Any positive semidefinite matrix $\mathbf{A}$ may be factored to obtain a matrix $\mathbf{B}$, where $\mathbf{A} = \mathbf{BB}'$. If the rank of matrix $\mathbf{A}$ is K, then $\mathbf{B}$ is a matrix of coordinates of n points on K dimensions in Euclidean space.

The above points indicate that if we have a data matrix with elements being real numbers, say $\mathbf{X}$, we can always find coordinates of the variables in Euclidean space. But, for some reasons or others, we often encounter negative eigenvalues in multivariate analysis. Why? This will be discussed later. There is another topic which may be too basic to look at, but it seems vitally important from the measurement point of view as seen in the next section.

3.2 Rectangular Coordinates

Consider that children were given two tests, one in mathematics (X_1) and the other in English (X_2). Typically, we construct a scatter plot using these variables as the two axes. As an educator, I always wonder why we have chosen these axes, treating X_1 and X_2 as if they were perpendicular. In multivariate analysis, we often consider a linear combination of variables such as

$$Y_i = w_1 X_{1i} + w_2 X_{2i}. \tag{3}$$

By imposing the condition that $w_1^2 + w_2^2 = 1$, the above linear combination can be expressed for non-zero constants a and b as

$$Y_i = \frac{a}{\sqrt{a^2 + b^2}} X_{1i} + \frac{b}{\sqrt{a^2 + b^2}} X_{2i} = X_{1i}cos\theta + X_{2i}sin\theta. \tag{4}$$

This means that all the composite scores, Y_i, are projections of data points (X_{1i}, X_{2i}) on the axis going through the origin and point (a, b), that is, the slope being $\frac{a}{b}$. In other words, *all linearly combined scores lie on a single axis.*

Consider a linear combination of n variables,

$$Y_i = w_1 X_{1i} + w_2 X_{2i} + w_3 X_{3i} + \cdots + w_n X_{ni}, \tag{5}$$

under the condition that the sum of the squared weights is 1. In principal component analysis, those weights are determined so as to make the variance of the composite scores Y_i be a maximum, and we know that this maximized variance is the eigenvalue λ and the corresponding axis is called the *principalaxis*. As is the case with two variables, the above composite is an axis going through the origin and a point with coordinates $(w_1, w_2, w_3, ..., w_n)$. Once we obtain n principal components, Y_i, $i = 1, 2, \cdots, n$, we can in turn express each variable as a linear combination of these orthogonal components as

$$X_{ji} = \sqrt{\lambda_1} w_1 Y_{1i} + \sqrt{\lambda_2} w_2 Y_{2i} + \cdots + \sqrt{\lambda_n} w_n Y_{ni} = u_1 Y_{1i} + u_2 Y_{2i} + \cdots + u_n Y_{ni}, \tag{6}$$

where the sum of squares of u_i is 1, the same condition as for w_i. Now, this final expression is important in that any variable X_j can be expressed as a linear combination of weights associated with the rectangular coordinates. In other words, any variable can be expressed as a single axis in multidimensional Euclidean space, going through the origin and a point with rectangular coordinates $(u_1, u_2, u_3, ..., u_n)$. Therefore, *all the scores on a variable lie on this single axis.* This is what one may call a correct representation of a variable in the rectangular coordinate system, in contrast to the general practice of using generally correlated variables (e.g., mathematics and English) as axes. Thus, if a variable can be expressed as a single axis in Euclidean space, it is easy to understand that the product-moment correlation is the cosine of the angle between the axes of the two variables, a statement that we often see in the literature on factor analysis and principal component analysis.

3.3 Multidimensional Space

In multivariate analysis, we commonly deal with multidimensional space. The concept of many dimensions is hard to visualize, but mathematically it is treated as a simple extension of one, two and three dimensions. Visually, however, we can understand three-dimensional space, but in the extension to the fourth dimension we typically lose the insight into the added dimension. Philosophically one often says that the fourth dimension is time. But typically we are dealing with the decomposition of a two-way table of data, where time is not involved. Pierce (1961) presented an interesting explanation of multidimensional space as follows. The area of a circle in 2-dimensional space of radius r is πr^2, and the area of a concentric circle of half the radius inside it is $\pi(0.5)^2 r$, which is just one-quarter. The volume of a sphere of half the "radius" is one-eighth of that of the radius r. As the dimension increases to n, one can safely say that the volume of a hyper-sphere of n dimensions is proportional to r^n. Then, with this preliminary, Pierce (1961) considers the volume of a hyper-sphere of radius r and that of radius $0.99r$. In two dimensional space, these two circles occupy almost the same amount of space. In three-dimensional space, the sphere of radius $0.99r$ occupies 97 percent of the sphere with radius r. For a 1000-dimensional hyper-sphere, a fraction of 0.00004 of the volume lies in a sphere of $0.99r$ the radius! Just raise 0.99r to the power of 1,000, and you will see this result. His conclusion, therefore, is that in the case of a hyper-sphere of a very high dimensionality, essentially all of the volume lies very near the surface. This is a remarkable revelation at least to some of the psychometricians.

Consider a point P in K-dimensional space with coordinates x_{pk}, where $k=1,2,...,K$. We can identify two fundamental properties of a hyper-space:

(1) The distance of P from the origin, d_p, is a monotonically increasing function of the number of dimensions, K, where

$$d_p = \sqrt{\sum_{k=1}^{K} x_{pk}^2}. \tag{7}$$

(2) Similarly, the distance between any two points, d_{pq} is a monotonically increasing function of the number of dimensions, where

$$d_{pq} = \sqrt{\sum_{k=1}^{K}(x_{pk} - x_{qk})^2}.$$ (8)

When we look at the graph of, for example, the first two principal components or factors, it is the general practice to identify only those clusters of closely located variables for an interpretation. Unless these two components account for nearly 100 per cent of the variance by the first two dimensions, however, we must admit that there is no guarantee that those closely located variables are indeed close to one another in a hyper-space. Only if two points are widely separated in a 2-dimensional graph, it is guaranteed that they are widely separated in a hyper-space. This point shows a great danger in interpreting those clusters of closely located variables in a 2-dimensional graph to draw a conclusion on their relationship. We must pay more attention to those points which are widely separated than to those which are closely located.

3.4 Principal Hyper-Space

Principal component analysis seeks the coordinate system expressed in terms of principal axes, and the space is then called principal hyper-space. This is probably the most preferred way to describe multivariate data. Its mathematics is provided by singular-value decomposition (SVD), due especially to Beltrami (1873), Jordan (1874), Schmidt (1907), and Eckart and Young (1936). The fact that the SVD transforms the data to another coordinate system means (i) that the data themselves have an exact geometric representation, and (ii) that the SVD offers only a special way of looking at the data, often said to be the most efficient way. As mentioned later, (i) applies not only to continuous data, but also to some types of categorical data, as described in the next section.

4. Measurement and Multivariate Analysis

One can look at many topics on the title of this section. However only two of them will be discussed here, that is, first on discrete versus continuous measurement for multivariate analysis, and second on a questionable assumption of measurement for multivariate analysis.

4.1 Continuous versus Discrete Measurement

Let us consider two mathematically equivalent procedures for multivariate analysis, one for continuous variables and the other for discrete variables. The former is principal component analysis (Pearson, 1901; Hotelling, 1933). abbreviated as PCA, and the latter is principal component analysis for categorical data (Torgerson, 1958), better known by many other such names as the method of reciprocal averages (Richardson and Kuder, 1933; Horst, 1935), simultaneous linear regressions (Hirschfeld, 1935; Lingoes, 1964), Hayashi's theory of quantification Type III (Hayashi, 1950), optimal scaling (Bock, 1960). analyse des correspondances (Escofier-Cordier, 1969), biplot (Gabriel,

1971), homogeneity analysis (de Leeuw, 1973), correspondence analysis (Hill, 1974) and dual scaling (Nishisato, 1980). The latter will be abbreviated as DS (dual scaling).

Suppose we have multiple-choice data where choices per question are ordered categories (e.g, never, sometimes, often, always). Researchers often assign integers 1, 2, 3 and 4 to these categories of each variable and subject thus 'quantified' data to PCA. In PCA, the variables (items) are used as units, and each variable is positioned in the principal space as a straight axis. In the current example, the four category points of the variable lie on the axis of the variable. The contribution of the variable to the graph is given by the distance from the origin to the point of the variable's coordinates. As may be conjectured from this linear representation of the variables and categories in hyper-space, PCA can neither capture non-monotone relations among categories nor non-linear relations between variables. Although this example used a 4-category variable treated as continuous, the same argument applies to truly continuous variables, where each variable is represented as an axis and scores on the variable all lie on this axis.

In contrast, DS captures both linear and nonlinear relations because the units of analysis are categories (response alternatives) of variables. If a variable has three categories, then each category will have its own coordinates, with the basis coordinates of the three categories being $(1,0,0)$, $(0,1,0)$ and $(0,0,1)$, and a single variable in two-dimensional space, for example, can be depicted by the three category points in such a way that when we connect the three points to form a triangle its area indicates the relative contribution of the variable to the two-dimensional space. When three category points lie on a straight line, an extremely rare case, it indicates the variable is correlated with other variables only linearly. Typically, however, we see that the three points are not on a straight line, but are scattered away from such a straight line. This means that variables are nonlinearly related. In relation to Pierce's conclusion, mentioned above, we should be aware that the area of the triangle becomes the largest in the total space, and that the three points *cannot* lie on a straight line in the total space. Only in a subspace, there exists a rare possibility that the three points lie on a straight line.

The above observation indicates how easy nonlinear relations can be analyzed when data are categorical: simply analyze a finite number of mutually exclusive categories of each variable as units, rather than variables as units, using singular value decomposition, and this does not require any prior specification of the form (shape) of non-linearity of pair-wise relations. This leads to the idea of subjecting categorized continuous variables to DS, in lieu of continuous variables to PCA. Most surprisingly, however, there does not seem to be any established knowledge available on how one can categorize continuous variables in a justifiable way, not to mention the optimal way. There does not seem to be an answer to even such a mundane question of how many categories we should introduce to this process. A modest preliminary study was done on optimal categorization (Nishisato, 1998, 2000), his problem being how to categorize continuous variables in such a way that the average inter-variable Tschuprow/Cramér coefficient, or its variant, be a maximum.

As for information contained in data, it is important to note that the sum of the eigenvalues of categorical data of the (1,0)response-pattern form is a function of the *average number of categories* of the variables (Nishisato, 1996). This implies that the total information contained in continuous variables, which can be regarded as having an extremely large number of categories, cannot be fully expressed simply by the sum of eigenvalues of the variance-covariance (or correlation) matrix. In other words, the trace of the variance-covariance matrix reflects only information captured by pair-wise linear relations of variables, and leaves out the rest of information unaccounted for.

Nishisato (1993) classified categorical data into incidence data (e.g., contingency tables and (1,0)pattern matrices) and dominance data (e.g., ranking data, paired comparison data), both of which are amenable to DS. However, only incidence data have exact geometric representations. Dominnace data for which we do not have the information of how much A is better than B, however, require something other than SVD. For the analysis of dominance data by DS, the readers are referred to Nishisato (1996).

4.2 Multivariate Normality and Multivariate Analysis

There are a number of cases in which measurement proceeds not necessarily in concert with multivariate analysis, or vice versa. We have already looked at one example in Section 2.1, in which the unbiased estimate of the variance becomes negative. The problem there was due to the fact that scaling of data was not carried out properly, that is, measurement was not appropriate for the computation of the variance. Let us extend this case to multivariate analysis.

In the social sciences, it is well known that when data are binary and researchers use tetracholic correlation for factor analysis or principal component analysis one would often observe negative eigenvalues, quantities that correspond to the variances of the individual components (e.g., Gorsuch, 1983; Yanai, Shigemasu, Maekawa and Ichikawa, 1990). In the structural equation modeling (e.g., Jöreskog and Sorbom, 1981; Muthén, 1987; Bentler, 1989; Bollen, 1989), often abbreviated as SEM, one often uses the Kendall-Stuart type canonical correlation (Kendall and Stuart, 1961) and polychoric correlation (e.g., Tallis, 1962) when data are polytomous. It is well known that the variance-covariance matrix calculated from these coefficients often yields a number of negative eigenvalues (e.g., Bollen and Long, 1993).

When negative eigenvalues are obtained, the Young-Householder theorem, mentioned earlier, asserts that the data cannot be mapped in Euclidean space. There is a serious measurement problem.

From the literature, it appears that the culprit behind the above problem lies in the assumption that the underlying distribution is normal for dichotomous variables in calculating tetracholic correlation (e.g., Gorsuch, 1983) and for polychomous variables in computing polychoric correlation and the Kendall-Stuart type canonical correlation (e.g., Nishisato and Hemsworth, 2001). In other words, the observed multi-category data are far from the one generated from multivariate normal variates. Many researchers look at only the first few components with large eigenvalues, ignoring the other components. But do they lead to interpretable results? Rather unlikely, or more precisely, we do not know what is happening if some other eigenvalues are negative.

There is a hint for a practical solution to the problem in the famous statement by Kendall and Stuart (1961) that if we manipulate the row spacing and the column spacing of a two way table so as to maximize the correlation we are basically transforming the data in such a way that the table of the adjusted row and column spacing would resemble that of a partitioned bivariate normal distribution. Thus, if DS is applied to the manifold incidence matrix as is done in multiple correspondence analysis and dual scaling of multiple-choice data, we are basically transforming the data into a multidimensional contingency table with the shape that resembles a partitioned multivariate normal distribution. This may be an alternative to the above problem of polychoric correlation and the Kendall-Stuart type canonical correlation because the use of DS definitely prevents us from seeing negative eigenvalues (Nishisato and Hemsworth, 2001; Hemsworth, 2002). If we were to maintain a weak order in the successive categories, DS under a weak order constraint (Nishisato & Arri, 1976; Nishisato, 1980, 1994) may be carried out. But, one may ask then if any order constraint is indeed necessary for data analysis. Should we not consider it more justified if we were open to such an order constraint and leave it to the structure of data in hand?

The message in this section is that such procedures as tetracholic and polychoric correlations are based on pair-wise normal distributions and that one should not use them if multivariate normality is a suspect. Regression of non-normal variables to the normal distribution cannot be used for the purpose of capturing information in data.

5. Concluding Remarks

Measurement is the very starting point for data analysis since the outcome of analysis depends on the quality of measurement. As Hand (1996) stated, however, the emphasis on measurement by the statistics community has not been that great. Yet, statistical theory, guarded by the scheme of random sampling and normal distribution, has pushed the frontiers of statistics into an independent scientific discipline. At the same time, we should note the resurgence of exploratory data analysis in many disciplines, which seems to tell us some very nature of statistics or the environment surrounding statistics, namely, the decision making under uncertainty. The more information we can collect through good designs for data collection the better.

From the measurement point of view, we cannot help but wonder why we have not made greater effort to develop many more innovative methods for data collection than we have. We should have learned from the old historical note that R. A. Fisher had experimented his experimental designs in his own garden. Surely we could have devised ways of collecting data such that the data are more informative, useful and easier to deal with, than what we see today. This is the exact point that our key-note speaker Dr. Chikio Hayashi has been advocating under the name of "data science." Valid measurement is essential for powerful statistical procedures to work properly.

From the statistics point of view, one problem was briefly touched on, that is, the use of assumptions to facilitate statistical reasoning. We often say that the assumptions do not have to be correct for a given situation so long as they are reasonable. So, for example, we state that the intelligence quotient, IQ, has a normal distribution, even though we know that it is hardly conceivable to observe an IQ of plus infinity or minus infinity. Yet this assumption is essential for statistical handling of IQ data. We often hear how robust a certain statistical procedure is over the violation of such assumptions. Our wishful thinking is that this applies to all the cases. However, we must remember that certain situations require the strict adherence to the assumptions. The multivariate normal assumption must be "correct" for the estimation of tetracholic and polychoric correlations. Otherwise, as we have seen earlier, the correlation matrix is likely to produce a number of negative eigenvalues.

Considering these aspects of both measurement and multivariate analysis, it seems that the adage "garbage in garbage out" cannot be solely attributed to the measurement side. This conference has given us an interdisciplinary forum for discussing measurement problems and multivariate analysis. Both are equally important partners.

Acknowledgements

The work was partially supported by a grant from the Natural Sciences and Engineering Council of Canada.

References:

Beltrami, E. (1873). Sulle funzioni bilineari [On the bilinear releations]. in G. Battagline and E. Fergola (Eds.) *Giornale di Mathematiche, 11*, 98-106.

Bentler, P.M. (1989). *EQS structural equation program manual.* Los Angeles: BMDP Statistical Software.

Bock, R. D. (1960). Methods and applications of optimal scaling. *The University of North Carolina Psychometric Laboratory Research Memorandum*, No.5.

Bollen, K.A. (1989). *Structural equations with latent variables.* New York: Wiley Interscience.

Bollen, K. and Long, S. (eds.) (1993) *Testing structural equation models.* Newbury Park: Sage.

Cronbach, L.J. (1951). Coefficient alpha and the internal structure of tests. *Psychometrika*, *16*, 297-334.

de Leeuw, J. (1973). *Canonical analysis of relational data*. Department of Data Theory, Leiden University, Report RN 007-68.

Eckart, C. and Young, G. (1936). The approximation of one matrix by another of lower rank. *Psychometrika*, *1*, 211-218.

Escofier-Cordier, E. (1969). L'analyse factorielle des correspondances. *Bureau Universitaire de Recherche Operationelle, Cahiers, Série Recherche*, **13**, 25-29.

Gabriel, K.R. (1971). The biplot graphical display of matrices with applications to principal component analysis. *Biometrika*, **58**, 453-457.

Gorsuch, R.L. (1983). *Factor analysis* (second edition). Hillsdale, N.J.: Lawrence Erlbaum.

Hamming, R.W. (1950). error detecting and error correcting codes. *The Bell System Technical Journal*, **26**, 147-160.

Hayashi, C. (1950). On the quantification of qualitative data from the mathematico-statistical point of view. *Annals of the Institute of Statistical Mathematics*, **2**, 35-47.

Hand. D. J. (1996). Statistics and the theory of measurement. *Journal of the Royal Statistical Society, A, 150*, Part 3, 445-492.

Hemsworth, D. (2002). The use of dual scaling for the production of correlation matrices for use in structural equation modeling. Unpublished Ph.D. thesis, University of Toronto.

Hill, M.O. (1974). Correspondence analysis: A neglected multivariate method. *Applied Statistics*, **23**, 340-354.

Hirschfeld, H.O. (1935). A connection between correlation and contingency. *Cambridge Philosophical Society Proceedings*, **31**, 520-524.

Horst, P. (1935). Measuring complex attitudes. *Journal of Social Psychology*, **6**, 369-374.

Hotelling, H. (1936). Relation between two sets of variables. *Biometrika*, **28**, 321-377.

Jordan, C. (1874). Mémoire sur les formes bilinieares [Note on bilinear forms]. *Journal de Mathématiques Pures et Appliquées, Deuxiéme Série, 19*, 35-54.

Jóreskog, K. and Sorbom, D. (1981). *Analysis of linear structural relationships by maximum likelihood and least squares methods*. (81-8). Uppsala: University of Uppsala.

Kendall, M.G. and Stuart, A. (1961). *The advanced theory of statistics*. Volume II. Longon: Griffin.

Minkowski, H. (1896). *Geometrie der Zahlen*. Leipzig: Teubner.

Muthén, B. O. (1987). *LISCOMP: Analysis of linear structure equations with a comprehensive measurement model*. Moorresville, Indiana: Scientific Software.

Nishisato, S. (1980). *Analysis of categorical data*. Toronto: University of Toronto Press.

Nishisato, S. (1993). On quantifying different types of categorical data. *Psychometrika, 58*, 617-629.

Nishisato, S. (1994). *Elements of dual scaling.* Hillsdale, N.J. :Lawrence Erlbaum.

Nishisato, S. (1996). Gleaning in the field of dual scaling. *Psychometrika, 61*, 559-599.

Nishisato, S. (1998). Unifying a spectrum of data types under a comprehensive framework for data analysis. A talk presented at a symposium at the Institute of Statistical Mathematics, Tokyo, Japan.

Nishisato, S. (2000). Data types and information: Beyond the current practice of data analysis. In Decker, R., and Gaul, W. (eds.), *Classification and Information Processing at the Turn of the Millennium.* Heidelberg: Springer-Verlag, 40-51.

Nishisato, S., and Arri, P. S. (1976). Nonlinear programming approach to optimal scaling of partially ordered categories. *Psychometrika, 40*, 525-548.

Nishisato, S. and Hemsworth, D. (2001, in press). Quantification of ordinal variables: A critical inquiry into polychoric and canonical correlation. To appear in Baba, Y. et al. (eds.), *Recent Advances in Statistical Research and Data Analysis.* Tokyo: Springer-Verlag.

Nishisato, S. and Sheu, W. J. (1980). Piecewise method of reciprocal averages for dual scaling of multiple-choice data. *Psychometrika, 45*, 467-478.

Pearson, K. (1901). On lines and planes of closest fit to systems of points in space. *Philosophical Magazine and Journal of Science, Series 6*, **2**, 559-572.

Pierce, J.R. (1961). *Symbols, signals and noise: The nature and process of communication.* New York Harper & Row.

Richardson, M. and Kuder, G.F. (1933). Making a rating scale that measures. *Personnel Journal,* **12**, 36-40.

Schmidt, E. (1907). Zür Teheorie der linearen und nichtlinearen Integralgleichungen. Erster Teil. Entwicklung willk—"urlicher Funktionen nach Systemen vorgeschriebener [On theory of linear and nonlinear integral equations. Part one. Development of arbitrary functions according to prescribed systems]. *Mathematische Annalen, 63*, 433-476.

Stevens, S. S. (1951). Mathematics, measurement, and psychophysics. In Stevens, S.S. (ed.), *Handbook of experimental psychology.* Wiley, Chapter 1, 1-49.

Tallis, G. (1962). The maximum likelihood estimation of correlation from contingency tables. *Biometrics, 18*, 342-353.

Torgerson, W.S. (1952). Multidimensional scaling. I. Theory and method. *Psychometrika, 17*, 401-419.

Torgerson, W. S. (1958). *Theory and methods of scaling.* New York: Wiley.

Yanai, H., Shigemasu, K., Maekawa, S. and Ichikawa, M. (1990). *Inshi bunseki (Factor analysis.* Tokyo: Asakura Shoten (in Japanese).

Young, G. and Householder, A.S. (1938). A note on multi-dimensional psychophysical analysis. *Psychometrika, 6*, 331-333.

An Intelligent Clustering Technique Based on Dual Scaling

Hans-Joachim Mucha

Weierstrass Institute for Applied Analysis and Stochastics (WIAS)
Mohrenstr. 39, D-10117 Berlin, Germany

Summary: Methods of cluster analysis (unsupervised classification) can help you in order to "Finding groups in data", so the suitable title of a book from Kaufman and Rousseeuw (1990). The intelligent clustering technique proposed here appears to be motivated by practical problems of analyzing mixed data. One way to deal with such problems is downgrading all data to the lowest scale level, that is, downgrading to categories without any information about their order. This new clustering technique based on dual scaling is presented in the simplest fashion possible for finding two groups (clusters) in data and for visualizing them, respectively. It is compared with well-known model-based clustering techniques. In the conclusion variations of improvement of the method are proposed.

1. Introduction

The guiding principle of this paper can be expressed by the following statement of D. Hand (1997): "…a key to intelligent data analysis is the ability to recognize what is important in a problem – what counts and what doesn't count". However, this is only one key among others to intelligent clustering. Anyway, this key "of knowing what to overlook" will be emphasized here extremely by the attempt of clustering of multivariate normal distributions based on dual scaling. The samples for both the examples and the simulation studies described later on consist of multivariate data, which is drawn from artificially generated multivariate distributions with known misclassification rate. In order to do dual scaling these data values are downgraded in a quite rough manner into categories (text data, characters, alphanumerical data) before applying the proposed iterative clustering technique based in the key part on univariate dual scaling of each axis separately. That means that almost all information in the data is lost beforehand. The new dual scaling clustering method is called *ScoreCutter*. It is compared with five well-known algorithms for model-based Gaussian clustering. These latter five methods work directly on the random generated metric data values.

2. The *ScoreCutter* Algorithm for Finding Two Clusters

The aim of *ScoreCutter* is finding two clusters in a two-way data table $X=(x_{ij})$ of I observations (row points) and J variables. It is only supposed for *ScoreCutter* that the number of categories of each variable is at least two. That means that this method can be applied in practice to many mixed data situation. For practical and interpretational reasons however, the number of categories should be limited to some few dozen at most. That suggests that metric variables should be downgraded into categories in some way in a ***data-preprocessing step***.

There is no information about classes known beforehand in cluster analysis. In order to do univariate dual scaling like in the credit scoring algorithm described by Mucha, Siegmund-Schultze, and Dübon (1998), here an artificially random generated variable setting up two categories is used as a starting point of the iterative algorithm *ScoreCutter*. This variable is called a cluster membership variable or partition $P^0(I,2)$ of I observations into two clusters. In that way also a hint about the prior probabilities of the clusters one is looking for are given.

2.1 Dual Scaling: *the upgrading step*

A cluster membership variable can be used in order to give categorical data a quantitative meaning (Nishisato 1980, 1994). Generally, we want to obtain a new variable z_j so as to make the derived scores within the given classes ss_w as similar as possible and the scores between classes ss_b as different as possible. The basis is a contingency table, which can be obtained by crossing a categorical variable j (K_j categories, where $K_j \geq 2$ is considered) with the cluster membership variable.

That is, regarding some constraints in the frame of a dual scaling approach, the squared correlation ratio has to be maximized:

$$\eta^2 = \frac{ss_h}{ss_h + ss_w} \left(= \frac{ss_h}{ss_t} \right) . \tag{1}$$

Because of the well-known variance decomposition

$$ss_t = ss_w + ss_h , \tag{2}$$

the squared correlation ratio lies in the interval $[0,1]$. Considering the special case of two classes, dual scaling can be applied without the calculation of orthogonal eigenvectors. Without loss of generality, a given category y_{kj} of a variable j is transformed into an optimal scaled one in the sense of maximal between classes variances by

$$z_{kj} = p_{kj}^{(1)} / (p_{kj}^{(1)} + p_{kj}^{(2)}) . \tag{3}$$

Here $p_{kj}^{(1)}$ is an estimate of the probability for being a member of cluster 1 when coming from category k of variable j, whereas on the other side $p_{kj}^{(2)}$ is an estimate of the probability for being a member of cluster 2 when coming from category k of variable j. The solution of correspondence analysis f_{kj} that depends only on the actual sample size of the two clusters is simply related to (3) by

$$z_{kj} = \frac{\sqrt{n_1 n_2}}{n_1 + n_2} f_{kj} + c , \qquad j=1,2,...,J, \ k=1,2,...,K_j \tag{4}$$

where n_1 and n_2 is the number of observations in cluster 1 and cluster 2, respectively. The constant c is responsible for the shift of the mean value of the scores f_{kj} to the origin. That is, every variable j, $j=1,2,...,J$, is scaled by the same value $b = \sqrt{n_1 n_2}/N$ and shifted by a constant c. Here the total number of observations is denoted by $N = n_1 + n_2$.

2.2 Distance Scores: *the clustering step*

After upgrading the categories in the described way all data are coming from the interval [0,1]. Without any loss of generality the hypothetical worst-case $z_w \equiv 0$ (that is $z_w \equiv (0,0,...,0)'$) is considered here. Otherwise one can look at the best-case model with $z_w \equiv 1$). Then the Manhattan distance t_{iw} between an observation z_i and the worst case is both suitable and really simple

$$t_{iw} = t(z_i, z_w) = \sum_{j=1}^{J} z_{ij} \, . \tag{5}$$

As a result distance scores are obtained. Now there are several ways possible to get a new partition $P^1(I,2)$. Here the simple way of looking for a suitable cut-off-point on the distance score axis is recommended, that preserves also the a priori probabilities of the clusters of the preceding partition $P^0(I,2)$ in some degree.

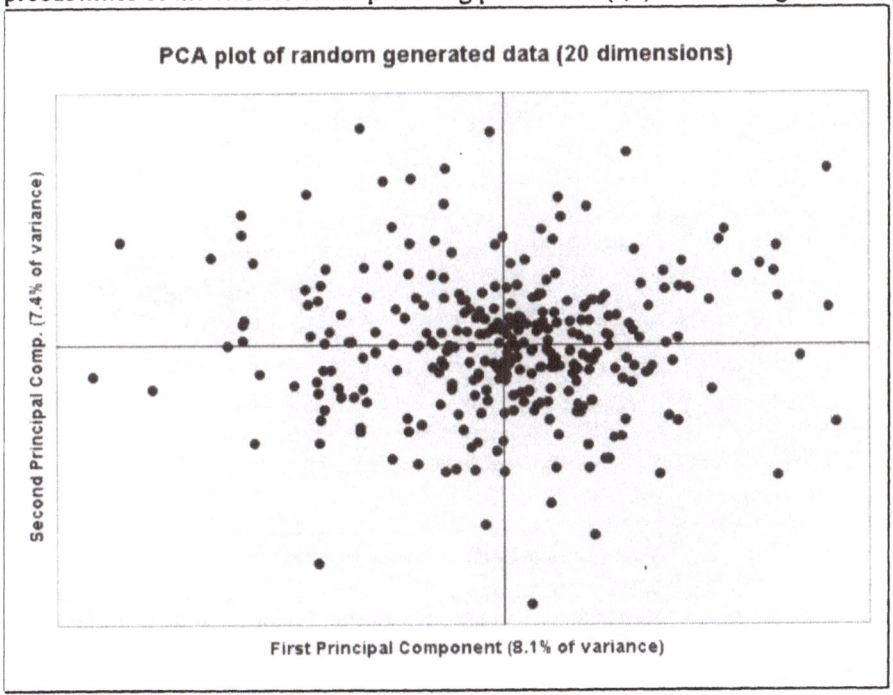

Figure 1: Principal components analysis plot of a "ringnorm" sample

2.3 Iteration

The iterative algorithm *ScoreCutter* repeats both the upgrading step and the clustering step until no change of two successive partitions $P^{m+1}(I,2)$ and $P^m(I,2)$ occur or until a fixed number of iterations are done.

2.3.1. Example "ringnorm" (Breiman (1996)):

The two class data consisting of $I=300$ observations is drawn from two normal distributions in $J=20$ dimensions with equal probabilities. Class 1 is multivariate normal with mean zero and covariance matrix 4 times the identity. Class 2 has unit covariance matrix and mean $(a,a,...,a)$ with $a=1/J^{1/2}$. Figure 1 shows a principal component analysis plot (PCA plot) of such a data set. Downgrading (or binning) of each variable is done here and later on simply by rounding the metric values to integers and making at the end a text value by a preceding "c" that stands for category. This is a rough discretization rule indeed. The upgrading step leads to a new geometry. In Figure 2 the PCA plot of the same data set as shown in Figure 1 is presented after convergence of *ScoreCutter*. Figure 3 and Figure 4 show the data of Figure 2 in a continuous bivariate and univariate representation, respectively.

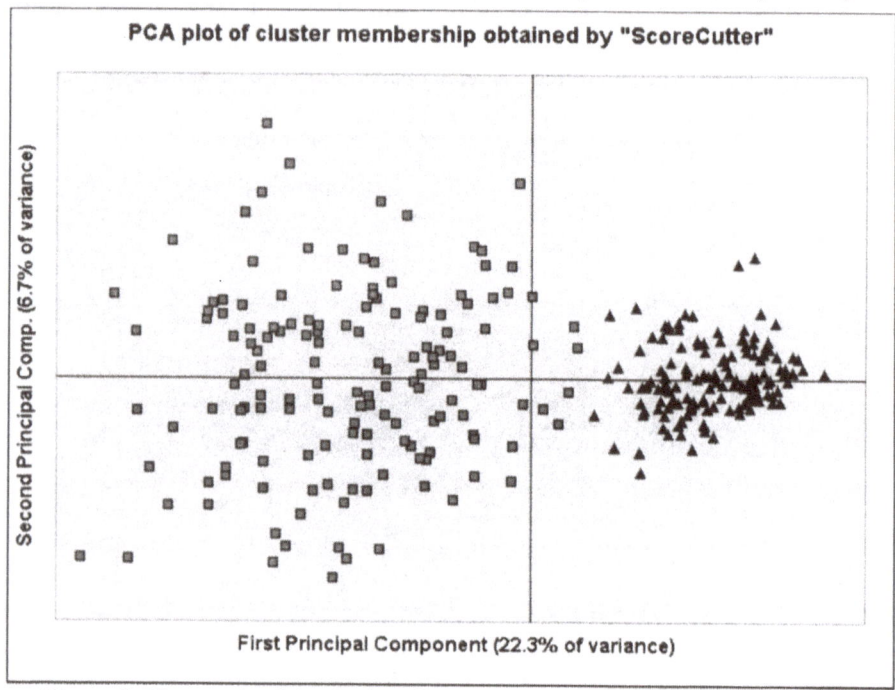

Figure 2: The result of the dual scaling clustering for the sample of Figure 1

2.4 Voting

Voting is beside univariate dual scaling another important key for *ScoreCutter*. Most generally speaking *ScoreCutter* looks for directions in the space and changes the geometry of the points (observations) in a more random than in a direct way. This is done partly by optimum univariate optimization on the one hand and on the other hand by some heuristics. Moreover, starting from a random initial partition a result is obtained in any case independently of the existence of a class structure. Here voting means finding the most typical partition (MTP) among a set of partitions that are obtained by *ScoreCutter* starting from different random initial partitions $P^0(1,2)$. Once more, there are several ways in defining a MTP.

Here in the two class data problem a MTP can be, for example, the one which sums over all simple matching similarities s_{mn} to the remaining partitions becomes a maximum. Taking into account the random permutation of cluster numbering the simple matching coefficient for two partitions P_m and P_n is defined as

$$s_{mn} = \text{Max}\left((a+d)/I, (b+c)/I\right) \quad , \qquad (6)$$

where (a + d) count the number of matches and (c+d) count the number of mismatches of all I observations, respectively.

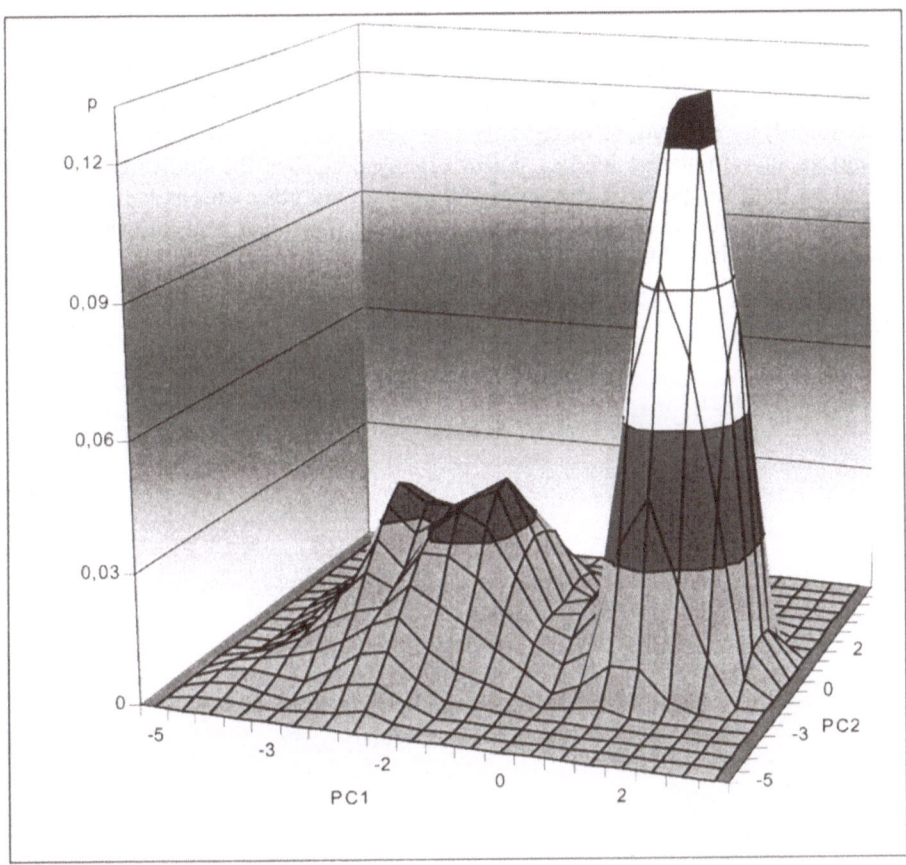

Figure 3: Bivariate nonparametric density estimation (Epanechnikow kernel) of the data set in Figure 2

3. Model-Based Gaussian Clustering

Well-known clustering algorithms have been derived from multivariate Gaussian models by Bock (1974), Jain and Dubes (1988), Mucha (1992), and Banfield and Raftery (1993). Here five methods are selected for a comparison with dual scaling clustering. Let a sample of I observations (matrix $\mathbf{X}=(x_{ij})$) be given. When the covariance matrix is constrained to be diagonal and uniform across all groups, the sum-of-squares criterion (left hand side of (7)) has to be minimized. The following formulation of this criterion is equivalent. At the right hand side it is formulated without using representatives (cluster centers):

$$tr(\sum_{k=1}^{K} W_k) = \sum_{k=1}^{K} \frac{1}{n_k} \sum_{r_i \in C_k} \sum_{\substack{r_j \in C_k \\ j > i}} d_{ij} \quad , \tag{7}$$

in which

$$W_k = \sum_{x_i \in C_k} (x_i - \bar{x}_k)(x_i - \bar{x}_k)' \tag{8}$$

is the sample cross-product matrix for the kth cluster and d_{ij} are the pairwise squared Euclidean distances. The well-known K-$Means$ method tries to minimize criterion (7) by exchanging observations between clusters. Here it starts with seed points in the simulation studies. Späth (1985) describes the equivalent method based on both the pairwise distances and exchanging observations between clusters. Here it is named $DistExch$. On the other side $Ward$ is a well-known agglomerative hierarchical clustering method minimizing criterion (7).

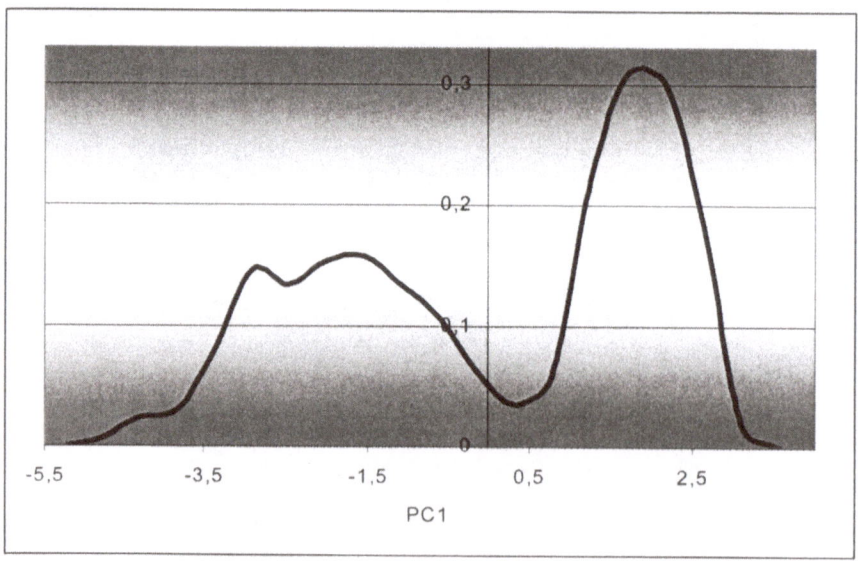

Figure 4: Univariate nonparametric density estimation (Epanechnikow kernel) of the first principal component of Figure 2

When the covariance matrix of each cluster is constrained to be diagonal, but otherwise allowed to vary between groups like in the example above, the logarithmic sum-of-squares criterion is minimized. Once more the following equivalent formulation holds:

$$\sum_{k=1}^{K} n_k \log tr({W_k}/{n_k}) = \sum_{k=1}^{K} n_k \log (\sum_{x_i \in C_k} \sum_{\substack{x_j \in C_k \\ j>i}} \frac{1}{n_k^2} d_{ij}) \quad , \qquad (9)$$

where n_k denotes the number of observation of the kth cluster. According to (9) the agglomerative hierarchical method proposed by Fraley (1996) is named here *Vary·Ward*. On the other hand a further distance based method tries to minimize criterion (9) by exchanging observations. Later on this last of the five comparative clustering methods considered here is named *VDistExch*.

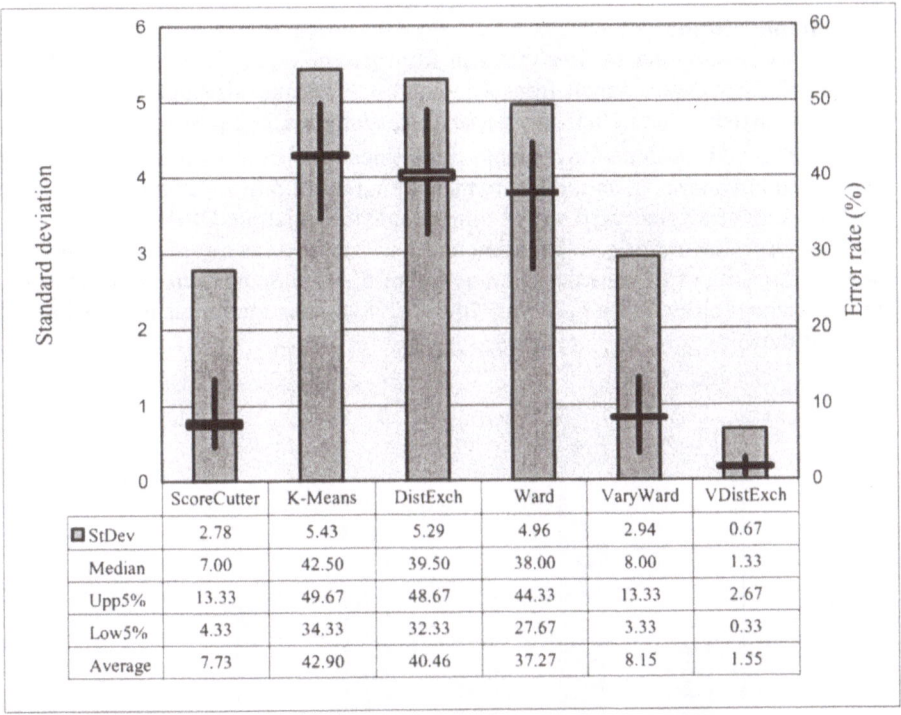

	ScoreCutter	K-Means	DistExch	Ward	VaryWard	VDistExch
■ StDev	2.78	5.43	5.29	4.96	2.94	0.67
Median	7.00	42.50	39.50	38.00	8.00	1.33
Upp5%	13.33	49.67	48.67	44.33	13.33	2.67
Low5%	4.33	34.33	32.33	27.67	3.33	0.33
Average	7.73	42.90	40.46	37.27	8.15	1.55

Figure 5: Univariate statistics of the misclassification errors of six methods for clustering the "ringnorm" data sets

4. Simulation Studies

Two examples are studied here. One of them is already described above. Generally there are 200 artificially generated Gaussian samples of size $I=300$ with equal class probabilities drawn. They are analysed in a parallel fashion by the six methods *ScoreCutter*, *K-Means*, *DistExch*, *Ward*, *VaryWard*, and *VDistExch*. Dual scaling clustering (*ScoreCutter*) is based on downgraded data values as described above. It has been applied here with 10 partitions used in the voting procedure.

4.1 Example "ringnorm"

Figure 5 shows both the most important numerical results concerning the mis-classification rate in percents as well as a corresponding graphical representation. The axis at the left hand side and the bars are assigned to the standard deviation ("StDev"), whereas the axis at the right hand side and the box-whisker-plots are linked to all other statistics. One can see that *ScoreCutter* is a stable method that outperforms all other clustering techniques except *VDistExch*. This partitioning method as well as the hierarchical method *VaryWard* are the correct and suitable model-based algorithms for this type of data.

4.2 Example "twonorm"

This two class data set is also taken from Breiman (1996), but it is slightly changed. Each class is drawn from a multivariate normal distribution with unit covariance matrix. Class 1 has mean $(a,a,...,a)$ and class 2 has mean $(-a,-a,...,-a)$ with $a=(2/J)^{1/2}$. In order to investigate the influence of outliers modifications of the original "twonorm" data model are made. First of all one out of the 300 observations is randomly generated with 4 times standard deviation. Often this observation lies far from the origin, but sometimes it may be not an outlier. Figure 6 shows the results in the same fashion as figure 5. Here *ScoreCutter* outperforms the hierarchical clustering algorithm *Ward*. The *K-Means* method is a quite instable algorithm.

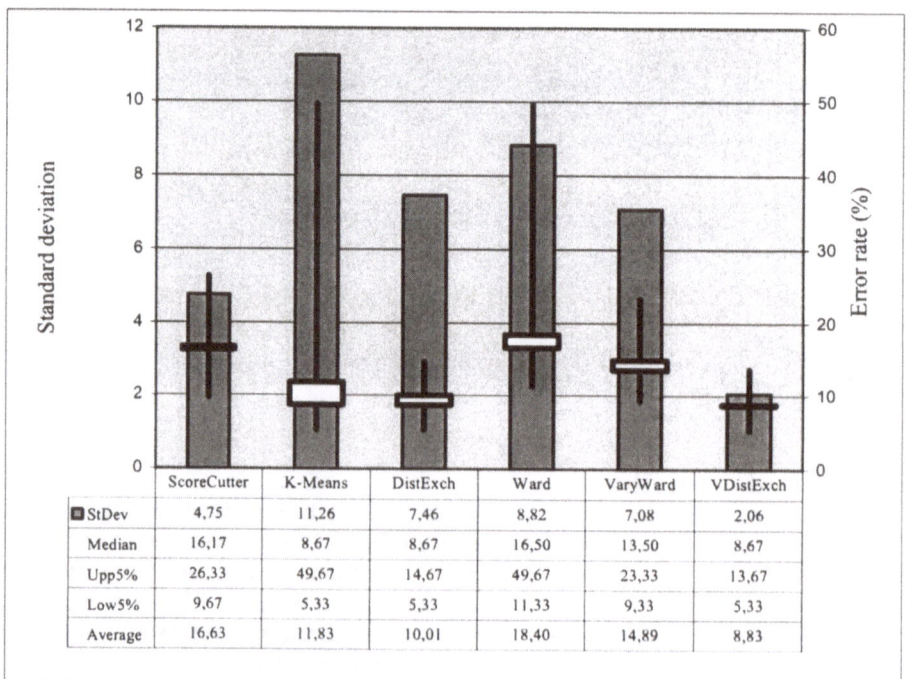

	ScoreCutter	K-Means	DistExch	Ward	VaryWard	VDistExch
▣ StDev	4,75	11,26	7,46	8,82	7,08	2,06
Median	16,17	8,67	8,67	16,50	13,50	8,67
Upp5%	26,33	49,67	14,67	49,67	23,33	13,67
Low5%	9,67	5,33	5,33	11,33	9,33	5,33
Average	16,63	11,83	10,01	18,40	14,89	8,83

Figure 6: Univariate statistics of the misclassification errors of six methods for clustering the "twonorm" data sets (one outlier)

Secondly, one out of the 150 observation of each class is randomly generated with 4 times standard deviation. As a consequence there is a high probability that the data contain at least one outlier. Figure 7 shows the results in the same fashion as figure 5. Here *ScoreCutter* is beside *DistExch* the most stable method. Once more it outperforms the hierarchical clustering method *Ward*. The K-Means method is the most instable algorithm.

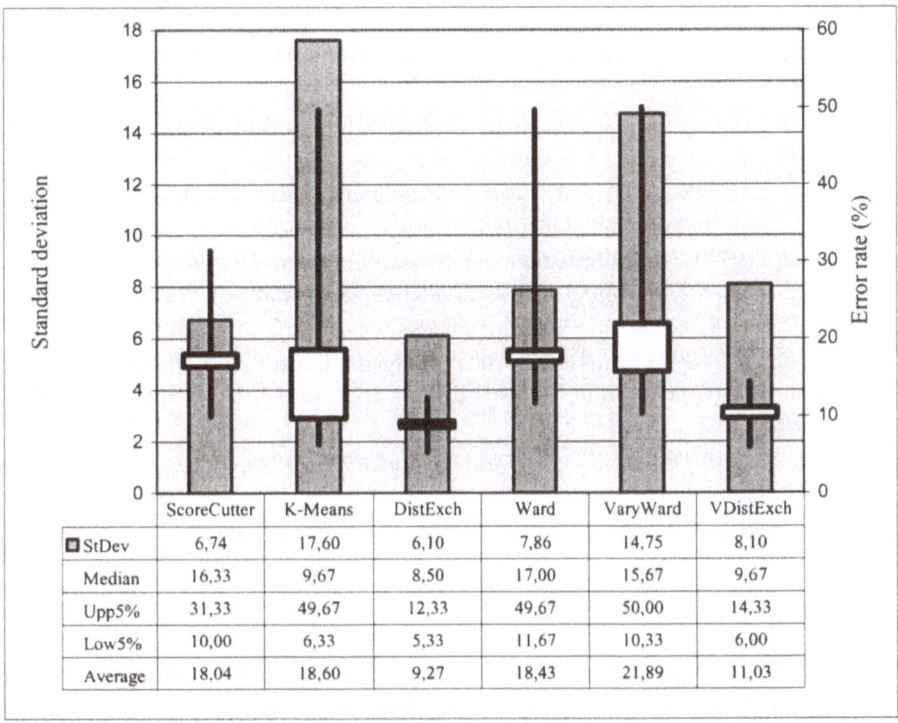

	ScoreCutter	K-Means	DistExch	Ward	VaryWard	VDistExch
☐ StDev	6,74	17,60	6,10	7,86	14,75	8,10
Median	16,33	9,67	8,50	17,00	15,67	9,67
Upp5%	31,33	49,67	12,33	49,67	50,00	14,33
Low5%	10,00	6,33	5,33	11,67	10,33	6,00
Average	18,04	18,60	9,27	18,43	21,89	11,03

Figure 7: Univariate statistics of the misclassification errors of six methods for clustering the "twonorm" data sets (two outliers)

5. Conclusion

The idea of dual scaling clustering is presented here in a quite simple fashion. *ScoreCutter* is a computationally efficient method because it iterates fast. In the simulation studies 9 iterations are needed in average. In a data-preprocessing step (discretization, binning) of *ScoreCutter*, outliers lay aside almost all of their influence on clustering. Missing values can be taken into consideration by using them as an additional category.

ScoreCutter can be improved by, for example,

- intelligent discretization methods,

- using rank data instead of categories as a result of discretization or binning,
- using more than 10 partitions in the voting procedure.

Unfortunately, the proposed method cannot be expanded in an easy manner to the case of more than two clusters.

References:

Banfield, J. D. and Raftery, A.E. (1993). Model-based Gaussian and non-Gaussian clustering. *Biometrics*, **49**, 803-821.

Bock, H. H. (1974). *Automatische Klassifikation*. Vandenhoeck & Ruprecht, Göttingen.

Breiman, L. (1996). Bias, variance, and arcing classifiers. *Technical Report 460*. Statistical Department, University of California, Berkeley.

Fraley C. (1996). Algorithms for model-based Gaussian Hierarchical Clustering. *Technical Report 311*. Department of Statistics, University of Washington, Seattle.

Hand, D. J. (1997). Intelligent data analysis: issues and opportunities. *In: Advances in Intelligent Data Analysis*. Liu, X. et al. (eds.), 1-14, Springer-Verlag, Heidelberg.

Jain, A. K. and Dubes, R. C. (1988). *Algorithms for clustering data*. Prentice Hall, New Jersey.

Kaufman, L. and Rousseeuw, P. J. (1990). *Finding groups in data*. Wiley, New York.

Mucha, H.-J. (1992). *Clusteranalyse mit Mikrocomputern*. Akademie Verlag, Berlin.

Mucha, H.-J., Siegmund-Schultze, R., and Dübon, K. (1998). Adaptive cluster analysis techniques - software and applications. *In: Data Science, Classification and Related Methods*, Hayashi, C. et al. (eds.), 231-238, Springer-Verlag Tokyo.

Nishisato, S. (1980). *Analysis of Categorical Data: Dual Scaling and its Applications*. University of Toronto Press, Toronto.

Nishisato, S. (1994). *Elements of Dual Scaling: An Introduction to Practical Data Analysis*. Lawrence Erlbaum Associates, Publishers, Hillsdale.

Späth, H. (1985). *Cluster dissection and analysis*. Ellis Horwood, Chichester.

Homogeneity and Smoothness Analysis for Quantifying a Longitudinal Categorical Variable

Kohei Adachi

Koshien University
10-1 Momijigaoka, Takarazuka, Hyogo 665-0006, Japan

Summary: A variant of homogeneity analysis is proposed to analyze longitudinal data that describe the categories chosen by individuals at each of time-points. In the proposed method, individuals are given the scores that are natural cubic spline functions of time, and categories are given time-invariant scores. Loss of homogeneity between individual and category scores is combined with loss of smoothness of individual scores, to form a penalized loss function. It is minimized using eigenvalue or singular value decomposition. The weight of smoothness relative to homogeneity is chosen by a cross-validation procedure. The resulting scores yield a graphical representation of individuals' trajectories with category points, which allows us to grasp longitudinal changes in individuals.

1. Introduction

Homogeneity analysis is a useful method for quantifying and graphically representing multivariate categorical data (Gifi, 1990; Meulman, Heiser, and SPSS, 1999). It is known that the resulting solutions are essentially equivalent to those of correspondence analysis (Greenacre, 1984), dual scaling (Nishisato, 1980) and the third type of quantification method (Hayashi, 1952).

Categorical data to be quantified are often observed longitudinally over time-points (Bijleveld and van der Burg, 1998). This paper focuses on the longitudinal data summarized into an $nq \times m$ indicator matrix $\mathbf{G} = [\mathbf{G}_1', \cdots, \mathbf{G}_n']'$. Here, $\mathbf{G}_i$ ($i = 1, \cdots, n$) is the q time-points by m categories matrix, whose (j, k)th element g_{ijk} indicates the category chosen by individual i at time-point j. That is, $g_{ijk} = 1$ if $k = K(i, j)$ and $g_{ijk} = 0$ otherwise, where $K(i, j)$ denotes the category chosen by i at j. Note $\mathbf{G}\mathbf{1}_m = \mathbf{1}_N$ with $N = nq$ and $\mathbf{1}_N$ the N-dimensional vector of ones. We consider that such data are collected with the purpose of finding longitudinal changes in individuals. This purpose would be achieved by quantifying $\mathbf{G}$, i.e., by assigning scores to its row and column entities. It is because each row corresponds to an individual at a time-point and thus the scores of rows express individuals' changes. A graphical representation of row scores with column ones would allow us to grasp the changes of the categories chosen by individuals. We consider a method of homogeneity analysis for performing such quantification.

Let $\mathbf{Y}$ ($m \times p$) and $\mathbf{X}_i$ ($q \times p$) contain the p-dimensional score vectors for m columns and q rows of $\mathbf{G}_i$, respectively. The (k, l)th element of $\mathbf{Y} = (y_{kl})$ expresses the score of category k on dimension l and the (j,l)th one of $\mathbf{X}_i = (x_{ijl})$ denotes the score on l for individual i at time-point j. Further, the (j,l)th element of $\mathbf{Y}_i^* = \mathbf{G}_i\mathbf{Y}$ indicates $y_{K(i,j),l}$, i.e., the score of category $K(i.j)$ on l. It is reasonably assumed that the scores for an individual should be similar (or homogeneous) to the scores of the categories chosen by that individual (Gifi, 1990). The departure from this assumption, which is called loss of homogeneity (LH), can be expressed as the sum of squares of $x_{ijl} - y_{K(i.j),l}$ over i, j and l;

$$LH(\mathbf{X},\mathbf{Y}) = \sum_{i=1}^{n} SSQ(\mathbf{X}_i - \mathbf{G}_i\mathbf{Y}) = SSQ(\mathbf{X} - \mathbf{GY}), \qquad (1)$$

where $\mathbf{X} = [\mathbf{X}_1', \cdots, \mathbf{X}_n']'$ and $SSQ(\mathbf{E}) = \mathrm{tr}\,\mathbf{E}'\mathbf{E}$ for any matrix or vector $\mathbf{E}$. However, $\mathbf{G}$ is univariate, and thus the $\mathbf{X}$ and $\mathbf{Y}$ minimizing (1) cannot be chosen uniquely, even if a normalization condition is imposed on $\mathbf{X}$ or $\mathbf{Y}$. For example, under a condition on $\mathbf{Y}$, we have an infinite number of trivial solutions, i.e., any $\mathbf{Y}$ meeting the condition and the $\mathbf{X}$ satisfying $\mathbf{X} = \mathbf{GY}$.

An approach for overcoming this difficulty is to introduce a penalty function, i.e., another loss function, besides the original loss (1). A penalty function is combined with the original one to form a new objective function, so that parameters to be obtained are constrained and uniquely chosen. Such approaches have been taken in some methods of data analysis (Hastie, Buja, and Tibshirani, 1995; Ramsay and Silverman, 1997). To introduce a penalty function into our problem, we assume that individual scores x_{ijl} change smoothly over time-point j. This assumption leads to loss of smoothness $LS(\mathbf{X})$. We propose to define an objective function, which is called loss of homogeneity and smoothness, as

$$LHS(\mathbf{X},\mathbf{Y}) = LH(\mathbf{X},\mathbf{Y}) + \alpha LS(\mathbf{X}), \qquad (2)$$

with $\alpha > 0$. This method is detailed in the following sections.

2. Method

2.1 Loss of Homogeneity and Smoothness

We treat individual scores as functions of continuous time t, and use $x_{il}(t)$ for a time series of the score of individual i on dimension l. Further we express the time for the j-th time-point as t_j. These expressions are used together with those in section 1. That is, the $(j.l)$ element of $\mathbf{X}_i$, i.e., x_{ijl}, equals the $x_{il}(t)$ at $t = t_j$

We require $x_{il}(t)$ to change smoothly with t. The roughness of $x_{il}(t)$, i.e., its departure from smoothness, can be expressed as $r_{il} = \int_{\beta}^{\gamma} x_{il}''^2(t)\, dt$ with $\beta < t_1$ and $\gamma > t_q$, where $x_{il}''(t)$ denotes the second derivative of $x_{il}(t)$. Roughness r_{il} takes a lower value when $x_{il}(t)$ is a natural cubic spline function of t with knots $t_1 < t_2 < \cdots < t_q$, than when $x_{il}(t)$ is any other function (Schoenberg, 1964; Green and Silverman, 1994). We thereby let $x_{il}(t)$ be that spline, which satisfies the following conditions: $x_{il}(t)$ is a cubic polynomial on each of the intervals (t_1, t_2), (t_2, t_3), $\cdots$, (t_{q-1}, t_q), it is linear outside (t_1, t_q), and $x_{il}(t)$, its derivative $x_{il}'(t)$ and $x_{il}''(t)$ are continuous at each t_j.

Although the use of splines for smoothing scores has been suggested by de Leeuw, van der Heijden, and Kreft (1985), they have not presented a concrete method.

Using the properties of natural cubic splines (Green and Silverman, 1994), the roughness is rewritten in the quadratic form

$$r_{il} = \tilde{\mathbf{x}}_{il}' \mathbf{A} \tilde{\mathbf{x}}_{il} , \tag{3}$$

where $\tilde{\mathbf{x}}_{il}$ is the l-th column of $\mathbf{X}_i$ and $\mathbf{A} = \mathbf{C} \mathbf{S}^{-1} \mathbf{C}'$ with $\mathbf{C}$ and $\mathbf{S}$ being $q \times (q-2)$ and $(q-2) \times (q-2)$ band matrices, respectively. Using $a_j = t_{j+1} - t_j$, the non-zero elements of $\mathbf{C} = (c_{ju})$ are defined as $c_{ij} = a_j^{-1}$, $c_{j+1,j} = -a_j^{-1} - a_{j+1}^{-1}$ and $c_{j+2,j} = a_{j+1}^{-1}$ for $j = 1, \cdots,$ $q-2$. The non-zero elements of $\mathbf{S} = (s_{ju})$ are defined as $s_{jj} = 3^{-1}(a_j + a_{j+1})$ for $j = 1, \cdots,$ $q-2$ and $s_{j-1,j} = s_{j,j-1} = 6^{-1} a_j$ for $j = 2, \cdots, q-2$. We can find $\mathbf{A} \mathbf{1}_q = \mathbf{0}$ with $\mathbf{0}$ the vector of zeros. Note that r_{jl} is expressed using x_{ijl}, not $x_{il}(t)$, in (3).

We use the sum of r_{il} for the loss of smoothness; $\mathrm{LS}(\mathbf{X}) = \sum_{i=1}^{n} \sum_{l=1}^{p} r_{il}$ $= \mathrm{tr}\, \mathbf{X}'(\mathbf{I}_n \otimes \mathbf{A})\mathbf{X}$. It is combined with $\mathrm{LH}(\mathbf{X}, \mathbf{Y})$ to form a loss function

$$\mathrm{LHS}(\mathbf{X}, \mathbf{Y}) = \mathrm{SSQ}(\mathbf{X} - \mathbf{G}\mathbf{Y}) + \alpha\, \mathrm{tr}\, \mathbf{X}'(\mathbf{I}_n \otimes \mathbf{A})\mathbf{X}. \tag{4}$$

Here, category scores are normalized such that the average of scores $y_{K(i,j),l}$ over i and j is zero and their covariance matrix is the $p \times p$ identity matrix $\mathbf{I}_p$:

$$\mathbf{1}_N' \mathbf{G}\mathbf{Y} = \mathbf{0}', \quad \text{equivalently,} \quad \mathbf{J}\mathbf{G}\mathbf{Y} = \mathbf{G}\mathbf{Y} \tag{5}$$

with $\mathbf{J} = \mathbf{I}_N - N^{-1} \mathbf{1}_N \mathbf{1}_N'$ (recall $N = nq$), and

$$\mathbf{Y}'\mathbf{D}\mathbf{Y} = N \mathbf{I}_p , \tag{6}$$

where $\mathbf{D} = \mathbf{G}'\mathbf{G}$ is the diagonal matrix having category frequencies $g_{\cdot \cdot k} = \sum_i \sum_j g_{ijk}$ in its diagonals, which are supposed to be positive. No explicit constraints are imposed on $\mathbf{X}$.

2.2 Homogeneity and Smoothness Analysis

We consider minimizing (4) over $\mathbf{X}$ and $\mathbf{Y}$ under normalization conditions (5) and (6), for a given $\alpha > 0$. Using (6) and $\mathbf{B} = \mathbf{I}_n \otimes \tilde{\mathbf{B}}$ with $\tilde{\mathbf{B}} = \mathbf{I}_q + \alpha \mathbf{A}$, (4) is rewritten as

$$\mathrm{LHS}(\mathbf{X}, \mathbf{Y}) = \mathrm{tr}\, \mathbf{X}'\mathbf{B}\mathbf{X} - 2\,\mathrm{tr}\, \mathbf{X}'\mathbf{G}\mathbf{Y} + Np$$
$$= \mathrm{tr}(\mathbf{B}\mathbf{X} - \mathbf{G}\mathbf{Y})'\mathbf{B}^{-1}(\mathbf{B}\mathbf{X} - \mathbf{G}\mathbf{Y}) - \mathrm{tr}\, \mathbf{Y}'\mathbf{G}'\mathbf{B}^{-1}\mathbf{G}\mathbf{Y} + Np . \tag{7}$$

Note that $\tilde{\mathbf{B}}$ and $\mathbf{B}$ are nonsingular, which follows from the nonnegative-definiteness of $\mathbf{A}$ and $\alpha > 0$. From (7), the optimal $\mathbf{X}$ is found to satisfy

$$\mathbf{X} = \mathbf{B}^{-1}\mathbf{G}\mathbf{Y} = (\mathbf{I}_n \otimes \tilde{\mathbf{B}}^{-1})\mathbf{G}\mathbf{Y} \tag{8}$$

where $\mathbf{B}^{-1} = \mathbf{I}_n \otimes \tilde{\mathbf{B}}^{-1}$ is used. Substituting (8) into (7) leads to

$$\text{LHS}(*, \mathbf{Y}) = Np - \text{tr } \mathbf{Y}'\mathbf{MY} \tag{9}$$

with $\mathbf{M} = \mathbf{G}'\mathbf{B}^{-1}\mathbf{G}$. The remaining problem is thus to maximize the quadratic form $F(\mathbf{Y}) = \text{tr } \mathbf{Y}'\mathbf{MY}$ over $\mathbf{Y}$ subject to (5) and (6). Considering (5), we have $F(\mathbf{Y}) = \text{tr } \mathbf{Y}'\tilde{\mathbf{M}}\mathbf{Y}$ with $\tilde{\mathbf{M}} = \mathbf{G}'\mathbf{JB}^{-1}\mathbf{JG}$. The optimal $\mathbf{Y}$ is obtained with the decomposition $\mathbf{D}^{-1/2}\tilde{\mathbf{M}}\mathbf{D}^{-1/2} = \tilde{\mathbf{W}}\tilde{\boldsymbol{\Delta}}^2\tilde{\mathbf{W}}'$, where $\tilde{\boldsymbol{\Delta}}^2$ is the diagonal matrix of order R with its l-th diagonal element the l-th largest eigenvalue of $\mathbf{D}^{-1/2}\tilde{\mathbf{M}}\mathbf{D}^{-1/2}$ and R its rank, and the columns of $\tilde{\mathbf{W}}$ are the normalized eigenvectors corresponding to $\tilde{\boldsymbol{\Delta}}^2$. The upper bound of $F(\mathbf{Y})$ is attained for

$$\mathbf{Y} = \sqrt{N}\mathbf{D}^{-\frac{1}{2}}\tilde{\mathbf{W}}_p, \tag{10}$$

where $\tilde{\mathbf{W}}_p$ is the $m \times p$ matrix whose columns are the first p ones of $\tilde{\mathbf{W}}$. It is easily proven that the $\mathbf{Y}$ given by (10) satisfies (5) and (6).

This solution of $\mathbf{Y}$ is also obtained using $\mathbf{M}$ instead of $\tilde{\mathbf{M}}$: $\tilde{\mathbf{W}}_p$ is given by the eigenvectors of $\mathbf{D}^{-1/2}\mathbf{MD}^{-1/2}$ related to its 2nd to $(p+1)$st largest eigenvalues, which is proven as follows. Noting that $\mathbf{A1}_q = \mathbf{0}$ leads to $\tilde{\mathbf{B}}\mathbf{1}_q = \mathbf{1}_q$ and $\tilde{\mathbf{B}}^{-1}\mathbf{1}_q = \mathbf{1}_q$, we find that $\mathbf{B}^{-1}\mathbf{1}_N = \mathbf{1}_N$ and thus $\mathbf{JB}^{-1}\mathbf{J} = \mathbf{B}^{-1} - N^{-1}\mathbf{1}_N\mathbf{1}_N'$. Using this, we have

$$\mathbf{D}^{-\frac{1}{2}}\mathbf{MD}^{-\frac{1}{2}} = \mathbf{D}^{-\frac{1}{2}}\tilde{\mathbf{M}}\mathbf{D}^{-\frac{1}{2}} + \mathbf{dd}' = \mathbf{W}\boldsymbol{\Delta}^2\mathbf{W}', \tag{11}$$

where $\mathbf{d} = N^{-1/2}\mathbf{D}^{-1/2}\mathbf{G}'\mathbf{1}_N = N^{-1/2}\mathbf{D}^{1/2}\mathbf{1}_m$ with $\mathbf{d}'\mathbf{d} = 1$, $\mathbf{W} = [\mathbf{d}, \tilde{\mathbf{W}}]$, and $\boldsymbol{\Delta}^2$ is the diagonal matrix with its first diagonal element 1 and the remaining diagonals equaling those of $\tilde{\boldsymbol{\Delta}}^2$. Next, we find that $\mathbf{d}'\tilde{\mathbf{W}} = \mathbf{0}'$ and (11) thus expresses the eigenvalue decomposition, by noting $\tilde{\mathbf{W}} = \mathbf{D}^{-1/2}\tilde{\mathbf{M}}\mathbf{D}^{-1/2}\tilde{\mathbf{W}}\boldsymbol{\Delta}^{-2}$ and $\mathbf{d}'\mathbf{D}^{-1/2}\mathbf{G}'\mathbf{J} = N^{-1/2}\mathbf{1}_N'\mathbf{J} = \mathbf{0}'$. Further, setting p at 1 and $\mathbf{Y}$ at $\mathbf{y}^* = N^{1/2}\mathbf{D}^{-1/2}\mathbf{w}$ in (9) with $\mathbf{w}$ a column of $\mathbf{W}$, we have $\text{LHS}(*, \mathbf{y}^*) = N(1-\lambda) \geq 0$, i.e., $\lambda \leq 1$, with λ the eigenvalue corresponding to $\mathbf{w}$. It implies that the largest eigenvalue of $\mathbf{D}^{-1/2}\mathbf{MD}^{-1/2}$ is 1 and the corresponding eigenvector is $\mathbf{d}$. This completes the proof.

As described above, solutions are obtained explicitly with the eigenvalue decomposition in (11). That is, (10) gives $\mathbf{Y}$ and the substitution of this $\mathbf{Y}$ into (8) gives $\mathbf{X}$. Note the following two properties of solutions. Firstly, they are nested. It means that the first p^* ($\leq p$) columns of the optimal $\mathbf{X}$ and $\mathbf{Y}$ are also the solution of dimensionality p^*. Secondly, solutions have rotational indeterminacy: (5) and (6) are satisfied and (4) remains unchanged, even if $\mathbf{X}$ and $\mathbf{Y}$ are replaced by $\mathbf{XT}$ and $\mathbf{YT}$, with $\mathbf{T}$ an arbitrary orthogonal matrix satisfying $\mathbf{T}'\mathbf{T} = \mathbf{TT}' = \mathbf{I}_p$. However, the rotation is not considered in this paper.

2.3 SVD Formulation of the Method
The solution is also obtained with SVD (singular value decomposition)

(Greenacre, 1984). From (11), we have $\mathbf{B}^{-1/2}\mathbf{GD}^{-1/2} = \mathbf{V}\Delta\mathbf{W}'$ with $\mathbf{V}'\mathbf{V} = \mathbf{W}'\mathbf{W} = \mathbf{I}_R$. These equations lead to the generalized SVD

$$\mathbf{B}^{-1}\mathbf{GD}^{-1} = \mathbf{K}\Delta\mathbf{L}' \tag{12}$$

with $\mathbf{K} = \mathbf{B}^{-1/2}\mathbf{V}$, $\mathbf{L} = \mathbf{D}^{-1/2}\mathbf{W}$ and $\mathbf{K}'\mathbf{BK} = \mathbf{L}'\mathbf{DL} = \mathbf{I}_R$. Comparing (12) with (8) and (10), we can find that the optimal $\mathbf{X}$ and $\mathbf{Y}$ are given by the 2nd to $(p+1)$st columns of $N^{1/2}\mathbf{K}\Delta$ and of $N^{1/2}\mathbf{L}$, respectively.

The matrix $\mathbf{H} = \mathbf{B}^{-1}\mathbf{GD}^{-1}$ decomposed as (12) is rewritten in the form

$$\mathbf{H} = (\mathbf{I}_n \otimes \tilde{\mathbf{B}}^{-1})[\mathbf{G}_1',\cdots,\mathbf{G}_n']'\mathbf{D}^{-1} = [\mathbf{H}_1',\cdots,\mathbf{H}_n']' \tag{13}$$

with $\mathbf{H}_i = \tilde{\mathbf{B}}^{-1}\mathbf{G}_i\mathbf{D}^{-1}$. Here, the (j,k)th element of $\mathbf{H}_i$ is $h_{ijk} = g_{i+k}^{-1}\sum_{u=1}^{q}b_{ju}g_{iuk}$, with b_{ju} the (j,u)th element of $\tilde{\mathbf{B}}^{-1}$. Recall $\tilde{\mathbf{B}}^{-1}\mathbf{1}_q = \mathbf{1}_q$, i.e., $\sum_{u=1}^{q}b_{ju} = 1$. These show that the matrix $\mathbf{H}$, whose generalized SVD provides the solution, consists of the weighted averages of the data divided by category frequencies. For example, if $q = 4$, $m = 3$, $\alpha = 1$ and $\mathbf{D} = e\mathbf{I}_3$ with some positive constant e,

$$\mathbf{H}_i = \begin{bmatrix} 0.776 & 0.353 & -0.129 \\ 0.319 & 0.647 & 0.034 \\ 0.034 & 0.647 & 0.319 \\ -0.129 & 0.353 & 0.776 \end{bmatrix} \times \frac{1}{e} \quad \text{for } \mathbf{G}_i = \begin{bmatrix} 1 & 0 & 0 \\ 0 & 1 & 0 \\ 0 & 1 & 0 \\ 0 & 0 & 1 \end{bmatrix},$$

$$\text{with } \mathbf{H}_i'\mathbf{H}_i = \begin{bmatrix} 0.722 & 0.457 & -0.179 \\ 0.457 & 1.086 & 0.457 \\ -0.179 & 0.457 & 0.722 \end{bmatrix} \times \frac{1}{e^2}.$$

It illustrates that the columns of the original data matrix $\mathbf{G}$ are independent of each other and do not show inter-category correlation, but that those of $\mathbf{H}$ have dependence which is brought by weight b_{ju}.

2.4 Graphical Representation Using Splines

The resulting $\mathbf{X}$ and $\mathbf{Y}$ provide a p-dimensional configuration of m categories and n individuals. In this configuration, category k is expressed as a point whose coordinates are given by the k-th row y_k' of the optimal $\mathbf{Y}$, and individual i is represented as a trajectory, i.e., as a vector time series $\mathbf{x}_i(t)=[x_{i1}(t),\cdots,x_{ip}(t)]'$. This series is given by substituting the optimal $\mathbf{X}=(x_{ijl})$ into the interpolation and extrapolation formula of natural cubic splines (Green and Silverman, 1994).

Using z_{ijl} for the j-th element of $\mathbf{z}_{il} = \mathbf{S}^{-1}\mathbf{C}'\tilde{\mathbf{x}}_{il}$, this formula is written as

$$x_{il}(t) = \begin{cases} x_{i1l} - (t_1 - t)\{a_1^{-1}(x_{i2l} - x_{i1l}) - 6^{-1}a_1 z_{i1l}\} & \text{for } t < t_1 \\ a_j^{-1}\{(t - t_j)x_{i,j+1,l} + (t_{j+1} - t)x_{ijl}\} - \eta_{ijl}(t) & \text{for } t_1 \le t < t_q \\ x_{iql} + (t - t_q)\{a_{q-1}^{-1}(x_{iql} - x_{i,q-1,l}) + 6^{-1}a_{q-1}z_{i,q-2,l}\} & \text{for } t \ge t_q \end{cases} \tag{14}$$

with $\eta_{yl}(t) = (6a_j)^{-1}(t-t_j)(t_{j-1}-t)\{z_{yl}(a_j+t-t_j)+z_{i,j-1,l}(a_j+t_{j+1}-t)\}$. A joint representation of $\mathbf{x}_i(t)$ and $\mathbf{y}_k$ in a low-dimensional space would allow us to visually interpret the outcomes of analysis.

2.5 Cross-Validated Selection of Penalty Weight

A correct classification rate (CCR) can be used as an index of the quality of a solution. The CCR is defined as the proportion of the cases where $\mathbf{y}_{K(i,j)}$ (the score vector of the category chosen by i at j) is closest to $\mathbf{x}_{ij} = [x_{ij1}, \cdots, x_{ijp}]'$ (the corresponding vector of i) among all category vectors, i.e., where $\mathbf{x}_{ij}$ is correctly classified into the category. It is written as

$$\text{CCR} = \frac{1}{N}\sum_{i=1}^{n}\sum_{j=1}^{q} o_{ij} \tag{15}$$

with

$$o_{ij} = \begin{cases} 1 & \textit{iff} \quad \text{SSQ}(\mathbf{x}_{ij} - \mathbf{y}_{K(i,j)}) = \min_{1 \le k \le m} \text{SSQ}(\mathbf{x}_{ij} - \mathbf{y}_k) \\ 0 & \text{otherwise} \end{cases} \tag{16}$$

The similarity between $\mathbf{x}_{ij}$ and $\mathbf{y}_{K(i,j)}$ is assumed in the homogeneity analysis, and CCR expresses the congruence of a solution to this assumption.

In the proposed method, the weight α for the penalty function must be chosen in advance, and CCR may be useful to that choice. However, it cannot be used straightforwardly, because the upper limit of CCR ($= 1$) is attained for $\alpha = 0$, which leads to trivial solutions as already stated. In order to handle this difficulty, we calculate CCR with the following cross-validation procedure.

For a given α, the proposed method is applied to the data set with the observations at time-point j deleted, $\mathbf{G}^{(j)} = [\mathbf{G}_1^{(j)}{}', \cdots, \mathbf{G}_n^{(j)}{}']'$, where $\mathbf{G}_i^{(j)}$ is the $(q-1) \times m$ matrix given by deleting the j-th row from $\mathbf{G}_i$. That is, we minimize

$$\text{LHS}(\mathbf{X}^{(j)}, \mathbf{Y}) = \text{SSQ}(\mathbf{X}^{(j)} - \mathbf{G}^{(j)}\mathbf{Y}) + \alpha \, \text{tr} \, \mathbf{X}^{(j)}{}'(\mathbf{I}_n \otimes \mathbf{A}^{(j)})\mathbf{X}^{(j)} \tag{17}$$

over $\mathbf{X}^{(j)}$ and $\mathbf{Y}$, under $\mathbf{1}'_{n(q-1)}\mathbf{G}^{(j)}\mathbf{Y} = \mathbf{0}'$ and $\mathbf{Y}'\mathbf{D}^{(j)}\mathbf{Y} = \mathbf{I}_p$, where $\mathbf{X}^{(j)}$ denotes $\mathbf{X}$ with the scores at j deleted, $\mathbf{A}^{(j)}$ is obtained by accommodating $\mathbf{A}$ to the deletion, and $\mathbf{D}^{(j)} = \mathbf{G}^{(j)}{}'\mathbf{G}^{(j)}$. Subsequently, the optimal $\mathbf{X}^{(j)}$ is used in (14) to *predict* the scores $\mathbf{x}_{ij}$ or the $\mathbf{x}_i(t)$ at $t=t_j$ associated with the deleted observations, and the resulting $\mathbf{x}_{ij}$ and $\mathbf{Y}$ provide o_{ij} with (16). These procedures are performed over $j=1,\cdots,q$ to yield a cross-validated CCR with (15). The CCR's are calculated for different values of α and the value yielding the highest CCR is chosen.

This method is expected to provide a reasonably large $\alpha > 0$ for the following reason. Loss of homogeneity $\text{SSQ}(\mathbf{X}^{(j)} - \mathbf{G}^{(j)}\mathbf{Y})$ is defined using data $\mathbf{G}^{(j)}$, while loss of smoothness is not. A too large weight for the former loss, i.e., an excessively small α, would *overfit* scores to the data, which leads to worse fitting (*prediction*) for the observations deleted from the data.

3. Example

To illustrate the proposed method, we use longitudinal preference data on seven categories of soft drinks (Adachi, 2000a). Those data describe the category that each of 117 individuals (Japanese subjects) liked best at each of five time-points, $j=1$ (in their first year of elementary school), $j = 2$ (fourth year of elementary school), $j = 3$ (first year of junior high school), $j = 4$ (first year of high school), and $j = 5$ (freshman year at university). Since the five time-points are approximately equally spaced, we set $t_j = j$ for $j=1,\cdots,5$. Only the two-dimensional ($p = 2$) results are reported here.

Cross-validated CCR's were obtained for two sets of values of α, i.e., for $\{0.1, 0.2, \cdots, 0.9\}$ and $\{1, 2, \cdots, 9\}$. The CCR's for the latter set, which were higher than for the former, are shown in Figure 1. There, CCR is found to achieve the highest for $\alpha = 3$. Adopting this α, we analyzed the data.

The resulting solution for $\alpha = 3$, whose CCR (full-data based CCR) was 0.674, is shown in Figure 2, where a circle represents a category point and a trajectory represents an individual time-series $x_i(t)$ for $t_1 \leq t \leq t_5$ with the tip of an allow denoting $[x_{i51}, x_{i52}]'$ at the final time-point. There, a lot of trajectories start from the left area where the points of milk and juice are located, showing that those categories are preferred at the younger ages. From this area, many trajectories extend in the right direction. We can thus interpret that the first (horizontal) dimension stands for a main trend in the preference changes with individuals' growth and that the left on this dimension corresponds to the younger ages. The second (vertical) dimension seems to divide individuals into two groups. A group of trajectories extends toward the black and green tea points located in the upper side. In contrast, the other group

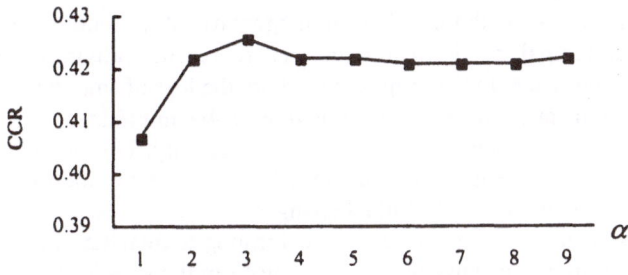

Figure 1: Cross-validataed CCR as a function of penalty weight α

extends toward the coffee point in the lower, by way of a cola point. We thus find that individuals are classified into *toward-tea* and *toward-coffee* clusters. Besides these two clusters, a group of trajectories remaining in the left area, though not major, is found by close inspection.

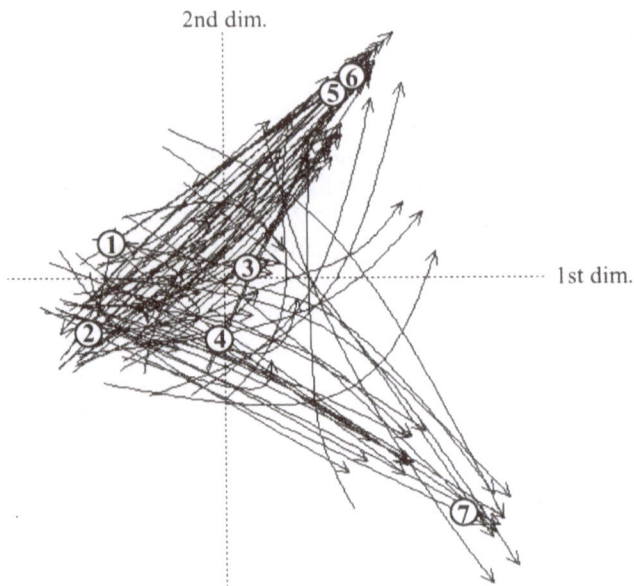

Figure 2: The configuration of individual trajectories and category points.
Numbers denote categories: 1 = milk, 2 = juice, 3 = refreshment
water, 4 = cola, 5 = black tea, 6 = green tea, and 7 = coffee.

4. Final Remarks

We proposed a method, called homogeneity and smoothness analysis, to quantify longitudinal categorical data and represent individual changes in a low-dimensional space. In the proposed method, the loss of smoothness is used for a penalty function, and the objective function to be minimized is defined as the weighted sum of the penalty and the loss of homogeneity. The weight is chosen with a cross validation procedure. As illustrated by the example, the method allows us easily to grasp the trends in individual changes.

Four approaches which have been made for longitudinal data analysis are related to our method, in that penalty functions are used in them. Adachi (2000b) and van Buuren (1990, pp. 64-71) have defined penalties using the differences of scores between time-points, which are similar to our approach. However, they have used first order differences treating individual scores as functions of discrete time-points, whereas we use the second order derivative of the scores treated as continuous time series. The other two are van Buuren's (1990, 1996) and Bijleveld and de Leeuw's (1991) approaches, in which penalty functions have been defined using dynamic time series models such as autoregression. These methods have been designed for the time series with large q, rather than for the data with small q such as that of $q = 5$

analyzed by ours in the example. Although those approaches differ from ours in some respects, it is interesting to compare them empirically and theoretically.

The cross-validation procedure in section 2.5 was to treat the observations at a time-point as missing. It suggests that our method is extended to handle the data with observations missing at some time-points for some individuals. Let $\mathbf{G}^{\#}$, $\mathbf{X}^{\#}$ and $\mathbf{A}^{\#}$ be defined by deleting the entries corresponding to missing values from $\mathbf{G}$, $\mathbf{X}$ and $\mathbf{A}$, respectively. Replacing the latter matrices by the former in (4), we can obtain $\mathbf{X}^{\#}$ and $\mathbf{Y}$. Further, using the resulting $\mathbf{X}^{\#}$ in the formula of natural cubic splines (14), we can interpolate or extrapolate the scores corresponding to the missing observations.

Another extension would be to generalize the method to deal with multiple-items data $\mathbf{G}^{\cdot} = [\mathbf{G}_{1}^{\cdot}, \cdots, \mathbf{G}_{u}^{\cdot}, \cdots, \mathbf{G}_{U}^{\cdot}]$, where each row of $\mathbf{G}_{u}^{\cdot}$ $(N \times K_{u})$ indicates the category of item u that an individual chose at a time-point. For this data set, loss of homogeneity is defined as the sum of $\mathrm{SSQ}(\mathbf{X} - \mathbf{G}_{u}^{\cdot}\mathbf{Y}_{u})$ over u, with $\mathbf{Y}_{u}$ $(K_{u} \times p)$ containing the scores of the K_{u} categories for item u. Then, a multi-items version of (4) is expressed as

$$\mathrm{LHS}(\mathbf{X}, \mathbf{Y}_{1}, \cdots, \mathbf{Y}_{U}) = \sum_{u=1}^{U} \mathrm{SSQ}(\mathbf{X} - \mathbf{G}_{u}^{\cdot}\mathbf{Y}_{u}) + \alpha \operatorname{tr} \mathbf{X}'(\mathbf{I}_{n} \otimes \mathbf{A})\mathbf{X}. \qquad (18)$$

The extended method thus amounts to minimizing (18) under a certain normalization condition on $\mathbf{Y}_{u}$.

The foundation of our approach is to define the objective function as (2). There, the weight of the loss of homogeneity is fixed at one, whereas that of the penalty, i.e., α, can be chosen by users. An alternative for the definition is such that

$$\mathrm{LHS}^{\cdot}(\mathbf{X}, \mathbf{Y}) = \beta\, \mathrm{LH}(\mathbf{X}, \mathbf{Y}) + \gamma\, \mathrm{LS}(\mathbf{X}) = \beta \left\{ \mathrm{LH}(\mathbf{X}, \mathbf{Y}) + \frac{\gamma}{\beta} \mathrm{LS}(\mathbf{X}) \right\} \qquad (19)$$

with $\beta + \gamma = 1$, $\beta > 0$ and $\gamma > 0$. For $\alpha = \gamma / \beta$, (2) and (19) lead to the same solution. If we are interested in the values of the function for different weights, (19) may be more useful (although those values were not dealt with in this paper). It is because β and γ are balanced with their sum constant, while the weights are not balanced in (2), and different values of α may yield non-comparable values of the objective function.

References:

Adachi, K. (2000a). Growth curve representation and clustering under optimal scaling of repeated choice data. *Behaviormetrika*, **27**, 15-32.

Adachi, K. (2000b). Optimal scaling of longitudinal choice variable with time-varying representation of individuals. *British Journal of Mathematical and Statistical Psychology*, **53**, 233-253.

Bijleveld, C.C.J.H. and de Leeuw, J. (1991). Fitting longitudinal reduced-rank regression models by alternating least squares. *Psychometrika*, **56**, 433-447.

Bijleveld, C. C. J. H. and van der Burg, E. (1998). Analysis of longitudinal categorical data using optimal scaling techniques. In C.C.J.H. Bijleveld and L.J.T. van der Kamp (Eds.), *Longitudinal data analysis: Designs, models and methods*. Sage, London.

de Leeuw, J., van der Heijden, P. G. M., and Kreft, I. (1985). Homogeneity analysis of event history data. *Methods of Operations Research*, **50**, 299-316.

Gifi, A. (1990). *Nonlinear multivariate analysis*. Wiley, Chichester.

Green, P. J. and Silverman, B. W. (1994). *Nonparametric regression and generalized linear models: A roughness penalty approach*. Chapman & Hall, London.

Greenacre, M. J. (1984). *Theory and applications of correspondence analysis*. Academic Press, London.

Hastie, T., Buja, A., and Tibshirani, R. (1995). Penalized discriminant analysis. *Annals of Statistics*, **23**, 73-102.

Hayashi, C. (1952). On the prediction of phenomena from qualitative data and the quantification of qualitative data from the mathematico-statistical point of view. *Annals of the Institute of Statistical Mathematics*, **3**, 69-98.

Meulman, J. J., Heiser, W. J., and SPSS Inc. (1999). *SPSS categories 10.0*. SPSS Inc., Chicago.

Nishisato, S. (1980). *Analysis of categorical data: Dual scaling and its applications*. University of Toronto Press, Toronto.

Ramsay, J. O. and Silverman, B. W. (1997). *Functional data analysis*. Springer, New York.

Schoenberg, I. J. (1964). Spline functions and the problem of graduation. *Proceedings of the National Academy of Sciences of the United States*, **52**, 947-950.

van Buuren, S. (1990). *Optimal scaling of time series*. DSWO Press, Leiden.

van Buuren, S. (1996). Optimal transformations for categorical autoregressive time series. *Statistica Neerlandica*, **51**, 90-106.

Partial Multiple Correspondence Analysis

Haruo Yanai[1] and Tadahiko Maeda[2]

[1] National Center for University Entrance Examinations
2-19-23 Komaba, Meguro-ku, Tokyo 153-8501, Japan
[2] The Institute of Statistical Mathematics
4-6-7 Minami-Azabu, Minato-ku, Tokyo 106-8569, Japan

Summary: The present paper proposes a method of multiple correspondence analysis (MCA), which we name partial multiple correspondence analysis(PMCA), where effects of an ancillary item are eliminated from the other items. The idea is a natural extension of partial correspondence analysis (PCA) introduced by the first author. While PCA analyses relationship between two items, the proposed method deals with more than two items. We begin by briefly reviewing the derivation of correspondence analysis, PCA, and MCA in terms of orthogonal projection operators. Using these formulations, extension of PCA to the multiple-item case (PMCA) is described. After introducing an expression of PMCA as a special case of constrained MCA, some properties of PMCA are demonstrated by a small numerical example. We will also refer to the relationship between PMCA and another method called conditional forced classification of dual scaling.

1. Introduction

Optimum quantification of two sets of categorical data has been investigated under the names of Correspondence Analysis (Bénzecri, 1977), Dual Scaling (Nishisato, 1980) and so on. Given G_1, G_2, i.e. matrices of dummy-coded variables, it is well known that canonical correlation analysis (CCA) between G_1 and G_2 leads to correspondence analysis or dual scaling. When there are more than two items, multiple correspondence analysis (MCA: Lebart, 1984; Greenacre, 1993) may duly be applied. MCA can be formalized as the generalized canonical correlation analysis (GCCA) by Yanai (1998), where m data sets of GCCA are replaced by m sets of dummy variables $G_1, G_2, \cdots, G_m$ ($m \geq 3$).

So far, Yanai (1986, 1987) has introduced a method of quantification of two sets of dummy variables G_1 and G_2 eliminating the effects of an ancillary variable G_0 such as sex, age group, and named it Partial Correspondence Analysis (PCA). Mathematically, PCA is equivalent to partial canonical correlation analysis between G_1 and G_2 where the effects of dummy variables

G_0 are eliminated.

In this paper, we extend the idea of PCA to the case where more than two sets of categorical variables $G_1, G_2, \cdots, G_m$ and an ancillary dummy-coded variable (G_0) are available and thus introduce a new method which we name Partial Multiple Correspondence Analysis (PMCA). We also discuss the relationship between PMCA and the forced classification of dual scaling introduced by Nishisato (1984).

It should be noted, however, that the optimal quantification of a set of dummy matrices $G_1, G_2, \cdots, G_m$ in a given subspace or a complimentary subspace, associated with G_i or a set of G_i has a long history and interested readers are referred to other relevant studies (e.g., Lawrence, 1985; Nishisato, 1972, 1980, 1984, 1986, 1988a, 1988b, 1994; Nishisato and Baba, 1999; Nishisato and Lawrence, 1989; Takane, 1995; Takane and Shibayama, 1991).

2. The method

Our approach in this paper is to formulate correspondence analysis, multiple correspondence analysis, partial correspondence analysis and partial multiple correspondence analysis by emphasizing extensive uses of projection operators which enable readers to see the mutual relationship among the methods easily.

2.1 Formulation of Correspondence Analysis

Let G_1, G_2 be matrices of orders $n \times c_1, n \times c_2$, which represent dummy variables corresponding to two items labelled as I_1, I_2. Here c_1 and c_2 are number of categories for I_1, I_2, respectively. Then $N_{12} = N_{21}' = G_1'G_2$ is the contingency table between I_1 and I_2, and $D_1 = G_1'G_1$ and $D_2 = G_2'G_2$ are diagonal matrices with the sums of responses to each categories. Further, let a_1 and a_2 be vectors that assign optimal weights to the categories of G_1 and G_2. Then maximizing

$$g_{CA} = \frac{a_1'G_1'G_2a_2}{\sqrt{a_1'G_1'G_1a_1}\sqrt{a_2'G_2'G_2a_2}} = \frac{a_1'N_{12}a_2}{\sqrt{a_1'D_1a_1}\sqrt{a_2'D_2a_2}}. \quad (1)$$

with respect to a_1 and a_2 under the normalization constraints $a_1'D_1a_1 = a_2'D_2a_2 = 1$ yields

$$N_{12}D_2^{-1}N_{21}a_1 = \lambda D_1a_1 \quad \text{and} \quad a_2 = \lambda^{-1/2}D_2^{-1}N_{21}a_1. \quad (2)$$

Let $P_1 = G_1(G_1'G_1)^{-1}G_1'$ and $P_2 = G_2(G_2'G_2)^{-1}G_2'$ be orthogonal projectors onto column subspaces of G_1 and G_2, respectively. Then (2) is equivalent to:

$$(P_1 P_2) G_1 a_1 = \lambda G_1 a_1 \quad \text{and} \quad (P_2 P_1) G_2 a_2 = \lambda G_2 a_2 . \qquad (3)$$

It follows that the maximum eigenvalue λ of (2) or (3) attains 1, since column spaces G_1 and G_2 have a unit vector of dimension n in common.

We show an alternative method to obtain optimal weights, using the principle of the homogeneity analysis (Gifi, 1989). With an unknown n-dimensional verctor f, let a minimizing criterion be

$$h_{CA} = \| f - G_1 a_1 \|^2 + \| f - G_2 a_2 \|^2 . \qquad (4)$$

Differentiating the above function with respect to a_1 and a_2, we have

$$a_1 = (G_1' G_1)^{-1} G_1' f \quad \text{and} \quad a_2 = (G_2' G_2)^{-1} G_2' f \qquad (5)$$

and with some derivations, we have

$$(P_1 + P_2) f = (\mu + 1) f . \qquad (6)$$

Multiplying P_1 from the left on both sides of (6) yields $P_1 f + P_1 P_2 f = (\mu + 1) P_1 f$, thus leading to

$$P_1 (P_2 f) = \mu (P_1 f). \qquad (7)$$

Similarly, multiplying P_2 from the left of (6) yields

$$P_2 (P_1 f) = \mu (P_2 f). \qquad (8)$$

Substituting (8) into (7) and using $\lambda = \mu^2$, we have

$$(P_1 P_2)(P_1 f) = \lambda (P_1 f) , \qquad (9)$$

which implies $\lambda_k(P_1 P_2) = \mu_k^2$, where $\lambda_k()$ denotes k-th eigenvalue of the matrix within parentheses. Accordingly we have

$$\lambda_k(P_1 P_2) = (\lambda_k(P_1 + P_2) - 1)^2 . \qquad (10)$$

Let $CC_k(G_1, G_2)$ be the k-th canonical correlation between G_1 and G_2. Then (10) implies

$$CC_k(G_1, G_2) = \lambda_k(P_1 + P_2) - 1 \quad (k = 1, \cdots, \min(c_1, c_2)). \qquad (11)$$

2.2 Formulation of Partial Correspondence Analysis

Next, we consider partial correspondence analysis in terms of projectors (Yanai, 1986, 1987). Suppose we have, in addition to the two items I_1 and

I_2 to be analyzed, the third variable I_0 such as sex or age group dividing n subjects into c_0 groups. Such an item is called an "criterion" item (variable). Our goal in PCA is to carry out correspondence analysis of I_1 and I_2 eliminating the effect of the criterion item I_0. G_0 is the matrix (of order $n \times c_0$) of dummy variables for I_0. Hereafter '|' is used to explicitly show the column concatenation of two matrices.

Let $G_{10} = [G_1|G_0]$ and $G_{20} = [G_2|G_0]$, and let $a_{10}' = [a_0'|a_1']$, $a_{20}' = [a_0'|a_2']$ be weight vectors. Then by maximizing

$$g_{PCA} = \frac{a_{10}'G_{10}'G_{20}a_{20}}{\sqrt{a_{10}'G_{10}'G_{10}a_{10}}\sqrt{a_{20}'G_{20}'G_{20}a_{20}}} \tag{12}$$

with respect to a_{10} and a_{20}, we obtain in terms of orthogonal projectors

$$P_{10}P_{20}G_{10}a_{10} = \lambda G_{10}a_{10} \quad \text{and} \quad P_{20}P_{10}G_{20}a_{20} = \lambda G_{20}a_{20}.$$

It is to be noted here that the maximum eigenvalue attains 1 with multiplicity $c_0 = \text{rank}(G_0) = \dim(R(G_0))$, since $R(G_{10}) \cap R(G_{20}) = R(G_0)$ where $R(G_\bullet)$ is the column space of $G_\bullet$.

2.3 Formulation of Multiple Correspondence Analysis

Next we give a formulation of Multiple Correspondence Analysis (MCA). Suppose that response of n subjects to m items were recorded in terms of the dummy variables $G = [G_1|G_2|\cdots|G_m]$. We look for an optimal weight vector $a' = [a_1'|a_2'|\cdots|a_m']$ considering the following criterion:

$$g_{MCA} = \frac{\sum_{j=1}^{m}\sum_{i=1}^{m}(G_i a_i, G_j a_j)}{\sum_{i=1}^{m}\|G_i a_i\|^2} = \frac{a'G'Ga}{a'D_{ss}a}. \tag{13}$$

where D_{ss} is the block diagnal matrix with $D_i = G_i'G_i$ at the i-th block diagonal. Maximizing (13) with respect to a yields

$$(G'G)a = \lambda D_{ss}a. \tag{14}$$

Putting $f = Ga$ and multiplying (14) from the left by GD_{ss}^{-1}, we have

$$(GD_{ss}^{-1}G')f = \lambda f \tag{15}$$

which is equivalent to

$$(P_1 + P_2 + \cdots + P_m)f = \lambda f. \tag{16}$$

Here $P_j = G_j(G_j'G_j)^{-1}G_j'$ is the orthogonal projector onto $R(P_j)$, i.e. the column space of G_j. Note that (15) follows also from (2), substituting $N_{21}=G, N_{12}=G', D_1 =D_{ss}$, and $D_2=mI$ into the left side of equation (2).

Observe that (16) can also be obtained from the homogeneity criterion:

$$h_{\text{MCA}} = \sum_{j=1}^{m} \| f - G_j a_j \|^2 . \qquad (17)$$

Minimizing (17) with respect to a_j leads to $a_j = (G_j'G_j)^{-1}G_j'f$ ($j = 1, \cdots, m$) and (16).

Following Theorem 1 of Yanai (1998), we have

$$0 \le \lambda_k(P_1 + P_2 + \cdots + P_m) \le m \quad (\text{for } k = 1, \cdots, \text{rank}(G)), \qquad (18)$$

and let

$$\text{GCC}_k(G_1, G_2, \cdots, G_m) = \frac{\lambda_k(P_1 + P_2 + \cdots + P_m) - 1}{m - 1} \qquad (19)$$

be (the k-th) Generalized Canonical Correlation (GCC). From (18) the range of GCC may be $-(m-1)^{-1} \le \text{GCC}_k \le 1$. It should be noted, however, that GCC_k is nonnegative, i.e. GCC cannot be defined for such $k \ge s$ that $\text{GCC}_s < 0$. Also, observe here that (11) is a special case of (19) when $m = 2$.

Thus the maximum value of (13) is given by λ_1, the maximum eigenvalue of (14). Then (14) implies nothing but the eigenequation of multiple correspondence analysis (MCA) (see, e.g. Lebart et al., 1984) or of the Quantification method of the third type of Hayashi (1952).

The following lemma will be useful.

Lemma 1: The following three statements are equivalent.

  (1) $\text{GCC}_k(G_1, G_2, \cdots, G_m) = 1$, for $k = 1, \cdots, r$;
  (2) $\dim(R(G_1) \cap R(G_2) \cap \cdots \cap R(G_m)) = r$,
  (3) $\lambda_k(P_1 + P_2 + \cdots + P_m) = m$, for $k = 1, \cdots, r$.

2.4 Formulation of Partial Multiple Correspondence Analysis

In this section, we propose a method for carrying out multiple correspondence analysis on items $I_1, \cdots, I_m$, eliminating the effects of a criterion item I_0. G_j is the $n \times c_j$ matrix of dummy variables for item I_j, as before. Further, put

$$G_{10} = [G_1|G_0] , \quad G_{20} = [G_2|G_0] , \quad \cdots , \quad G_{m0} = [G_m|G_0] , \qquad (20)$$

and let the vectors of weights for G_{j0} be

$$a_{10}' = [a_1'|a_0'] \ , \ a_{20}' = [a_2'|a_0'] \ , \ \cdots \ , \ a_{m0}' = [a_m'|a_0'] \ . \qquad (21)$$

In line with the formulation of MCA, we look for weight vectors a_{j0} $(j = 1, \cdots, m)$ which maximize the following L_1 criterion,

$$g_{\text{PMCA}}(\equiv L_1) = \sum_{i=1}^{m} \sum_{j=1}^{m} (G_{i0}a_{i0} , \ G_{j0}a_{j0}) / \sum_{j=1}^{m} \| \ G_{j0}a_{j0} \ \|^2 , \qquad (22)$$

or equivalently, weight vectors which minimize the following S_1 criterion,

$$h_{\text{PMCA}}(\equiv S_1) = \sum_{j=1}^{m} \| \ f_0 - G_{j0}a_{j0} \ \|^2 \ . \qquad (23)$$

Then f_0 can be obtained as the eigenvector of the following equation:

$$(P_{10} + P_{20} + \cdots + P_{m0}) f_0 = \lambda f_0 \qquad (24)$$

where $P_{j0} = G_{j0}(G_{j0}'G_{j0})^- G_{j0}'$. For the reason explained later, g-inverse '$-$' is used here. Finally the weight vectors a_{j0} $(j = 1, \ldots, m)$ is obtained from f_0 as follows:

$$a_{j0} = (G_{j0}'G_{j0})^- G_{j0}' f_0 \ . \qquad (25)$$

As was the case with MCA, writing $G_z = [G_{10}|G_{20}|\cdots|G_{m0}]$, $a_z = [a_{10}|a_{20}|\cdots|a_{m0}]$ and after some algebraic manipulations, it can be shown that (24) is equivalent to the following equation which can also be obtained directly from the maximization of L_1:

$$(G_z'G_z)a_z = \lambda D_{zz}a_z \ . \qquad (26)$$

Blockwise expression of (26) is as follows:

$$\begin{bmatrix} D_{110} & N_{120} & \cdots & N_{1m0} \\ N_{210} & D_{220} & \cdots & N_{2m0} \\ \vdots & \cdots & \ddots & \vdots \\ N_{m10} & D_{m20} & \cdots & D_{mm0} \end{bmatrix} \begin{bmatrix} a_{10} \\ a_{20} \\ \vdots \\ a_{m0} \end{bmatrix} = \lambda \begin{bmatrix} D_{110} & O & \cdots & O \\ O & D_{220} & \cdots & O \\ \vdots & \cdots & \ddots & \vdots \\ O & O & \cdots & D_{mm0} \end{bmatrix} \begin{bmatrix} a_{10} \\ a_{20} \\ \vdots \\ a_{m0} \end{bmatrix} \qquad (27)$$

where

$$N_{ij0} = G_{i0}'G_{j0} \ , \quad D_{jj0} = G_{j0}'G_{j0} \ , \quad N_{ji0}' = N_{ij0} \ . \qquad (28)$$

Equation (26) is convenient for calculations and we will use this in practice.

Note that matrices D_{jj0} $(j = 1, \cdots, m)$ are singular, because both submatrices in G_{j0}, i.e. G_j and G_0, have an n-dimensional unit vector with all elements being unity as a common subspace. Therefore we need some tricks in the calculation of (27). One way is to exclude an arbitrary column vector from G_0, thus reducing G_0 to G_{00}. Then let $G_{j00} = [G_j|G_{00}]$ and substitute

$$N_{ij0} = G_{i00}'G_{j00}, \quad D_{ii0} = G_{i00}'G_{i00}, \quad N_{ji0}' = N_{ij0}$$

for (27) instead of (28).

We will call the above method Partial Multiple Correspondence Analysis(PMCA).

2.5 Some Properties of PMCA

We give some properties of the solution of PMCA.

Property 1: The maximum eigenvalue of (24) is m with multiplicity of c_0 [= rank(G_0)], i.e. (24) has c_0 trivial solutions. The weight vector a_z given by the eigenvector f_0 corresponding to the $(c_0 + 1)$-th eigenvalue of (24) maximizes L_1 in (22) and minimizes S_1 in (23).

Proof of Property: The proof of property 1 about the maximum eigenvalues can be obtained using the following lemma.

Lemma 2: Let $Q_0 = I_n - P_0 = I_n - G_0(G_0'G_0)^{-1}G_0'$ be the orthogonal projector onto the orthocomplement subspace of $R(P_0)$. Then

$$\lambda_k(P_{10} + P_{20} + \cdots + P_{0m}) = \lambda_{k-c_0}(P_{Q_0G_1} + P_{Q_0G_2} + \cdots + P_{Q_0G_m}).$$

$$\left[\begin{array}{l} \text{for } k = c_0 + 1, c_0 + 2, \cdots, r; \ r = c_0 + \text{rank}([G_1|G_2|\cdots|G_m]) - 1 \\ \qquad\qquad\qquad\qquad = c_0 + \sum_{s=1}^{m} c_s - m \end{array} \right]$$

For the proof, observe that $R(G_{10}) \cap R(G_{20}) \cap \cdots \cap R(G_{m0}) = R(G_0)$, thus in view of Lemma 1, it follows that the maximum eigenvalue of (24) is m with multiplicity of $c_0 = \text{rank}(G_0)$. Further, using Theorem 5 of Rao & Yanai (1979), we have $P_{0j} = P_{G_0} + P_{Q_0G}$, $(j = 1, \cdots, m)$, thus establishing

$$\sum_{j=1}^{m} P_{0j} = \sum_{j=1}^{m}(P_{G_0} + P_{Q_0G_j}) = mP_{G_0} + \sum_{j=1}^{m} P_{Q_0G_j}. \qquad (29)$$

Observe that $\lambda_k(\sum_{j=1}^{m} P_{0j}) = \lambda_1(mP_{G_0}) = m\lambda_1(P_{G_0}) = m \ (k = 1, \cdots, c_0)$, thus the proof is established. (Q.E.D.)

In terms of GCC as defined by (19), Lemma 2 can be stated as

$$\text{GCC}_k(G_{10}, G_{20}, \cdots, G_{m0}) = \text{GCC}_{k-c_0}(Q_0G_1, Q_0G_2, \cdots, Q_0G_m).$$

From the above lemma, we can prove the following property.

Property 2: f_0 and a_0 that maximize L_1 of (22) and minimize S_1 in (23) also maximize L_2 and minimize S_2 in the following equations:

$$S_2 = \sum_{j=1}^{m} \| Q_0 f_0 - Q_0 G_j a_j \|^2 . \tag{30}$$

$$L_2 = \sum_{i=1}^{m}\sum_{j=1}^{m} (Q_0 G_i a_i , Q_0 G_j a_j) / \sum_{j=1}^{m} \| Q_0 G_j a_j \|^2 . \tag{31}$$

2.6 Relationship of PMCA with MCA with Linear Constraints

Following Theorem 2 of Yanai (1998), maximization of (22) subject to the linear constraints of the forms $C_j' a_{j0} = 0$ for $j = 1, \cdots, m$ yields

$$\sum_{j=1}^{m} (P_{j0} - P_{G_{j0}(G_{j0}'G_{j0})^- C_j}) f = \lambda f \tag{32}$$

which is called GCCA with linear constraints(GCCA-LC). We show that PMCA is a special case of GCCA-LC, more specifically, a special case of MCA with linear constraints(MCA-LC).

For the purpose, put $C_j = (G_{j0}'G_0)$ and noting $G_{j0}(G_{j0}'G_{j0})^- G_{j0}'G_0 = P_{j0}G_0 = G_0$ since $R(G_0)$ is a subspace of $R([G_j|G_0])$, we have

$$P_{j0} - P_{G_{j0}(G_{j0}'G_{j0})^- C_j} = P_{j0} - P_{G_0} = P_{Q_0 G_j}$$

Thus, it follows that the equation (32) reduces into $(\sum_{j=1}^{m} P_{Q_0 G_j}) f = \lambda f$ which is exactly MCA based on $Q_0 G_j$ ($j = 1, \cdots, m$). This implies PMCA also can be considered as a special case of MCA-LC. It is interesting to note that substituting $C_j = G_{j0}'G_0$ into $C_j' a_{j0} = 0$ yields $G_0'(G_{j0} a_{j0}) = 0$.

2.7 Relationships of PMCA with the Forced Classification Methods

The idea of eliminating the effect of an ancillary item in the quantification of categorical data has been already discussed by other authors. Among others, Nishisato's method of forced classification (hereafter abbreviated to FC), which Nishisato (1984, 1988a) proposed in the context of his continuing work on Dual Scaling (DS) is without doubt the most relevant to our proposal of PMCA. To be more specific, our methods may correspond to what Nishisato and Baba (1999) call conditional FC, as is pointed out in the paper. Unfortunately, however, we have so far not succeeded in deriving the mathematical relationships between conditional FC and our methods. Therefore, we will only briefly refer to our conjecture that some relationships exist between our methods and conditional FC of DS.

According to the papers cited above, given $(m + 1)$ dummy matrices G_0 and $G = [G_1|G_2|\cdots|G_m]$, FC and relevant methods are classified into:

(M1) DS of $[G|G_0\,G_0\cdots G_0]$, where G_0 is repeated ℓ times,
(M2) DS of $[G|\ell \times G_0]$,
(M3) DS of P_0G (P method) and DS of Q_0G (Q method),
(M4) DS of $G_0'G$. (M method).

Principle of equal partitioning given by Nishisato (1980) guarantees that (M1) and (M2) are equivalent with any positive integer ℓ. When ℓ is taken large enough, then (M2) is called the method of FC and is asymptotically equivalent to (M3). Here 'asymptotic' is understood as ℓ approaching infinity and 'equivalence' is understood as follows: It is known that this method of FC yields both $(c_0 - 1)$ proper solutions which correspond to those of P method in (M3) and subsequent (at least $(c_0 - 1)$) solutions which correspond to those of Q method in (M3). The latter case which corresponds to Q method is called conditional FC, where effects of the criterion item G_0 is partialled out. Further, P method is equivalent to (M4) (Nishisato, 1988a, 1994).

In this connection, our method called PMCA is described as

(M5) MCA of $[(G_1|G_{00})|(G_2|G_{00})|\cdots|(G_m|G_{00})]$.

In case of $\ell = m$, (M1) is seemingly very similar to (M5), but the solution is not identical since the order of the dummy matrices G_0 in (M1) and G_{00} in (M5) are different.

Rather, our conjecture is that there are some relations between (M5) and Q method of (M3). But up to now we are not sure whether or not these two methods are equivalent, since we have not succeeded in giving a mathematical proof of the equivalence. We will further examine more detailed mathematical relathionship that may exist between the FC method and our method in the near future.

3. A Numerical Example

In this section we will give a numerical example with a small artificial data set. Consider that responses by eight subjects to three items I_1, I_2, I_3 and a criterion item I_0, each having two categories, are given by the matrices of dummy variables G_1, G_2, G_3, and G_0 as follows:

$$
G_1 = \begin{bmatrix} 1 & 0 \\ 0 & 1 \\ 1 & 0 \\ 1 & 0 \\ 0 & 1 \\ 1 & 0 \\ 1 & 0 \\ 1 & 0 \end{bmatrix}, G_2 = \begin{bmatrix} 1 & 0 \\ 0 & 1 \\ 0 & 1 \\ 0 & 1 \\ 1 & 0 \\ 0 & 1 \\ 0 & 1 \\ 0 & 1 \end{bmatrix}, G_3 = \begin{bmatrix} 0 & 1 \\ 1 & 0 \\ 0 & 1 \\ 0 & 1 \\ 0 & 1 \\ 1 & 0 \\ 0 & 1 \\ 1 & 0 \end{bmatrix}, G_0 = \begin{bmatrix} 1 & 0 \\ 1 & 0 \\ 1 & 0 \\ 1 & 0 \\ 0 & 1 \\ 0 & 1 \\ 0 & 1 \\ 0 & 1 \end{bmatrix}.
$$

Table 1: Results of PMCA on an artificial dataset

Item-Category	D0-1	D0-2	D1	D2	D3
1-1	0.057	0.497	−0.105	−0.274	0.159
1-2	0.057	0.497	0.314	0.821	−0.478
2-1	0.057	0.497	0.749	0.098	0.661
2-2	0.057	0.497	−0.250	−0.033	−0.220
3-1	0.057	0.497	−0.365	0.346	0.361
3-2	0.057	0.497	0.365	−0.346	−0.361
λ_k	3.000	3.000	1.505	1.145	0.350
GCC	1.000	1.000	0.502	0.382	0.117

We take the first column of G_0 as G_{00} and concatenate it to G_i to obtain G_{i00} ($i = 1, 2, 3$).

The result of PMCA where effects of I_0 are eliminated is shown in Table 1. From property 1 in section 2.5, we have the maximum eigenvalue 3 with multiplicity 2(=rank(G_0)), which is shown in the columns labeled as D0-1 and D0-2. These are so-called trivial solutions. Remaining three columns contain eigenvalues (λ_k), generalized canonical correlation (GCC), and optimal weights for categories as given by the rescaled eigenvectors. Interpretation of dimensions should be based on weights in these non-trivial solutions.

With regard to the conjecture in the section 2.7, for this particular example, it is easily verified that results of PMCA and conditional FC with large enough ℓ are quite similar. The correlation coefficients between the derived component (so-called total scores) given by the weight vertors were very close to 1 for each of the three dimensions. But this is not always the case. According to our experience on other data, similarity of components tends to decrease when the size of data bocomes larger. In such cases, however, more or less similar components can be found after proper selection of dimensions.

We need furthur examinations on the question of under which conditions the results of two methods coincide exactly. Our current opinion is that both methods and their formulation have their own virtues and intuitive appeals.

4. Some Additional Comments

From a practical point of view, we need more knowledge on when and how PMCA are useful to reveal the structure of categorical data. The knowlede should be based on the experiences of analysis of empirical data sets. We are also interested in extending PMCA to situations where multiple criterion items effects of which should be eliminated from the analysis are available. One of such generalization of FC is is already discussed in Nishisato and

Lawrence (1989) in a general form. We believe that there are still other directions of extension, which remains to be investigated in the future.

References:

Bénzecri, J.P. (1977). Histoire et préhistorie de l'analyse des données: L'analyse des correspondances. *Les Cahiers de l'Analyse des Donneé*, **2**, 9–40.

Gifi, A. (1989). *Nonlinear Multivariate Analysis*, John Wiley & Sons, New York.

Greenacre, M. (1993). Multivariate generalisations of correspondence analysis, *In Multivariate Analysis: Future Directions 2*, Cuadras, C.M. et al. (eds.), 327–340, North Holland, Amsterdam.

Hayashi, C. (1952). On the prediction of phenomena from qualitative data from the mathematico-statistical point of view, *Annals of the Institute of Statistical Mathematics*, **3**, 69–96.

Lawrence, D.R. (1985). Dual scaling of multidimensional data structures: An extended comparison of three methods. *Doctoral Dissertation*, University of Toronto.

Lebart, L., Morineau, A.K. and Warwick, M. (1984). *Multivariate Descriptive Statistics*, John Wiley & Sons, New York.

Nishisato, S. (1972). Analysis of variance through optimal scaling. *Proceedings of the First Canadian Conference in Applied Statistics*, 306–317, Sir George Williams University Press, Montreal.

Nishisato, S. (1980). *Analysis of Categorical Data: Dual Scaling and its Applications*, University of Toronto Press, Toronto.

Nishisato, S. (1984). Forced classification: A simple application of a quantification method, *Psychometrika*, **49**, 25–36.

Nishisato, S. (1986). Generalized forced classification for quantifying categorical data, *Data Analysis and Informatics, IV*, Diday, E. et al. (eds.), 351–362, North-Holland, Amsterdam.

Nishisato, S. (1988a). Forced Classification Procedure of Dual Scaling: Its Mathematical Properties. In H. H. Bock (Ed.), *Classification and Related Methods of Data Analysis*, 523–532, North-Holland, Amsterdam.

Nishisato, S. (1988b). Market segmentation by dual scaling through generalized forced classification. In Gaul, W. and Schader, M. (eds.), *Data, Expert Knowledge and Decisions*, 268–278. Springer-Verlag, Berlin.

Nishisato, S. (1994) *Elements of Dual Scaling: An Introduction to Practical Data Analysis*. Hillsdale, New Jersey: Lawrence-Erlbaum Associates.

Nishisato, S. and Baba, Y. (1999) On contingency, projection and forced classification of dual scaling, *Behaviormetrika*, **26**, 207–219.

Nishisato, S. and Lawrence, D. R. (1989). Dual Scaling of Multiway Data Matrices: Several Variants. In R. Coppi and S. Bolasco (Eds.), *Multiway Data Analysis*, 317–326. Amsterdam, The Netherlands: North-Holland.

Rao, C.R. and Yanai, H. (1979). General definition of a projector, its decomposition, and application to statistical problems, *Journal of Statistical Planning and Inference*, **3**, 1–17.

Takane, Y. (1995). *Seiyaku Tsuki Shuseibun Bunsekiho (Constrained Principal Component Analysis)*. Asakura Shoten, Tokyo.

Takane, Y. and Shibayama, T. (1991). Principal component analysis with external information on both subjects and variables. *Psychometrika, 56*, 97–120.

Yanai, H. (1986). Some generalization of correspondence analysis in terms of projection operators, *In Data Analysis and Informatics IV*. Diday, E.L. et al. (eds.), 193–207, North Holland, Amsterdam.

Yanai, H. (1987). Partial Correspondence Analysis and its properties. *In Recent Development in Clustering and Data Analysis, Proceedings of the Japanese-French Scientific Seminar*, Hayashi, C. et al. (eds.), 259–266, Academic Press.

Yanai, H. (1998). Generalized canonical correlation analysis with linear constraints, *In Data Science, Classification and Related Methods*, Hayashi, C. et al.(eds.), 539–546, Springer, Tokyo.

Studying Triadic Distance Models Under a Likelihood Function

Mark de Rooij

Leiden University
Department of Psychology

Summary: Triadic distance models are relatively new. Their merits and demerits are fairly unknown. In the present paper we will study triadic distance models and bring the understanding of those models to a next level. Therefore, the models are studied under a Multinomial sampling scheme and a detailed investigation of the likelihood function results in relationships with multiple correspondence analysis and three-way quasi-symmetry models.

1. Introduction

The analysis of three-way tables has received an enormous amount of attention in the last few decades. In the case of distance models, the attention was initially focused on three-way two-mode data, but recently the attention shifted towards three-way one-mode and three-way three-mode data. The latter data types require a different modeling strategy: we are not after a graphical representation of one mode that is afterwards transformed for each specific instance of the second mode, but instead we are looking for a graphical representation of the entire three-way table in a single Euclidean space. The entries of the three-way table can in some way be viewed as (dis)similarities that relate three categories of (different) variables. Triadic distance models try to represent the three categories as points in a Euclidean space, such that a measure of distance for these three points as closely as possible approximates the dissimilarities in the three-way table. A popular class of triadic distance models is formed by the L_p-transform, where a triadic distance is defined on dyadic distances:

$$d_{ijk} = \left[d_{ij}^p + d_{jk}^p + d_{ik}^p \right]^{1/p} . \tag{1}$$

Here d_{ijk} is the triadic distance between the three points i, j, and k; d_{ij} are the usual (dyadic) distances between two points i and j. The precise definition of the dyadic distances is left for the next section.

In the present paper we will study triadic distances, defined by the L_2-transformation. Therefore, we apply them to contingency tables assuming a Multinomial sampling scheme, and study the likelihood function in more detail. This procedure establishes relationships of triadic distance models with Multiple Correspondence Analysis and Quasi-Symmetry models for three-way tables.

2. Triadic Distance Models

The study of triadic distance models was initiated by Hayashi (1972). A number of papers have been written since: Cox, Cox, and Branco (1991), Pan and Harris (1991), Joly and Le Calvé (1995), Daws (1996), Heiser and Bennani (1997), De Rooij and Heiser (2000), De Rooij (2001, submitted). Two papers (Joly and Le Calvé, 1995 and Heiser and Bennani, 1997) propose an axiomatic framework for the study of triadic distance models. All these papers assumed a three-way square, i.e., a $K \times K \times K$, proximity matrix.

We will focus on the Generalized Euclidean Model, where $p = 2$ in the L_p-transform. The triadic distance in this case is defined as the square root of the sum of squared dyadic Euclidean distances. The squared triadic distance is given by

$$d_{ijk}^2(\mathbf{X}) = d_{ij}^2(\mathbf{X}) + d_{jk}^2(\mathbf{X}) + d_{ik}^2(\mathbf{X}), \tag{2}$$

where $d_{ij}(\mathbf{X})$ is the usual Euclidean distance between points i and j, i.e.,

$$d_{ij}(\mathbf{X}) = \left[\sum_m (x_{im} - x_{jm})^2 \right]^{1/2}. \tag{3}$$

Joly and Le Calvé (1995) show this triadic distance is equal to the square root of the inertia, the sum of squared distances of each of the three points towards their center of gravity. In a one dimensional representation the triadic distance is then equal to the standard deviation of the three points. In a multidimensional representation we can consider the triadic distance as a natural generalization of the standard deviation, and so the squared triadic distance as a natural generalization of the variance of the three points.

For a further development later, it is good to note here that Heiser and Bennani (1997) showed this model (2) can be rewritten as

$$\begin{aligned} d_{ijk}^2(\mathbf{X}) &= tr\mathbf{X}^T \mathbf{A}_{ij}\mathbf{X} + tr\mathbf{X}^T \mathbf{A}_{jk}\mathbf{X} + tr\mathbf{X}^T \mathbf{A}_{ik}\mathbf{X} \\ &= tr\mathbf{X}^T \mathbf{A}_{ijk}\mathbf{X}, \end{aligned} \tag{4}$$

where tr denotes $trace$, the sum of the diagonal elements of a matrix, $\mathbf{A}_{ij} = (\mathbf{e}_i - \mathbf{e}_j)(\mathbf{e}_i - \mathbf{e}_j)^T$, and $\mathbf{e}_i$ is the i-th column of a identity matrix of order K.

For the Generalized Euclidean Model we give an example of a graphical representation in Figure 1. In this figure we see 5 points a, b, c, d and e. The triadic distance is defined as the square root of the sum of squared dyadic distances. Comparing some distances we find that $d_{abe} < d_{ade}$ since $d_{ab} < d_{ad}$ and the other dyadic distances are equal; $d_{abd} < d_{acd}$, since $\sqrt{(2^2 + 2^2 + 4^2)} < \sqrt{(3^2 + 1^2 + 4^2)}$.

De Rooij and Heiser (2000) extend the work of Heiser and Bennani to the unfolding situation, and discuss restrictions on triadic unfolding models to visualize trends in longitudinal studies. For the triadic unfolding model any $I \times J \times K$-matrix with proximity data can be used. The restricted unfolding

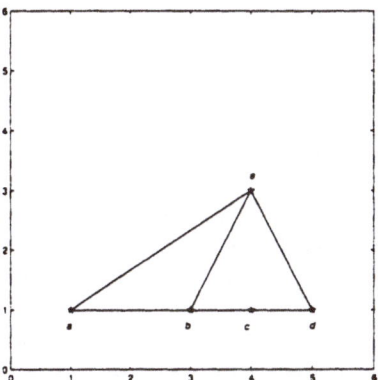

Figure 1: An configuration with triadic distances.

models, proposed by De Rooij and Heiser, can only be applied to $K \times K \times K$-tables. In the present paper we will study the (restricted) triadic unfolding models in more detail.

In the triadic unfolding model we estimate a coordinate matrix for each way. For three-way square tables the triadic unfolding model allows for three-way asymmetry. Still the triadic distance is defined as the square root of the sum of dyadic distances, but the dyadic distances are defined by

$$d_{ij}(\mathbf{X}; \mathbf{Y}) = \left[\sum_m (x_{im} - y_{jm})^2 \right]^{1/2}. \tag{5}$$

The squared triadic unfolding distance is then defined as

$$d_{ijk}^2(\mathbf{X}; \mathbf{Y}; \mathbf{Z}) = d_{ij}^2(\mathbf{X}; \mathbf{Y}) + d_{jk}^2(\mathbf{Y}; \mathbf{Z}) + d_{ik}^2(\mathbf{X}; \mathbf{Z}). \tag{6}$$

In the triadic unfolding model the subscript i is attached to the first way, and thus to the coordinates x; j is attached to the second way, and thus to the coordinates y; k is attached to the coordinates for the third way, z. The triadic unfolding model has the same interpretation as shown in Figure 1, but in the triadic unfolding model distances between points representing categories of one way are not related to observations.

We can rewrite the triadic unfolding model also in matrix terms. Therefore, first define the matrix $\mathbf{S}$ with $\mathbf{X}$, $\mathbf{Y}$, and $\mathbf{Z}$ concatenated vertically, that is, $\mathbf{S}$ is defined as

$$\mathbf{S} = \begin{pmatrix} \mathbf{X} \\ \mathbf{Y} \\ \mathbf{Z} \end{pmatrix}. \tag{7}$$

Equation 6 can then be rewritten as

$$
\begin{aligned}
d_{ijk}^2(\mathbf{X};\mathbf{Y};\mathbf{Z}) &= tr\mathbf{S}^T\mathbf{A}_{ij}\mathbf{S} + tr\mathbf{S}^T\mathbf{A}_{jk}\mathbf{S} + tr\mathbf{S}^T\mathbf{A}_{ik}\mathbf{S} \\
&= tr\mathbf{S}^T\mathbf{A}_{ijk}\mathbf{S},
\end{aligned}
\tag{8}
$$

where in this case $\mathbf{A}_{ij} = (\mathbf{e}_i - \mathbf{e}_{I+j})(\mathbf{e}_i - \mathbf{e}_{I+j})^T$; $\mathbf{A}_{ik} = (\mathbf{e}_i - \mathbf{e}_{I+J+k})(\mathbf{e}_i - \mathbf{e}_{I+J+k})^T$; $\mathbf{A}_{jk} = (\mathbf{e}_{I+j} - \mathbf{e}_{I+J+k})(\mathbf{e}_{I+j} - \mathbf{e}_{I+J+k})^T$ and $\mathbf{e}_i$ is the i-th column of a identity matrix of order $I + J + K$.

The generalized slide vector models proposed by De Rooij and Heiser are restricted triadic unfolding models. These restricted models can only be applied to three-way tables where each way refers to the same variable. So the three-way matrix needs to be square. The imposed restrictions are $z_{jm} = y_{jm} - v_m$, and $y_{jm} = x_{jm} - u_m$, for the slide-2 model. Consequently, $z_{jm} = x_{jm} - u_m - v_m$. The square slide-2 distance model is then defined by

$$
d_{ijk}^2(\mathbf{X};\mathbf{u};\mathbf{v}) = d_{ij}^2(\mathbf{X};\mathbf{u}) + d_{jk}^2(\mathbf{X};\mathbf{v}) + d_{ik}^2(\mathbf{X};\mathbf{u};\mathbf{v}),
\tag{9}
$$

where $d_{ij}^2(\mathbf{X};\mathbf{u}) = \sum_m (x_{im} - x_{jm} + u_m)^2$, the slide vector model as defined by Zielman and Heiser (1993). A further restriction is $v_m = u_m$, for the slide-1 model. We obtain the triadic distance model by imposing the restriction $v_m = u_m = 0$.

In matrix terms the constraints can be written as $\mathbf{S} = \mathbf{ER}$, where $\mathbf{E}$ is a known design matrix, and $\mathbf{R}$ is a matrix with the coordinates $\mathbf{X}$ and possibly the slide vectors $\mathbf{u}$ and $\mathbf{v}$ concatenated vertically. Define the three arrays $\mathbf{E}$ for the slide-2 model, the slide-1 model and the triadic distance model, respectively, with identity matrices of order K ($\mathbf{I}$), and $K \times 1$ vectors with ones ($\mathbf{1}$) and zeros ($\mathbf{0}$):

$$
\mathbf{E}_2 = \begin{pmatrix} \mathbf{I} & \mathbf{1} & \mathbf{0} \\ \mathbf{I} & \mathbf{0} & \mathbf{0} \\ \mathbf{I} & \mathbf{0} & -\mathbf{1} \end{pmatrix}, \quad \mathbf{E}_1 = \begin{pmatrix} \mathbf{I} & \mathbf{1} \\ \mathbf{I} & \mathbf{0} \\ \mathbf{I} & -\mathbf{1} \end{pmatrix}, \quad \mathbf{E}_0 = \begin{pmatrix} \mathbf{I} \\ \mathbf{I} \\ \mathbf{I} \end{pmatrix}.
\tag{10}
$$

We will make use of these design matrices in the next section to further analyze the generalized slide vector models.

3. Application to Contingency Tables

De Rooij and Heiser (2000) apply their triadic distance model to a three-way contingency table, assuming that the frequencies are a measure of similarity. Here we will also apply the models to contingency tables, but in the present paper we will assume a Multinomial sampling scheme for the observed frequencies. In general, the kernel of the log-likelihood function for a model under Multinomial sampling can be written

$$
\mathcal{L} = \sum_{ijk} f_{ijk} \log(\pi_{ijk}),
\tag{11}
$$

where π_{ijk} are the expected probabilities under a specified model. The f_{ijk} are the observed frequencies. The simple model we will study is that the expected probabilities are given by

$$\pi_{ijk} = \exp(-d_{ijk}^2), \qquad (12)$$

that is the expected probabilities are related to distances in Euclidean space by the Gaussian transform. The larger the distance the smaller expected probability; the smaller the distance the larger the expected probability. If a combination of categories often occur the categories are close in Euclidean space, which is in line with our assumption that frequencies are a measure of similarity.

If we insert our model into the likelihood function we obtain

$$\mathcal{L} = -\sum_{ijk} f_{ijk} d_{ijk}^2. \qquad (13)$$

We will use different distance models and develop the likelihood function. This will give a more detailed view on triadic distance models as developed by now.

3.1 Triadic Unfolding Models

For unfolding models we saw in the previous section that the distance can be written as

$$d_{ijk}^2(\mathbf{X};\mathbf{Y};\mathbf{Z}) = tr\mathbf{S}^T\mathbf{A}_{ijk}\mathbf{S}. \qquad (14)$$

Inserting this definition in the log-likelihood function (13), and developing we obtain

$$
\begin{aligned}
\mathcal{L}_u &= -\sum_{ijk} f_{ijk} \times tr\mathbf{S}^T\mathbf{A}_{ijk}\mathbf{S} \\
&= tr\mathbf{S}^T\mathbf{C}_u\mathbf{S},
\end{aligned} \qquad (15)
$$

where

$$
\mathbf{C}_u = \begin{pmatrix} -2\mathbf{F}_i & \mathbf{F}_{ij} & \mathbf{F}_{ik} \\ \mathbf{F}_{ij}^T & -2\mathbf{F}_j & \mathbf{F}_{jk} \\ \mathbf{F}_{ik}^T & \mathbf{F}_{jk}^T & -2\mathbf{F}_k \end{pmatrix}. \qquad (16)
$$

Here $\mathbf{F}_{ij}$ is a two-way matrix obtained by summing the three-way table over the third way k, and similarly for the other marginal arrays, $\mathbf{F}_i$ denotes the diagonal matrix with elements f_{i++}.

The matrix $\mathbf{C}_u$ has the same form as the Burt matrix that is decomposed in Multiple Correspondence Analysis (MCA) (Greenacre, 1984; Gifi, 1990). For the triadic unfolding model the diagonal blocks have matrices defined by minus two times the univariate margin, where in MCA the margin itself is used. However, there has been a discussion lately about these diagonal blocks (Greenacre, 1988: Boik, 1996; Tateneni and Brown, 2000). A new method called Joint Correspondence Analysis (JCA) is devised to reduce the influence of the diagonal blocks on the result of MCA. Further research could be done in the field of correspondence analysis whether the replacement of the diagonal blocks as proposed here would add to the understanding of the method. At least then the interpretation of MCA can be done in terms of triadic Euclidean distances.

3.2 Triadic Slide Vector Models

The triadic unfolding model could be written as a trace function. The slide vector models and the symmetric model are restricted unfolding models, where the restriction has the form $\mathbf{S} = \mathbf{ER}$. We will use these restrictions in our development of the likelihood function.

Inserting $\mathbf{S} = \mathbf{E}_2\mathbf{R}$ in the likelihood function for the triadic unfolding model (15) we obtain the likelihood function for the slide-2 model, that is

$$
\begin{aligned}
\mathcal{L}_{sv2} &= tr\mathbf{R}^T\mathbf{E}_2^T\mathbf{C}_u\mathbf{E}_2\mathbf{R} \\
&= tr\mathbf{R}^T\mathbf{C}_{sv2}\mathbf{R},
\end{aligned} \tag{17}
$$

where $\mathbf{C}_{sv2}$ has the form

$$
\mathbf{C}_{sv2} = \begin{pmatrix} \mathbf{C}_s & \mathbf{n}_1 & \mathbf{n}_2 \\ \mathbf{n}_1^T & -2f_{+++} & -f_{+++} \\ \mathbf{n}_2^T & -f_{-++} & -2f_{+++} \end{pmatrix}, \tag{18}
$$

in which the vector $\mathbf{n}_1$ has elements $\{n_i^1\}$ defined by $n_i^1 = f_{++i}+f_{+i+}-2f_{i++}$, and the vector $\mathbf{n}_2$ has elements $\{n_i^2\}$ defined by $n_i^2 = 2f_{++i} - f_{+i+} - f_{i++}$. The matrix $\mathbf{C}_s$ has elements $\{c_{ij}^s\}$ defined by $c_{ij}^s = \sum_k g_{ijk}$ if $i \neq j$, else $c_{ij}^s = \sum_j c_{ij}^s$, and the g_{ijk} are given by $g_{ijk} = \frac{1}{6}(f_{ijk}+f_{ikj}+f_{jik}+f_{jki}+f_{kij}+f_{kji})$.

Using the same steps as above, the likelihood function for the slide-1 model is given by

$$
\begin{aligned}
\mathcal{L}_{sv1} &= tr\mathbf{R}^T\mathbf{E}_1^T\mathbf{C}_u\mathbf{E}_1\mathbf{R} \\
&= tr\mathbf{R}^T\mathbf{C}_{sv1}\mathbf{R},
\end{aligned} \tag{19}
$$

where $\mathbf{C}_{sv1}$ has the form

$$
\mathbf{C}_{sv1} = \begin{pmatrix} \mathbf{C}_s & \mathbf{n}_3 \\ \mathbf{n}_3^T & -6f_{+++} \end{pmatrix}, \tag{20}
$$

in which $\mathbf{n}_3$ has elements $\{n_i^3\}$ defined by $n_i^3 = 3(f_{++i} - f_{i++})$.

For the symmetric generalized Euclidean model the design matrix $\mathbf{E}_0$ can be used and the likelihood function obtains the following form

$$
\begin{aligned}
\mathcal{L}_s &= tr\mathbf{R}^T\mathbf{E}_0^T\mathbf{C}_u\mathbf{E}_0\mathbf{R} \\
&= tr\mathbf{R}^T\mathbf{C}_s\mathbf{R}.
\end{aligned} \tag{21}
$$

The matrix C_s is the same as above in the likelihood functions for the triadic slide vector models. The matrix R in this case is equal to the coordinate matrix X.

Both the slide-1 and the slide-2 model will fit the same coordinate matrix as the symmetric triadic distance model, corresponding to the matrix C_s. The slide-1 model represents in addition the difference between the third margin and the first margin. The slide-2 model represents two differences: (1) The difference of the first margin compared with the second and the third margin; (2) The difference of the third margin compared with the first and the second margin.

We can compare these models to quasi-symmetry models for three-way tables. The quasi-symmetry model for three-way tables is given by

$$\log(\pi_{ijk}) = \lambda + \lambda_i^R + \lambda_j^C + \lambda_k^P + \lambda_{ijk}, \tag{22}$$

where $\lambda_{ijk} = \lambda_{ikj} = \lambda_{jik} = \lambda_{jki} = \lambda_{kij} = \lambda_{kji}$, i.e., the interaction term is three-way symmetric. In the triadic slide vector models the symmetric part is modeled by the distances between the points in Euclidean space given in the coordinate matrix X. The main effect terms λ_i^R, λ_j^C, and λ_k^P represent the occurences of each of the categories. In the generalized slide vector models the differences of these main effect parameters are represented by vectors, which are attached to the dimensions of the Euclidean space. This does give us a nice representation of, for example, the trends in longitudinal research. For the symmetric generalized Euclidean model we find no marginal differences, i.e., the model is equal to the model of three-way symmetry with metric constraints.

4. Conclusions

We studied triadic distance models under a Multinomial sampling scheme. For the triadic unfolding model we found an interesting relationship with MCA. In MCA the distances between points are defined via the underlying subject points. In triadic unfolding models we have a direct distance definition between the three points. In both models no three-way information is represented (see De Rooij, 2001, submitted), which is clear from the Burt matrix and the matrix obtained in (16). This might be a disadvantage of both procedures in case three way relations are of specific interest. However, many multivariate data anayses techniques only consider bivariate relationships and these bivariate relationship are often more interesting compared to the trivariate or higher order relationships. An advantage of looking only at bivariate relationships is that large tables can be handled without many complications. In log-linear analysis, for example, one should always be careful analyzing large sparse tables.

Distance models for three-way two-mode data, as the well known INDSCAL model (Carroll and Chang, 1970) do represent three-way relationships (see De Rooij, 2001). In the INDSCAL model, however, only one of the three bivariate relationships has a distance representation. The other two bivariate relationships are hard to grasp from the results of the model. Moreover, the INDSCAL model has a very specific interpretation, which is only nice when there is a specific interest in differences between individuals or groups of individuals. Many three-way contingency tables do not have that nature, and in that case our triadic distance models have a more natural interpretation. It would be interesting to combine the two approaches, that is a topic of current research.

References:

Boik, R. J. (1996), "An efficient algorithm for joint correspondence analysis," *Psychometrika*, **61**, 255-269.

Carroll, J. D., and Chang, J. J. (1970), "Analysis of individual differences in multidimensional scaling via an N-way generalization of 'Eckart-Young' decomposition," *Psychometrika*, **35**, 283-319.

Cox, T. F., Cox, M. A., and Branco, J. A. (1991), "Multidimensional scaling for n-tuples," *British Journal of Mathematical and Statistical Psychology*, **44**, 195-206.

Daws, J. T., (1996), "The analysis of free-sorting data: Beyond pairwise cooccurences," *Journal of Classification*, **13**, 57-80.

De Rooij, M. (2001, submitted), "Distance models for three-way tables and three-way information: a theoretical note,"

De Rooij, M., and Heiser, W. J. (2000), "Triadic distance models for the analysis of asymmetric three-way proximity data," *British Journal of Mathematical and Statistical Psychology*, **53**, 99-119.

Gifi, A. (1990), *Nonlinear multivariate analysis*. Chichester, England: Wiley.

Greenacre, M. J. (1984), *Theory and applications of correspondence analysis*. New York: Academic Press.

Greenacre, M. J. (1988), "Correspondence analysis of multivariate categorical data by weighted least squares," *Biometrika*, **75**, 457-467.

Hayashi, C. (1972), "Two dimensional quantifications based on a measure of dissimilarity among three elements," *Annals of the Institute of Statistical Mathematics*, **25**, 251-257.

Heiser, W. J., and Bennani, M. (1997), "Triadic distance models: Axiomatization and least squares representation," *Journal of Mathematical Psychology*, **41**, 189-206.

Joly, S., and Le Calvé, G. (1995), "Three-way distances," *Journal of Classification*, **12**, 191-205.

Pan, G., and Harris, D. P. (1991), "A new multidimensional scaling technique based upon association of triple objects pijk and its application to the analysis of geochemical data," *Mathematical Geology*, **23**, 861-886.

Tateneni, K., and Browne, M. W. (2000), "A noniterative method of joint correspondence analysis," *Psychometrika*, **65**, 157-165.

Zielman, B., and Heiser, W. J. (1993), "The analysis of asymmetry by a slide-vector," *Psychometrika*, **58**, 101-114.

A Generalized Modification of Scheffé's Paired Comparisons:
A theoretical approach to decrease the number of experiments

Masaya IIZUKA[1] and Katsumi UJIIE[2]

[1] Faculty of Law, Okayama University
3-1-1, Tsushima-naka, Okayama 700-8530, Japan
[2] Department of Mathematics, School of Science, Tokai University
1117, Kitakaname, Hiratsuka 259-1292 Japan

Summary: In the sensory evaluation, Scheffé's paired comparisons are theoretically interesting and also practically used many times. However, the number of its experiments is so large that there are a few cases in which its experiments can not be carried out. On those occasions, Scheffé's method can not be used. The current study proposes a generalized modification of Scheffé's method and demonstrates the parameter estimation and testing, with a manageable number of experiments by keeping the minimum structure of Scheffé's model. To do this, we suppose the following assumption for all experiments. (1) The scores do not have order effects. (2) Combination effect may exists in a few combination, and the value is constant γ. (3) The observation is considered a random sample drawn from the same population. It is of interest to see if the number of experiments can be reduced without sacrificing the estimation process.

1. Introduction

Consider a set of n objects, T_i, $i = 1, 2, \cdots, n$, and let us indicate by x_{ijk} the score given to the paired comparison of (T_i, T_j) by judge k, $k = 1, 2, \cdots, r$.

The score x_{ijk} is now considered to have the following structure,

$$x_{ijk} = \alpha_i - \alpha_j + \gamma_{ij} + e_{ijk}. \tag{1}$$

where $\sum_{i=1}^{n} \alpha_i = 0$, and e_{ijk} are identically distributed as an independent normal distribution $N(0, \sigma^2)$.

Table 1 shows the combinations of objects and judges. This means that combination effects exist only for the case of $i = 1$ and $j = 2, \cdots, h$ $(h < n)$, and other cases have no combination effects.

Table 1: The structure of (i,j,k)

i	j	k
1	2	1
		2
		:
		r
	3	1
		2
		:
		r
	:	:
	h	1
		2
		:
		r

i	j	k
1	$h+1$	1
	$h+2$	1
	:	:
	n	1
2	3	1
	4	1
	:	:
	n	1
:	:	:
$n-1$	n	1

Then,

(i) $E(x_{ijk}) = \begin{cases} \alpha_i - \alpha_j + \gamma_{ij} & (i = 1, j = 2, \cdots, h) \\ \alpha_i - \alpha_j \ (that\ is, \gamma_{ij} = 0) & (otherwise) \end{cases}$,

$$\sum_{i=1}^{n} \alpha_i = 0$$

(ii) $Var(x_{ijk}) = \sigma^2$.

(iii) x_{ijk} are distributed according to $N(0, \sigma^2)$ and are independent random variables.

A method of analyzing paired comparison experiments is developed in this paper for experiments in which preferences are expressed on a scale of several ordered points.

In a 5-point scoring system the judge presented with the ordered pair (T_i, T_j) makes one of the following 5 statements.

score   (2) I prefer i to j moderately.   (-1) I prefer j to i slightly.
        (1) I prefer i to j slightly.   (-2) I prefer j to i moderately.
        (0) No preference.

The corresponding values of the scores x_{ijk} might be those shown in parentheses. In any event it is assumed that the numerical scores reflect the order relations with 0 as the no-preference point.

We therefore do not need a new symbol for the preference of the judges for i over j when presented in the order (j,i), since it would be $-x_{jik}$ for the k-th judge presented with the pair in this order.

2. Analysis

When there are n objects and r judges, the formula (1) can be expressed as follows;

$$
\left\{
\begin{aligned}
z_{121} &= \alpha_1 - \alpha_2 + \gamma_{12} + \epsilon_{121}\\
z_{122} &= \alpha_1 - \alpha_2 + \gamma_{12} + \epsilon_{122}\\
&\ \ \vdots\\
z_{12r} &= \alpha_1 - \alpha_2 + \gamma_{12} + \epsilon_{12r}\\
z_{131} &= \alpha_1 - \alpha_3 + \gamma_{13} + \epsilon_{131}\\
z_{132} &= \alpha_1 - \alpha_3 + \gamma_{13} + \epsilon_{132}\\
&\ \ \vdots\\
z_{13r} &= \alpha_1 - \alpha_3 + \gamma_{13} + \epsilon_{13r}\\
&\ \ \vdots\\
z_{1h1} &= \alpha_1 - \alpha_h + \gamma_{1h} + \epsilon_{1h1}\\
z_{1h2} &= \alpha_1 - \alpha_h + \gamma_{1h} + \epsilon_{1h2}\\
&\ \ \vdots\\
z_{1hr} &= \alpha_1 - \alpha_h + \gamma_{1h} + \epsilon_{1hr}\\
z_{1,h+1,1} &= \alpha_1 - \alpha_{h+1} + \epsilon_{1,h+1,1}\\
z_{1,h+2,1} &= \alpha_1 - \alpha_{h+1} + \epsilon_{1,h+2,1}\\
&\ \ \vdots\\
z_{1n1} &= \alpha_1 - \alpha_n + \epsilon_{1n1}\\
z_{231} &= \alpha_2 - \alpha_3 + \epsilon_{231}\\
z_{241} &= \alpha_2 - \alpha_4 + \epsilon_{241}\\
&\ \ \vdots\\
z_{2n1} &= \alpha_2 - \alpha_n + \epsilon_{2n1}\\
z_{341} &= \alpha_3 - \alpha_4 + \epsilon_{341}\\
z_{351} &= \alpha_3 - \alpha_5 + \epsilon_{351}\\
&\ \ \vdots\\
z_{3,h-1,1} &= \alpha_3 - \alpha_{h-1} + \epsilon_{3,h-1,1}\\
z_{3h1} &= \alpha_3 - \alpha_h + \epsilon_{3h1}\\
z_{3,h+1,1} &= \alpha_3 - \alpha_{h+1} + \epsilon_{3,h+1,1}\\
&\ \ \vdots\\
z_{3n1} &= \alpha_3 - \alpha_n + \epsilon_{3n1}\\
&\ \ \vdots\\
z_{h-1,h,1} &= \alpha_{h-1} - \alpha_h + \epsilon_{h-1,h,1}\\
z_{h-1,h+1,1} &= \alpha_{h-1} - \alpha_{h+1} + \epsilon_{h-1,h+1,1}\\
&\ \ \vdots\\
z_{h-1,n,1} &= \alpha_{h-1} - \alpha_n + \epsilon_{h-1,n,1}\\
z_{h,h+1,1} &= \alpha_h - \alpha_{h+1} + \epsilon_{h,h+1,1}\\
z_{h,h+2,1} &= \alpha_h - \alpha_{h+2} + \epsilon_{h,h+2,1}\\
&\ \ \vdots\\
z_{hn1} &= \alpha_h - \alpha_n + \epsilon_{hn1}\\
z_{h+1,h+2,1} &= \alpha_{h+1} - \alpha_{h+2} + \epsilon_{h+1,h+2,1}\\
&\ \ \vdots\\
z_{h+1,n,1} &= \alpha_{h+1} - \alpha_n + \epsilon_{h+1,n,1}\\
&\ \ \vdots\\
z_{n-1,n,1} &= \alpha_{n-1} - \alpha_n + \epsilon_{n-1,n,1}
\end{aligned}
\right.
\tag{2}
$$

where, $\sum_{i=1} a_i = 0$, e_{ijk} are each identically distributed as an independent normal distribution $N(0, \sigma^2)$.

The current model can be expressed in matrix notation as follows:

$$
\begin{aligned}
\mathbf{x} &= \mathbf{A\Theta} + \mathbf{e} \\
&= (\mathbf{A_1}, \mathbf{A_2}) \begin{pmatrix} \mathbf{\Theta_1} \\ \mathbf{\Theta_2} \end{pmatrix} + \mathbf{e} \\
&= \mathbf{A_1\Theta_1} + \mathbf{A_2\Theta_2} + \mathbf{e}
\end{aligned}
\tag{3}
$$

Where $\mathbf{x}$ is an observation vector, $\mathbf{A}$ a design matrix, $\mathbf{\Theta}$ the matrix of parameters, and $\mathbf{e}$ an error vector. The design matrix and the corresponding parameter matrix are each partitioned in the following way: $\mathbf{A}$ partitioned into $\mathbf{A_1}$ and $\mathbf{A_2}$ the part associated with $\alpha_1, \cdots, \alpha_n$ and the part associated with $\gamma_{12}, \cdots, \gamma_{1h}$. $\mathbf{\Theta}$ partitioned into $\mathbf{\Theta_1}$ and $\mathbf{\Theta_2}$ the part associated with $\alpha_1, \cdots, \alpha_n$ and the part associated with $\gamma_{12}, \cdots, \gamma_{1h}$.

${}^t\mathbf{AA}$ is a singular matrix. Thus any solution matrix $\hat{\mathbf{\Theta}}$ from the normal equation ${}^t\mathbf{AA\Theta} = {}^t\mathbf{Ax}$ is the least square estimator.

For N-dimensional vector $\mathbf{R}^n$, if the subspace of $\mathbf{R}^n$ spanned by $\mathbf{A}$ column vector to $\mathbf{R}^n$ of N-dimensional vector is shown $L(\mathbf{A})$, $L_{\mathbf{A}}^{\perp}$ of all the vertical vectors to $L(\mathbf{A})$ makes the subspace of $\mathbf{R}^n$. When $L(\mathbf{A})$ has a subspace $L(\mathbf{A_1})$, in $L(\mathbf{A})$ all the vertical vectors vertical to $L(\mathbf{A_1})$ makes a subspace $L_{\mathbf{A_1}}^{\perp}(\mathbf{A})$. That is to say, $L(\mathbf{A})$'s orthogonal complement in the relation of $L(\mathbf{A_1})$ is $L_{\mathbf{A_1}}^{\perp}(\mathbf{A})$ and the orthogonal complement in the relation of $L(\mathbf{A})$ is $L_{\mathbf{A}}^{\perp}$.

$\mathbf{R}^n$ is divided into a direct sum as follows:

$$
\mathbf{R}^n = L(\mathbf{A_1}) \ominus L_{\mathbf{A_1}}^{\cdot}(\mathbf{A}) \ominus L_{\mathbf{A}}^{\perp}
$$

If the projection matrix to $L(\mathbf{A})$ is $\mathbf{G}$, the projection matrix to each subspace are $\mathbf{G_1}$, $\mathbf{G} - \mathbf{G_1}$ and $\mathbf{I} - \mathbf{G}$.

where

$$
\mathbf{I}_n = \mathbf{G_1} + (\mathbf{G} - \mathbf{G_1}) + (\mathbf{I} - \mathbf{G})
\tag{4}
$$

Multiply (4) by ${}^t\mathbf{x}$ from the left and by $\mathbf{x}$ from the right, then we obtained the following formula (5).

$$
\begin{aligned}
{}^t\mathbf{xx} &= {}^t\mathbf{xG_1x} + {}^t\mathbf{x(G - G_1)x} + {}^t\mathbf{x(I - G)x} \\
&= S_1 + (S - S_1) + S_e \quad (\text{where } S = {}^t\mathbf{xGx}, \; S_1 = {}^t\mathbf{xG_1x})
\end{aligned}
\tag{5}
$$

Next, seek $\mathbf{G}$ and $\mathbf{G_1}$. Find out $\mathbf{G} = \mathbf{AB}$ which is the projection matrix from $\mathbf{B}$ to $L(\mathbf{A})$ which satisfies ${}^t\mathbf{AAB} = {}^t\mathbf{A}$. And find out the projection matrix $\mathbf{G_1} = \mathbf{A_1B_1}$ from $\mathbf{B_1}$ to $L(\mathbf{A_1})$, which satisfies ${}^t\mathbf{A_1A_1B_1} = {}^t\mathbf{A_1}$.

$\mathbf{G}$ is the projection matrix to $L(\mathbf{A})$ and $\mathbf{G_1}$ is the projection matrix to $L(\mathbf{A_1})$. The ranks of the projection matrices satisfy the following identities:

$$
rank\mathbf{G} = tr\mathbf{G}, rank\mathbf{G_1} = tr\mathbf{G_1}, rank(\mathbf{G} - \mathbf{G_1}) = tr(\mathbf{G} - \mathbf{G_1})
$$

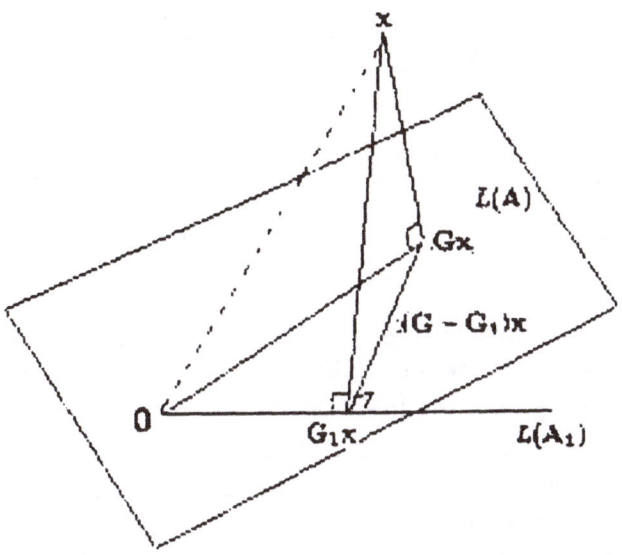

Figure 1: graphical expression of analysis

The projection matrix to $L_{\mathbf{A}_1}^{\perp}(\mathbf{A})$ is $\mathbf{G} - \mathbf{G}_1$.
    The graphic expression is shown in Figure 1.

3. Calculation of the Projection Matrices $\mathbf{G}$ and $\mathbf{G}_1$

Consider the matrices $\mathbf{G}$ and $\mathbf{G}_1$. To obtain these matrices, we find out $\mathbf{B}$ which satisfies the following.

$$^t\mathbf{AAB} = {}^t\mathbf{A}$$

By the matrix $\mathbf{B}$, we have the projection matrix from $\mathbf{B}$ to $L(\mathbf{A})$.

$$\mathbf{G} = \mathbf{AB}$$

$\mathbf{G}$ can be written as

$$\mathbf{G} = \begin{pmatrix} \mathbf{Y} & & \\ & \mathbf{Z} & {}^t\mathbf{T}' \\ & \mathbf{T}' & \mathbf{T} \end{pmatrix}$$

where $\mathbf{G}$ is $\left\{ (h-1)r + \left(\frac{-h^2-3h+2nh}{2} + 1 \right) + \frac{(n-h-1)(n-h-2)}{2} \right\} \times$
$\left\{ (h-1)r + \left(\frac{-h^2-3h+2nh}{2} + 1 \right) + \frac{(n-h-1)(n-h-2)}{2} \right\}$ square matrix.

Table 2: Analysis of variance for α

space	sum of squares	d.f	Mean Square
$L(\mathbf{A_1})$	$S_1 = {}^t\mathbf{x}\mathbf{G_1}\mathbf{x}$	$n-1$	${}^t\mathbf{x}\mathbf{G_1}\mathbf{x}/(n-1)$
$L_{\bar{\mathbf{A}}_1}^{\perp}(\mathbf{A})$	$S - S_1$	$(h-1)r - 2h$ $+\frac{n^2-5n+8}{2}$	
$L_{\bar{\mathbf{A}}}^{\perp}$	$S_e = {}^t\mathbf{x}(\mathbf{I} - \mathbf{G})\mathbf{x}$	$n + h - 2$	${}^t\mathbf{x}(\mathbf{I} - \mathbf{G})\mathbf{x}/(n + h - 2)$
$\mathbf{R^n}$	${}^t\mathbf{x}\mathbf{x}$	$(h-1)r - h$ $+\frac{n^2-n+2}{2}$	

The $(h-1)r \times (h-1)r$ square matrix $\mathbf{Y}$ has the element $\mathbf{Y_1}$ in its main diagonal positions and zero in all other locations and $r \times r$ matrix $\mathbf{Y_1}$ has the all elements $\frac{1}{r}$. $\mathbf{Z}$ is the $\left(\frac{-h^2-3h+2nh}{2} + 1\right) \times \left(\frac{-h^2-3h+2nh}{2} + 1\right)$ matrix and $\mathbf{T}$ is the $\frac{(n-h-1)(n-h-2)}{2} \times \frac{(n-h-1)(n-h-2)}{2}$ matrix.

And then seek $\mathbf{G_1} = \mathbf{A_1}\mathbf{B_1}$ which is projection matrix from $\mathbf{B_1}$ to $L(\mathbf{A_1})$, which satisfies ${}^t\mathbf{A_1}\mathbf{A_1}\mathbf{B_1} = {}^t\mathbf{A_1}$.

Matrix $\mathbf{G_1}$ can be written as

$$\mathbf{G_1} = \begin{pmatrix} \mathbf{U} & \mathbf{R} & \\ \mathbf{V} & \mathbf{S} & {}^t\mathbf{T'} \\ & \mathbf{T'} & \mathbf{T} \end{pmatrix}$$

where $\mathbf{G_1}$ is the matrix of $\left\{(h-1)r + \left(\frac{-h^2-3h+2nh}{2} + 1\right) + \frac{(n-h-1)(n-h-2)}{2}\right\} \times \left\{(h-1)r + \left(\frac{-h^2-3h+2nh}{2} + 1\right) + \frac{(n-h-1)(n-h-2)}{2}\right\}$. $\mathbf{U}$ is $(h-1)r \times (h-1)r$ matrix. $\mathbf{S}$ is the $\left(\frac{-h^2-3h+2nh}{2} + 1\right) \times \left(\frac{-h^2-3h+2nh}{2} + 1\right)$ matrix and ${}^t\mathbf{V} = \mathbf{R}$.

From the result mentioned above, the analysis of variance for α can be shown as in Table 2.

4. Matrix $\mathbf{G_2}$

In the same way, the analysis of variance for γ can be obtained by the division of $\mathbf{R^n}$. $\mathbf{R^n}$ is divided into a direct sum as follows:

$$\mathbf{R^n} = L(\mathbf{A_2}) \ominus L_{\bar{\mathbf{A}}_2}^{\perp}(\mathbf{A}) \ominus L_{\bar{\mathbf{A}}}^{\perp}$$

If the projection matrix to $L(\mathbf{A})$ is $\mathbf{G}$, the projection matrices to subspaces are $\mathbf{G_2}$, $\mathbf{G} - \mathbf{G_2}$ and $\mathbf{I} - \mathbf{G}$, where

$$\mathbf{I_n} = \mathbf{G_2} + (\mathbf{G} - \mathbf{G_2}) + (\mathbf{I} - \mathbf{G}) \tag{6}$$

Multiply (6) by ${}^t\mathbf{x}$ from the left and by $\mathbf{x}$ from the right, and then the following is obtained.

$${}^t\mathbf{x}\mathbf{x} = {}^t\mathbf{x}\mathbf{G_2}\mathbf{x} + {}^t\mathbf{x}(\mathbf{G} - \mathbf{G_2})\mathbf{x} + {}^t\mathbf{x}(\mathbf{I} - \mathbf{G})\mathbf{x}$$

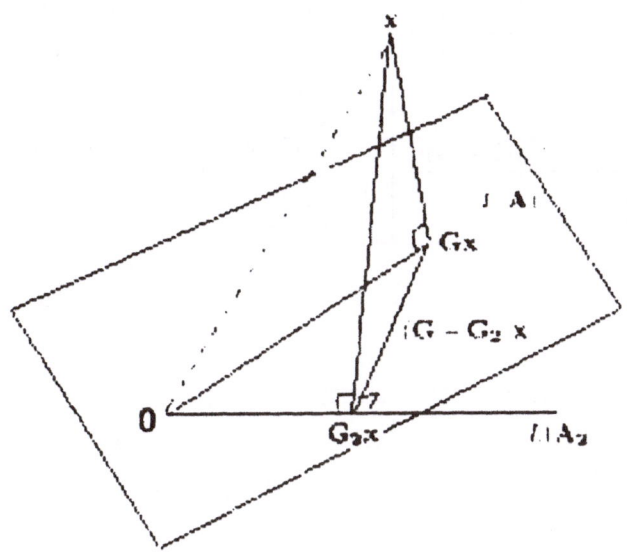

Figure 2: graphical expression of analysis

$$= \quad S_2 + (S - S_2) + S_e \quad (where\ S = {}^t\mathbf{x}\mathbf{G}\mathbf{x},\ S_2 = {}^t\mathbf{x}\mathbf{G}_2\mathbf{x})$$

The graphic expression of the analysis above is shown in Figure 2.
The projection matrix to $\mathbf{L}(\mathbf{A_2})$ is $\mathbf{G_2}$.
Matrix $\mathbf{G_2}$ can be written as

$$\mathbf{G}_2 = \begin{pmatrix} \mathbf{Y} & \mathbf{O} \\ \mathbf{O} & \mathbf{O} \end{pmatrix}$$

We have ${}^t\mathbf{x}\mathbf{G_2}\mathbf{x}$. Thus the analysis of variance for γ is as shown in Table 3.

5. Estimation of Parameters α, γ and σ^2

Solve the simultaneous equation ${}^t\mathbf{A}\mathbf{A}\Theta = {}^t\mathbf{A}\mathbf{x}$, $\displaystyle\sum_{i=1}^{n} \alpha_i = 0$.

Then, we have

Table 3: Analysis of variance for γ

space	sum of squares	d.f	Mean Square
$L(\mathbf{A_2})$	$S_2 = {}^t\mathbf{x}\mathbf{G_2}\mathbf{x}$	$h-1$	${}^t\mathbf{x}\mathbf{G_2}\mathbf{x}/(h-1)$
$L_{\mathbf{A_2}}^{\perp}(\mathbf{A})$	$S - S_2$	$(h-1)r - 3h$ $+\frac{n^2-3n+8}{2}$	
$L_{\mathbf{A}}^{\perp}$	$S_e = {}^t\mathbf{x}(\mathbf{I}-\mathbf{G})\mathbf{x}$	$n+h-2$	${}^t\mathbf{x}(\mathbf{I}-\mathbf{G})\mathbf{x}/(n+h-2)$
$\mathbf{R}^n$	${}^t\mathbf{x}\mathbf{x}$	$(h-1)r - h$ $+\frac{n^2-n+2}{2}$	

$$
\begin{cases}
\hat{\alpha}_1 = \frac{(n-1)x_1-(x_2+x_3+\cdots+x_h)-n(y_1+y_2+\cdots+y_{h-1})}{n(n-h)} \\[4pt]
\hat{\alpha}_2 = \frac{-(n-1)x_1+(n^2-nh+1)x_2+(x_3+x_4+\cdots+x_h)+n(n-h-1)y_1-n(y_2+y_3+\cdots+y_{h-1})}{n(n-1)(n-h)} \\[4pt]
\hat{\alpha}_3 = \frac{-(n-1)x_1+(n^2-nh+1)x_3+(x_2+x_4+\cdots+x_h)+n(n-h-1)y_2+n(y_1+y_3+\cdots+y_{h-1})}{n(n-1)(n-h)} \\
\vdots \\
\hat{\alpha}_h = \frac{-(n-1)x_1+(n^2-nh+1)x_h+(x_2+x_3+\cdots+x_{h-1})+n(n-h-1)y_{h-1}+n(y_1+y_2+\cdots+y_{h-2})}{n(n-1)(n-h)} \\[4pt]
\hat{\alpha}_{h+1} = \frac{x_{h+1}}{n} \\[2pt]
\hat{\alpha}_{h+2} = \frac{x_{h+2}}{n} \\
\vdots \\
\hat{\alpha}_n = \frac{x_n}{n} \\[2pt]
\hat{\gamma}_{12} = \frac{y_1}{r} \\
\quad + \frac{-n(n-1)x_1+n(n-h-1)x_2+n(x_3+x_4+\cdots+x_h)+n(2n-h-2)y_1+n^2(y_2+y_3+\cdots+y_{h-1})}{n(n-1)(n-h)} \\[4pt]
\hat{\gamma}_{13} = \frac{y_2}{r} \\
\quad + \frac{-n(n-1)x_1+n(n-h-1)x_3+n(x_2+x_4+\cdots+x_h)+n(2n-h-2)y_2+n^2(y_1+y_3+\cdots+y_{h-1})}{n(n-1)(n-h)} \\
\vdots \\
\hat{\gamma}_{1h} = \frac{y_1}{r} \\
\quad + \frac{-n(n-1)x_1+n(n-h+1)x_h+n(x_2+x_3+\cdots+x_{h-1})+n(2n-h-2)y_{h-1}+n^2(y_1+y_2+\cdots+y_{h-2})}{n(n-1)(n-h)}
\end{cases}
\tag{7}
$$

where, x_i $(i = 1, 2, \cdots, h, \cdots, n)$ and y_j $(j = 1, 2, \cdots, h-1)$ indicate the elements of ${}^t\mathbf{A}\mathbf{x}$.

Here, each elements of ${}^{t}\mathbf{Ax}$ are

$$
\begin{cases}
x_1 = x_{121} + \cdots + x_{12r} + x_{131} + \cdots + x_{13r} + \cdots + x_{1h1} \\
\qquad + \cdots + x_{1hr} + x_{1,h+1,1} + \cdots + x_{1,n,1} \\
\quad = \sum_{i=1}^{r} x_{12i} + \sum_{i=1}^{r} x_{13i} + \cdots + \sum_{i=1}^{r} x_{1hi} + x_{1,h+1,1} + \cdots + x_{1,n,1} \\
\quad = x_{12\bullet} + x_{13\bullet} + \cdots + x_{1h\bullet} + x_{1,h+1,1} + \cdots + x_{1,n,1} \\
x_2 = -(x_{121} + \cdots + x_{12r}) + x_{231} + x_{241} + \cdots + x_{2n1} \\
\quad = -\sum_{i=1}^{r} x_{12i} + \sum_{j=3}^{n} x_{2j1} \\
\quad = -x_{12\bullet} + x_{2\bullet 1} \\
x_3 = -(x_{131} + \cdots + x_{13r}) - x_{231} + x_{341} + x_{351} + \cdots + x_{3n1} \\
\quad = -\sum_{i=1}^{r} x_{13i} - x_{231} + \sum_{j=4}^{n} x_{3j1} \\
\quad = -x_{13\bullet} - x_{231} + x_{3\bullet 1} \\
\vdots \\
x_h = -(x_{1h1} + \cdots + x_{1hr}) - x_{2h1} - \cdots - x_{h-1,h,1} + x_{h,h+1,1} + \cdots + x_{hn1} \\
\quad = -\sum_{j=1}^{r} x_{1hj} - \sum_{i=2}^{h-1} x_{ih1} + \sum_{k=h+1}^{n} x_{hk1} = -x_{1h\bullet} - x_{\bullet h1} + x_{h\bullet 1} \\
x_{h+1} = -x_{1,h+1,1} - x_{2,h+1,1} - \cdots - x_{h,h+1,1} + x_{h+1,h+2,1} + \cdots + x_{hn1} \\
\quad = -\sum_{j=1}^{h} x_{j,h+1,1} + \sum_{i=h+2}^{n} x_{h+1,i,1} = x_{h+1,\bullet,1} \\
\vdots \\
x_n = -x_{1n1} - x_{2n1} - \cdots - x_{n-1,n,1} = -\sum_{i=1}^{n-1} x_{in1} \\
\quad = -x_{\bullet n1} = x_{n\bullet 1} \\
y_1 = x_{12\bullet} \\
y_2 = x_{13\bullet} \\
\vdots \\
y_{h-1} = x_{1n\bullet}
\end{cases}
$$

$$
\hat{\sigma}^2 = \frac{S_e}{n + h - 2} \tag{8}
$$

6. Conclusions

Although we have a few assumption, we obtain the number of experiment $\frac{n(n-1)}{2} + (h-1)(r-1)$. Based upon the above, an analysis on "estimation" and "testing" can be obtained for the cases of smaller sample sizes than $\frac{rn(n-1)}{2}$.

Based on this analysis, it also can be proved that in accordance with our objectives, an analysis can be obtainable for the number of experiments $\frac{n(n-1)}{2} + (h-1)(r-1)$.

References:

Ujiie. K and Iizuka. M (2000). A Generalized Modification of Scheffé's Paired Comparisons -Theoretical Approach to Decrease the Number of Experiments-. *Proceedings of the International Conference on Measurement and Multivariate Analysis*, **2**, 234–237. Banff, Alberta, Canada

Ujiie. K, Iizuka. M and Nonaka. T (1996). Modified Scheffé's paired comparisons. *Proceeding s of the 10th Symposium of the Japanese Society of Computational Statistics.* 1-8-111

Ujiie.K and Nonaka. T (1996). Modified Scheffé's paired comparisons -Theoretical Approach to Decrease the Number of Experiments-. Proceedings of the 26th symposium on the sensory analysis. *Union of Japanese Scientists and Engineers (JUSE),* 189-196 (in Japanese)

Scheffé,H. (1952). An analysis of variance for paired comparisons. *Journal of American Statistical Assosiation,* **47**. 381-400

Analysing Dependence in Large Contingency Tables: NonSymmetric Correspondence Analysis and Regression with Optimal Scaling

Pieter M. Kroonenberg

Department of Education, Leiden University
Wassenaarseweg 52, Leiden, The Netherlands

Summary: In this presentation a brief survey is presented of the relative merits of two alternative approaches to the problem of modelling dependence for categorical variables when they have more than a few categories. The first approach is categorical regression with optimal scaling. The other technique is nonsymmetric(al) correspondence analysis. On a very general level, it will be shown to what an extent the techniques are similar and different.

1. Introduction

When the number of categories for both the predictors and the dependent variable are fairly large, standard multivariate techniques using dummy coding can become cumbersome, and these techniques often use nearly as many parameters as there are original data points. An alternative is to define transformations for the variables either a priori or empirically. One example of the latter approach is categorical regression with optimal scaling (see Meulman, Heiser, and SPSS, 2000, for a general overview), and another approach is to use nonsymmetric (also referred to as nonsymmetrical) correspondence analysis (see Lauro and D'Ambra, 1984, for the first non-Italian publication, Lauro and Balbi, 1995, for a survey, and Kroonenberg and Lombardo, 1999, for a tutorial). As both techniques deal explicitly with categorical variables and a dependence structure, it is of some interest to evaluate their similarities and differences. The comparison will be made at a fairly general level and for more detailed mathematical descriptions and considerations, one should go to the original publications.

The basic type of data considered in this paper are two-way contingency tables of relative frequencies $P = (p_{ij})(i = 1, \cdots, I; j = 1, \cdots, J)$ with I response categories of row variable Y and J predictor categories of column variable X. In one section, a three-way table with one criterion and two predictors is briefly considered as well, i.e we have a table $P = (p_{ijk})(i = 1, \cdots, I; j = 1, \cdots, J; k = 1, \cdots, K)$ with I categories of the response variable Y and J categories of the predictor variable X and K categories of the predictor variable Z.

2. Theory

An important distinction needs to be made between measuring the size of the dependence and the structure of the dependence. The size of the variability of the criterion accounted for by the predictors is expressed by the multiple correlation coefficient in both standard and optimal scaling regression and by Goodman and Kruskal's (1954)' τ in the nonsymmetric correspondence analysis. In standard regression, the structure of the dependence is contained in the explicit (e.g. linear or quadratic) regression function and the estimated coefficient(s). Because of the categorical nature of the variables, in the nominal case even without any numeric relationships between the categories themselves, no explicit function exists and the structure of the dependence has to be modelled in a more detailed fashion. As an example for optimal scaling regression, next to the regression function for the quantified variables, the analysis employs empirical quantification functions or transformations which are not explicit, but empirically defined. An interpretation of the dependence is therefore not complete without an interpretation of the quantifications and their relationships with the original categories.

In this paper, we will pay attention both to the question of measuring dependence and to the question of modelling and portraying the dependence. Unfortunately, the required length of the paper does not enable an extensive, substantive treatment of a large example.

2.1 Measuring Dependence

The aim of most dependence techniques is to predict the criterion variable Y from one or more predictors X and to measure the increase in predictability. Restricting ourselves for the moment to one predictor, in ordinary regression this is done with the squared multiple correlation R^2, the squared correlation between the criterion Y and the predicted values $\hat{Y}$, which is interpreted as the increase in predictability of Y due to knowledge of X expressed through the predicted values $\hat{Y}$, where Y and $\hat{Y}$ both are numerical. More formally,

$$R^2 = \frac{\sum_{i=1}^{N}\left(\hat{Y}_i - \bar{Y}\right)^2}{\sum_{i=1}^{N}\left(Y_i - \bar{Y}\right)^2} = \frac{SS_{\text{between}}}{SS_{\text{total}}}. \tag{1}$$

On the other hand, Goodman & Kruskal's (1954) τ can be used to measure the increase in predictability if both variables are categorical and it is the difference between prediction of Y based on its marginal distribution, i.e. $p_{i.} = Pr[i]$ and the conditional distribution of Y given column categories of X, $p_{ij}/p_{.j} = Pr[i|j]$. Following Light and Margolin (1971) and Margolin and Light (1974), between and total mean squares can be defined as

$$MS_{\text{between}} = \frac{1}{2}\sum_i\sum_j p_{.j}\left[\frac{p_{ij}}{p_{.j}} - p_{i.}\right]^2 = \frac{1}{2}\sum_i\sum_j p_{.j}\Pi_{ij}^2 \tag{2}$$

$$MS_{\text{total}} = \frac{1}{2}(1 - \sum_i p_{i.}^2). \tag{3}$$

Analogous to the numerical case, we get a measure (i.e. τ) for the increase in predictability by dividing these mean squares,

$$\tau = \frac{MS_{\text{between}}}{MS_{\text{total}}} \tag{4}$$

Note that for two variables τ is asymmetric and measures nominal predictability, while R^2 is symmetric and measures numerical predictability. With more than two variables, also R^2 is no longer symmetric.

2.2 Handling Different Measurement Levels

When some of the variables have nonmetric and possibly different measurement levels, one of the options is to resort to explicit functional transformations of the criterion variable or predictor variables or both. More recently, tools have become available to use optimal scaling for finding empirical transformation functions, such as monotone functions for ordinal measurement. In essence, optimal scaling aims to transform categorical (nominal or ordinal or categorised numerical) measurements into numeric quantifications in accordance with the model that is being fitted. If all variables are nominal, nonsymmetric correspondence analysis, which is also a quantification technique, can also be used as will be explained below.

If standard regression is defined as

$$\hat{\mathbf{y}} = \sum_j \hat{b}_j \mathbf{x}_j \tag{5}$$

then categorical regression with optimal scaling is defined as

$$\mathbf{G}_y \hat{\gamma}_y = \sum_j \hat{\beta}_j \mathbf{G}_j \hat{\xi}_j. \tag{6}$$

The matrix $\mathbf{G}_y$ is an indicator matrix linking the original categories of Y with the predictor quantifications γ_y and the indicator matrices $\mathbf{G}_j$ link the original categories of the predictors X_j with the predictor quantifications ξ_j. Figure 1 gives a pictorial impression of the process of quantification.

Due to the lack of intrinsic standardisation, the quantified variables are always standardised so that the regression coefficients, β_j are always standardised as well. The measure of dependence for optimal scaling regression is still R^2 but the multiple correlation now refers to the quantified predictor variable $\mathcal{Y} = \mathbf{G}_y \gamma_y$ and the quantified predictor variables $\mathcal{X}_j = \mathbf{G}_j \xi_j$.

2.3 Nonsymmetric Correspondence Analysis

Whereas the primary focus of (optimal scaling) regression is on measuring the increase in predictability, nonsymmetric correspondence analysis is far

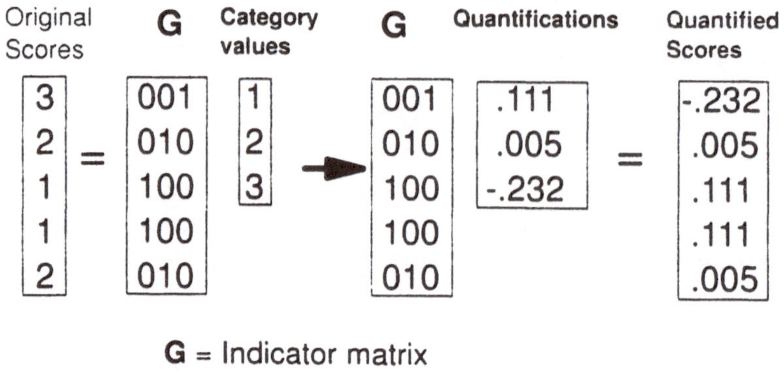

Original Scores	**G**	Category values	**G**	Quantifications		Quantified Scores
$\begin{vmatrix} 3 \\ 2 \\ 1 \\ 1 \\ 2 \end{vmatrix}$	$= \begin{vmatrix} 001 \\ 010 \\ 100 \\ 100 \\ 010 \end{vmatrix}$	$\begin{vmatrix} 1 \\ 2 \\ 3 \end{vmatrix}$	$\rightarrow \begin{vmatrix} 001 \\ 010 \\ 100 \\ 100 \\ 010 \end{vmatrix}$	$\begin{vmatrix} .111 \\ .005 \\ -.232 \end{vmatrix}$	$=$	$\begin{vmatrix} -.232 \\ .005 \\ .111 \\ .111 \\ .005 \end{vmatrix}$

G = Indicator matrix

Figure 1: The process of quantification

more focused on analysing the structure of the dependence. Typically with several categories of the criterion variable, the τ coefficient itself is small and it seems that there is little increase in predictability from knowing the column category an observation is in. This is rather misleading if one looks at the generally much larger associated, and highly siginificant, chi-square, but it is still a somewhat puzzling phenomenon, which τ shares with the inertia in ordinary correspondence analysis (for a discussion, see Kroonenberg and Lombardo, 1999). Nonsymmetric correspondence analysis has, as its name suggests, considerable similarity to that of standard correspondence analysis (see e.g. Greenacre, 1984), but the former technique takes into account that there is a dependence structure between the rows and columns of the contingency table. To uncover the structure of the dependence, the (numerator of) τ is decomposed via quantifications for rows and columns such that the increase in predictability measured by τ due to these quantification is explained as well as possible. Most of the theory has been developed by Lauro, D'Ambra and their co-workers at the University of Naples, see e.g. Lauro and D'Ambra (1984); Siciliano (1992); Lauro and Balbi (1995).

A two-dimensional decomposition of the numerator of τ has the following form:

$$\hat{\Pi}_{ij}^{(2)} = \sum_{s=1}^{2}(a_{is})(\lambda_s b_{js}) = \sum_{s=1}^{2}\gamma_{is}\xi_{js}, \qquad (7)$$

where the rows (belonging to categories of the criterion variable), γ_{is}, are in standard co-ordinates (lengths equal to 1), and the columns (belonging to the categories of the predictor variable), ξ_{js}, are in principal co-ordinates (length equal to the eigenvalues). To view the structure of the dependence we can make use of biplots (Tucker, 1960; Gabriel, 1971) with co-ordinates for rows point $\Gamma_i = (a_{i1}, a_{i2})$ $(\sum_i a_{is}a_{is'} = \delta_{ss'})$ and co-ordinates for the column points $\Xi_j = (\lambda_1 b_{j1}, \lambda_2 b_{j2})$ $(\sum_j p_{.j}b_{js}b_{js'} = \delta_{ss'})$.

From the above it can be seen that also nonsymmetric correspondence analysis quantifies categorical variables, but that independent quantifications exist for each dimension. For the first dimension, $\gamma_i = a_{i1}$, i.e. category i has quantification γ_i and category j has quantification $\xi_i = \lambda_1 b_{j1}$. The original dependent (row) variable y with categories $i = 1, \cdots, I$ is changed into a quantified variable $\mathcal{Y} = \mathbf{G}_y \gamma_y$, and the original predictor variable x with categories $j = 1, \cdots, J$ is changed into a quantified variables $\mathcal{X}_j = \mathbf{G}_x \xi_j$, where the indicator matrices $\mathbf{G}$ link the subjects to the categories. A difference with optimal scaling regression is that there are several independent quantifications rather than only one.

2.4 Nonsymmetric and Standard Correspondence Analysis

Nonsymmetric correspondence analysis and standard correspondence analysis are different in their handling of the relationship between the row and columns variables as can most easily seen from the definitions of the measures of dependence. In standard correspondence analysis the deviation from independence is measured:

$$X_{ij} = \frac{p_{ij}}{p_{i.}p_{.j}} - 1 = \frac{p_{ij} - p_{i.}p_{.j}}{p_{i.}p_{.j}}, \tag{8}$$

while in nonsymmetric correspondence analysis the deviation from the row margin is measured:

$$\Pi_{ij} = \frac{p_{ij}}{p_{.j}} - p_{i.} = \frac{p_{ij} - p_{i.}p_{.j}}{p_{.j}}, \tag{9}$$

This difference has several consequences for interpretation and sensitivity for small row margins (for details, see D'Ambra and Lauro, 1992).

2.5 Increase in Predictability in Three-Way Tables

Without going into much detail, it is instructive to show that the dependence measure τ on which nonsymmetric correspondence analysis is based, behaves exactly like its counterpart the multiple correlation coefficient in ordinary regression. We have seen for a two-way table that Goodman and Kruskal's τ can be used as a measure for the increase in predictability, in particular the increase in knowledge of the row categories given the column categories, while the multiple correlation coefficient measures the increase in predictability of Y given X. This analogy between the two measures extends to three-way and multiway tables as well. In particular, the hierarchical partitioning of the multiple regression coefficient, r^2_{mul}, here illustrated with three variables, and the partial regression coefficient have their parallels in the τ-domain (Gray and Williams, 1975, published in 1981; D'Ambra and Lauro, 1989, 1992; Siciliano, 1992).

Again we take Y to be the criterion variable, but now we have X and Z as predictor variables. The hierarchical formula for multiple regression is

$$r^2_{mul} = r^2_{Y|X \& Z} = r^2_{Y \cdot X} + r^2_{YZ|X}(1 - r^2_{Y \cdot X}) \qquad (10)$$

where $r^2_{YZ|X}$ is squared partial correlation between Y and Z given X, indicating which proportion of the variance of Y which has not yet been explained by X is explained by Z. Furthermore, $r^2_{Y \cdot X}$ is the squared marginal correlation of Y and X, and $r^2_{YZ|X}(1 - r^2_{Y \cdot X})$ is the increase in predictability due to Z given X.

The analogous situation holds for τ. The hierarchical formula for multiple categorical regression based on τ is

$$\tau^2_{mul} = \tau^2_{Y|X \& Z} = \tau^2_{Y \cdot X} + \tau^2_{YZ|X}(1 - \tau^2_{Y \cdot X}) \qquad (11)$$

where $\tau^2_{YZ|X}$ is squared partial τ between Y and Z given X, indicating which proportion of τ of Y which has not yet been explained by X, is explained by Z. Furthermore, $\tau^2_{Y \cdot X}$ is the squared marginal τ of Y and X, and $\tau^2_{YZ|X}(1 - \tau^2_{Y \cdot X})$ is the increase in predictability by Z given X (see also the appendix of Kroonenberg and Lombardo, 1999). The form of the multiple τ is such that for a multiple regression like nonsymmetric correspondence analysis, a juxta-positioning of the slices of the three-way table suffices.

2.6 Optimal Scaling Regression and τ Regression are Different

As indicated above, the (multiple) correlation as measure of increase in prediction is intrinsically symmetric in regression analysis with one predictor. The only asymmetric part is the regression coefficient, but even that asymmetry disappears for standardised variables. In this respect, the τ coefficient is fundamentally different due to its asymmetry, and thus also the analysis of the structure of dependence is fundamentally different whether the row variable or the column variable is seen as the dependent variable. As in optimal scaling regression all quantified variables are standardised, it is completely symmetric given one predictor.

With a single quantification, the quantified scores in correspondence analysis and those of optimal scaling regression are identical . Above we have already seen that the measure of relationship between row and column categories is different between standard correspondence analysis and nonsymmetric correspondence analysis. As optimal scaling with one criterion and one predictor is equal to one-dimensional correspondence analysis, and as correspondence analysis is not the same as nonsymmetric correspondence analysis, optimal scaling regression is not the same as τ regression.

There is an interesting side-effect of the realisation that optimal scaling regression is equal to one-dimensional correspondence analysis. The former procedure as implemented in the *CatReg* module of SPSS (Meulman, Heiser, and SPSS, 2000) has facilities for handling ordinal variables, and thus if one has two ordinal variables, the optimal scaling solution of the regression for these two variables is the same as ordinal correspondence analysis in one dimension. Thus one does not need special programs or algorithms such as Ritov and Gilula (1993).

3. Dream Disturbance Example

In Ritov and Gilula (1993, Table 1) a table is reproduced from Maxwell (1961) in which the severity of dream disturbance of 223 boys is cross-classified with their age. They use these data to demonstrate their procedure of correspondence analysis with order constraints. To illustrate some of the points made above these data have been reanalysed. Also the biplot from a two-dimensional nonsymmetric correspondence analysis is added and discussed.

All correspondence analyses and optimal scaling regressions were carried out with the program *CatReg* contained in the *Categories* module from SPSS, version 10 (Meulman, Heiser and SPSS, 2000[1]) and the nonsymmetric correspondence analysis with the author's freeware program AsymTab available from The Three-Mode Company[2].

3.1 Different Quantifications

In Table 1 the quantifications of all one-dimensional analyses are reported. There are some differences between our correspondence analysis result and that reported in Ritov and Gilula, but the cause is unclear. Because of the small differences in the correlations between quantified variables in the various analyses, that particular aspect is a bit difficult to compare across authors. For the analyses where this was available, the log-likelihood ratio is also given (L^2).

Overall the agreement between comparable analysis is satisfactory, showing that ordinal correspondence analysis via optimal scaling regression leads (at least in this case) to the same results as Ritov and Gilula's EM-algorithm. That nonsymmetric correspondence analysis is different, can be seen in the last column of Table 1. As the predictor variable is treated in the analysis as it would be in standard correspondence analysis, these coefficients are comparable. However, those of the row criterion variable are clearly different.

3.2 Nonsymmetric Correspondence Analysis in Two-Dimensions

To give a flavour of the kind of analysis one gets with nonsymmetric correspondence analysis, we present the biplot for the Dream disturbance data based on a two-dimensional analysis (Figure 1) with dream disturbance as criterion variable and age group as predictor. This figure represents 98 % of the increase in predictability as measured by τ ($\tau = 0.063$ with an approxi-

---

[1] http://www.spss.com/spss10/categories/cat95-98.htm
[2] http://www.fsw.leidenuniv.nl/~kroonenb/genprogs/programs.html

Table 1: Disturbed dreams data: Quantifications of One-dimensional Analyses

	Propor-tion	Unrestricted CatReg	R&G	Ordinal CatReg	R&G	Nonsymmetric Corr.An.
Severity Disturbance						
Very light	.45	1.09	1.10	1.09	1.10	0.86
Light	.19	-0.93	-0.84	-0.70	-0.76	-0.31
Severe	.18	-0.56	-0.68	-0.70	-0.76	-0.19
Very Severe	.18	-1.18	-1.16	-1.26	-1.18	-0.37
Age Groups						
5-7	.09	-0.81	-0.74	-1.23	-1.23	-0.74
8-9	.22	-1.41	-1.42	-1.23	-1.23	-1.42
10-11	.22	0.13	0.12	0.12	0.12	0.10
12-13	.27	0.18	0.16	0.16	0.15	0.16
14-15	.20	1.57	1.58	1.61	1.62	1.59
r		.353	.349	.347	.342	.352
L^2			3.50		4.67	3.65

mate chi-square of 31.5 with 12 df). Of the increase in predictability, 93% is contained in the first dimension and the second dimension only accounts for an additional 5%. The figure clearly shows this largely one-dimensionality in the age groups, and the log-likelihood ratio confirms the one-dimensionality of the solution: for one dimension $L^2 = 3.65$ with 6 df for two dimensions $L^2 = 1.34$ with 2 df.

The origin represents the prediction on the basis of the marginal distribution of dream disturbance. The co-ordinates of dream disturbance are in standard co-ordinates and those of age groups in principal co-ordinates so that we may represent the age groups as vectors and project the disturbance points onto these vector for interpretation. (A few examples can be seen in the graph.) For the 14-15 age group, one predicts a relative larger proportion of very light disturbances compared with the average (i.e. compared with the marginal distribution). This follows from the high projection of Very light on the positive side of the 14-15 group arrow. Young adolescents between 10 and 13 are located near the origin, so that their conditional distributions follow the marginal probability of dream disturbance.The 5-7 and 8-9 year old children are comparable in that they are predicted to be somewhat more often in all dream disturbance categories but very seldom in the very low disturbance one. The extra five percent explained variability from the second dimension arises out of an increased risk for very serious disturbances for the youngest group compared to an increase in serious disturbances for the 8-9 year olds. This would seem to indicate that, if the youngest children are disturbed in their dreams, it tends be more serious.

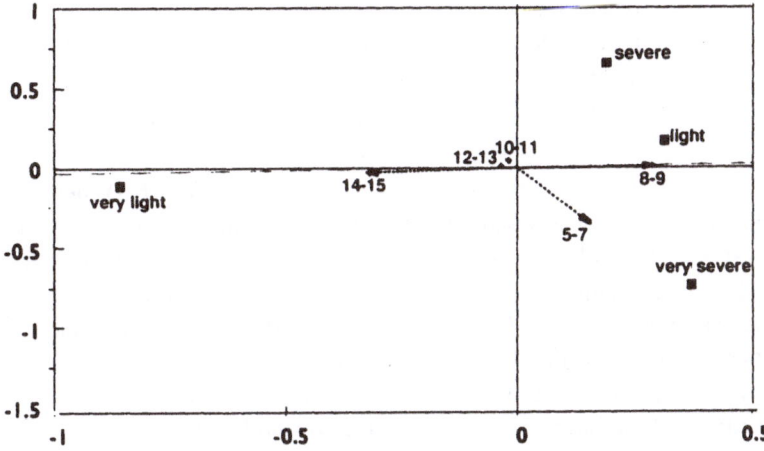

Figure 2: Nonsymmetric Correspondence Analysis of Dream Disturbance Data: Biplot

4. Conclusion

In this paper we have given an exposé of two techniques to handle the regression problem given two categorical variables with a little excursion to three-way tables. It was indicated that nonsymmetric correspondence analysis and optimal scaling regression are fundamentally different techniques even though they both are based on quantifying the original categories of the variables. However, categorical regression with optimal scaling can handle far more different situations than nonsymmetrical correspondence analysis. In particular, it can analyse much larger numbers of variables and as implemented by Meulman, Heiser and SPSS (2000), it can handle many more different types of transformations. On the other hand, nonsymmetric correspondence analysis seems particularly suited for those situations with two or three categorical variables with a fair amount of categories, because then the biplot displays can assist in unravelling the (possibly complex) patterns of increase in predictability due to the predictor variable. An advantage of nonsymmetric correspondence analysis with respect to standard correspondence analysis, is that the interpretation of increase and decrease in predictability due to the knowledge of the category of the predictor, is a much clearer concept than the deviance from independence (see e.g. Kroonenberg and Lombardo, 1999).

References:

D'Ambra, L. and Lauro, N. C. (1989). Non symmetrical analysis of three-way contingency tables. In R. Coppi and S. Bolasco (Eds.), *Multiway data analysis* (pp.

301–315). Amsterdam: Elsevier.

D'Ambra, L. and Lauro, N. C. (1992). Non symmetrical exploratory data analysis. *Statistica Applicata*, 4, 511–529.

Gabriel. K. R. (1971). The biplot graphic display with application to principal component analysis, *Biometrika*, 58, 453–467.

Gifi, A. (1990). *Nonlinear multivariate analysis*. Chicester, UK: Wiley.

Goodman, L. A., and Kruskal, W. H. (1954). Measures of association for cross classifications, *Journal of the American Statistical Association*, 49, 732–764.

Gray, L. N. and Williams, J. S. (1975). Goodman and Kruskal's *tau_b*. Multiple and partial analogs. *Proceedings of the Social Statistics Section of the American Statistical Association* (pp. 444–448). Washington, DC: ASA

Gray, L. N. and Williams, J. S. (1981). Goodman and Kruskal's *tau_b*. Multiple and partial analogs *Sociological Methods and Research*, 10, 50–62.

Greenacre, M. J. (1984). *Theory and applications of correspondence analysis*. London: Academic Press.

Kroonenberg, P. M. and Lombardo, R. (1998). Nonsymmetric correspondence analysis: A tool for analysing contingency tables with a dependence structure. *Multivariate Behavioral Research*. 34, 367–396.

Lauro, N. C. and Balbi, S. (1995). The analysis of structured qualitative data. In A. Rizzi (Ed.), *Some relations between matrices and structures of multidimensional data analysis* (pp. 53–92). Pisa, Italy: CNR.

Lauro, N. C. and D'Ambra, L. (1984). L'Analyse non symétrique des correspondances [Nonsymmetric correspondence analysis]. In E. Diday et al. (Eds.), *Data analysis and Informatics III* (pp. 433–446). Amsterdam: Elsevier.

Light, R. J. and Margolin, B. H. (1971). An analysis of variance for categorical data, *Journal of the American Statistical Association*, 66, 534–544.

Margolin, B. H. and Light, R. J. (1974). An analysis of variance for categorical data, II: Small sample comparisons with Chi Square and other competitors, *Journal of the American Statistical Association*, 69, 755–764.

Maxwell, A. E. (1961). *Analyzing qualitative data*. London: Methuen

Meulman, J. J., Heiser, W. J., and SPSS Inc (2000) *Categories 10.0*. Chicago, IL: SPSS Inc.

Ritov, Y. and Gilula, Z. (1993). Analysis of contingency tables by correspondence models subject to order constraints. *Journal of the American Statistical Association*, 88, 1380–1387.

Siciliano, R. (1992). Reduced-rank models for dependence analysis of contingency tables. *Statistica Applicata*, 4, 481–501.

Tucker, L.R. (1960). Intra-individual and inter-individual multidimensionality. In H. Gulliksen, & S. Messick (Eds.), *Psychological Scaling: Theory and applications* (pp. 155-167). New York: Wiley.

Multidimensional Scaling with Different Orientations of Dimensions for Symmetric and Asymmetric Relationships

Akinori Okada[1] and Tadashi Imaizumi[2]

[1] Department of Industrial Relations, School of Social Relations
Rikkyo (St. Paul's) University
3-34-1 Nishi Ikebukuro, Toshima-ku, Tokyo, Japan 171-8501
[2] Department of Management and Information Sciences
Tama University
4-4-1 Hijirigaoka, Tama City, Tokyo, Japan 206-0022

Summary: A model and an associated nonmetric algorithm for analyzing two-mode three-way asymmetric proximities are presented. The model represents proximity relationships among objects which are common to all sources, the salience of symmetric proximity relationships along dimensions for each source, and the salience of asymmetric proximity relationships along dimensions. The salience of asymmetric proximity relationships is represented by a set of dimensions, which have different orientations from that for the symmetric relationships.

1. Introduction

Two-mode three-way proximities (object × object × source) consist of a set of proximity matrices among a set of objects, where each proximity matrix comes from a different source of data and represents proximities among a set of objects for the source. When each of the matrices is not necessarily symmetric, the matrices form two-mode three-way asymmetric proximities.

Several multidimensional scaling (MDS) procedures for analyzing two-mode three-way asymmetric proximities (two-mode three-way asymmetric MDS) have already been presented (Chino, Grorud, and Yoshino, 1996; DeSarbo, Johnson, Manrai, Manrai, and Edwards, 1992; Okada and Imaizumi, 1997; Zielman,1991; Zielman and Heiser, 1993). Zielman (1991) allows representing the asymmetric component of proximity relationships by using different dimensions from those for the symmetric component. But the other procedures use the same set of dimensions to represent symmetric and asymmetric components. The two components might reflect different backgrounds of proximity relationships among objects, and might show different aspects of proximity relationships, suggesting that they are represented by two different sets of dimensions not by the same set of dimensions.

A hypothetical example, where a different set of dimensions is needed to represent each of symmetric and asymmetric components of relationships among objects, is shown in Figure 1. In the two-dimensional object configuration of Figure 1, four objects (plane figures) are represented. Two sets

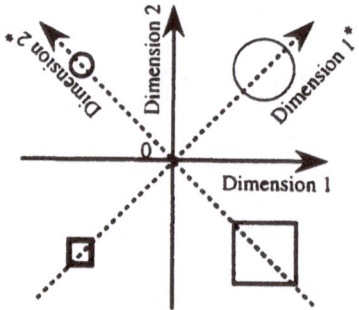

Figure 1: Two sets of dimensions.

of dimensions; dimensions 1 and 2, and dimensions 1* and 2*, are shown. Let us suppose that the symmetric component is represented by dimensions 1 and 2, and that the asymmetric component is represented by dimensions 1* and 2*.

Dimension 1 corresponds to the differences between large and small figures. Dimension 2 corresponds to the differences between squares and circles. Thus, the symmetric component is represented by the size and the shape of the figures. Dimension 1* corresponds to the differences of thickness of the frame of the figures. Dimension 2* corresponds to the differences of darkness of the figures. Thus, the asymmetric component is represented by the thickness of the frame and the darkness of the figures.

The purpose of the present paper is to present a model and an associated nonmetric algorithm for analyzing two-mode three-way asymmetric proximities which allow for different orientations of dimensions to represent symmetric and asymmetric components of proximity relationships among a set of objects (Okada and Imaizumi, 2000b).

2. The Model

The present model is an extended version of the predecessor (Okada and Imaizumi, 1997), and consists of the common object configuration, the symmetry weight configuration, and the set of asymmetry weights. Figure 2 shows the two sets of dimensions; dimensions 1 and 2 represent the symmetric component of proximity relationships among objects, and dimensions 1* and 2*, which are derived by orthogonally rotating dimensions 1 and 2, represent the asymmetric component. In the common object configuration, object j $(j = 1, 2, \cdots, n)$ is represented as a point $(x_{j1}, x_{j2}, \cdots, x_{jt}, \cdots, x_{jp})$ and a circle (sphere, hypersphere) having radius r_j centered at that point in a multidimensional Euclidean space, where x_{jt} is the coordinate of object j

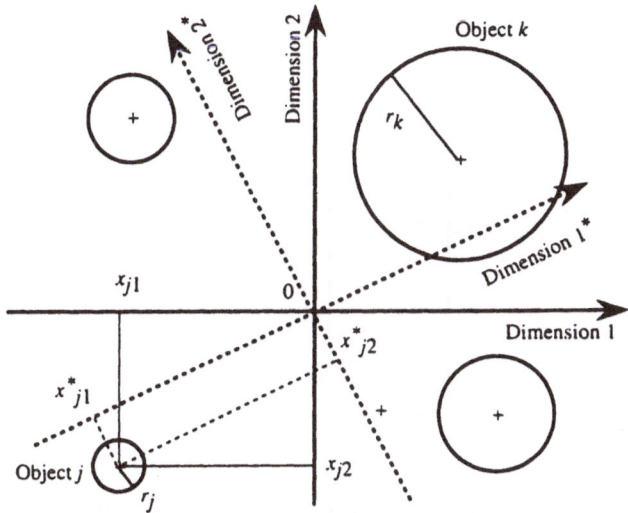

Figure 2: Common object configuration.

on dimension t, n is the number of objects, and p is the dimensionality of the space. Radii r_j are normalized so that the smallest one is zero. The common object configuration is exactly the same as that of the predecessor except for dimensions 1^* and 2^*. x_{js}^* $(s = 1, 2, \cdots, p)$ is the coordinate of object j on dimension s;

$$x_{js}^* = \sum_{t=1}^{p} x_{jt} t_{ts},\qquad(1)$$

where $\mathbf{T}=[t_{ts}]$ is a $p \times p$ orthogonal matrix.

In the symmetry weight configuration, source i $(i = 1, 2, \cdots, N)$ is represented as a point $(w_{i1}, w_{i2}, \cdots, w_{it}, \cdots, w_{ip})$ in a multidimensional space like the subject space of the INDSCAL model (Carroll and Chang, 1970), where $w_{it}(w_{it} > 0)$ shows the salience of the symmetric component of proximity relationships along dimension t for source i, and N is the number of sources (Figure 3). The symmetry weight configuration represents differences among sources in symmetric proximity relationships, and has a uniquely determined orientation of dimensions up to permutations.

The set of asymmetry weights $(u_1^*, u_2^*, \cdots, u_s^*, \cdots, u_p^*)$ consists of weights along p dimensions which are obtained by orthogonally rotating the dimensions of the symmetry weight configuration. The set of asymmetry weights is common to all sources, i.e., the orthogonal rotation is common to all sources. $u_s^*(u_s^* > 0)$ represents the salience of the asymmetric component of proxim-

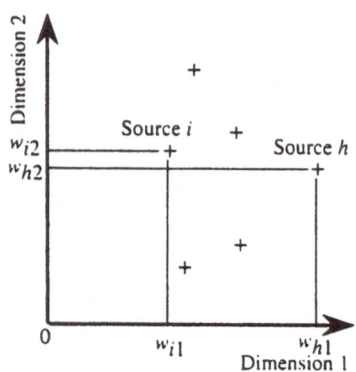

Figure 3: Symmetry weight configuration.

ity relationships along dimension s of the orthogonally rotated dimensions shown by dashed arrows in Figure 2.

In the present model, source i has symmetry weights $(w_{i1}, w_{i2}, \cdots, w_{it}, \cdots, w_{ip})$. which means that each source has its own salience of the symmetric component along each dimension, while in the model of the predecessor, source i has symmetry weight w_i, which means that each source has its own salience of the symmetric component but all dimensions have the same salience. In the present model, all sources have the same asymmetry weights $(u_1^*, u_2^*, \cdots, u_s^*, \cdots, u_p^*)$, which means that each dimension has its own salience of the asymmetric component but each source does not have its own salience of the asymmetric component, while in the model of the predecessor, source i has asymmetry weight $(u_{i1}, u_{i2}, \cdots, u_{it}, \cdots, u_{ip})$, which means that each source has its own salience of the asymmetric component along each dimension.

Each source has its own configuration of objects derived by transforming the common object configuration. In a configuration of objects for a source, each object is represented as a point and an ellipse (ellipsoid, hyperellipsoid) centered at that point in a multidimensional Euclidean space (Figure 4). The object configuration for source i is derived (a) by applying symmetry weight w_{it} onto dimension t of the common object configuration in order to stretch or shrink the configuration of points in the common object configuration along dimension t, and (b) by applying asymmetry weight u_s^* onto the radius along dimension s of the common object configuration in order to transform a circle in the common object configuration to an ellipse. A circle is transformed to an ellipse by stretching or shrinking it along dimensions shown by the dashed arrows in Figures 2 and 4.

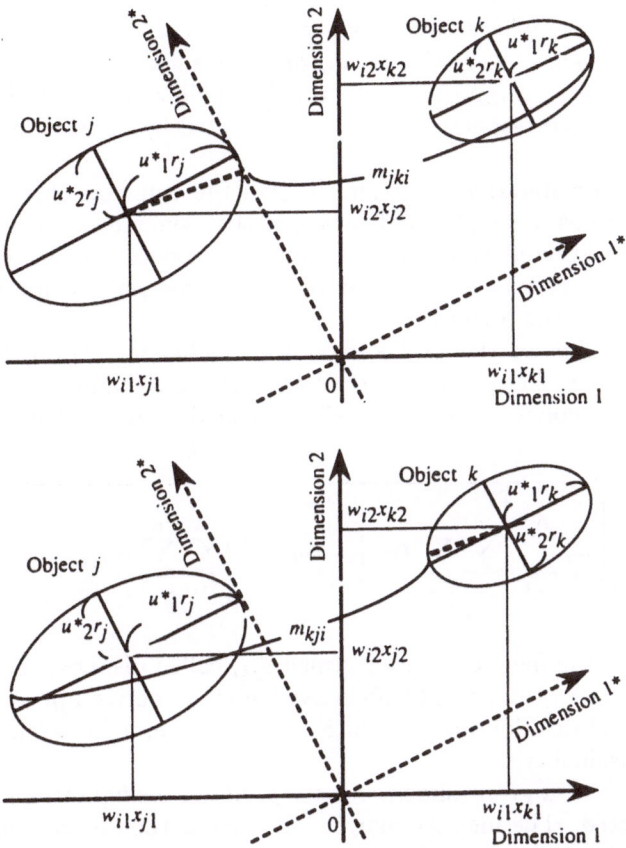

Figure 4: m_{jki} and m_{kji} in the configuration of objects for source i.

3. The Algorithm

Let s_{jki} be the observed similarity from objects j to k for source i. s_{jki} is assumed to be monotonically increasingly related to m_{jki} defined as

$$m_{jki} = d_{jki} - \frac{d_{jki} r_j}{\sqrt{\sum_{s=1}^{p} \left[\frac{x_{js}^* - x_{ks}^*}{u_s^*} \right]^2}} + \frac{d_{kji} r_k}{\sqrt{\sum_{s=1}^{p} \left[\frac{x_{ks}^* - x_{js}^*}{u_s^*} \right]^2}}, \tag{2}$$

where d_{jki} is the distance between two points representing objects j and k in the configuration of objects for source i, and is defined as

$$d_{jki} = \sqrt{\sum_{t=1}^{p} w_{it}^2 (x_{jt} - x_{kt})^2}. \tag{3}$$

In each of upper and lower panels of Figure 4, objects j and k are represented in the two-dimensional configuration of objects for source i. m_{jki} which corresponds to s_{jki} is represented by the gray bold arrow in the upper panel, and m_{kji} which corresponds to s_{kji} is represented by the gray bold arrow in the lower panel.

A nonmetric iterative algorithm derives the common object configuration, the symmetry weight configuration, the asymmetry weights and the orthogonal rotation matrix from those of the symmetry weight configuration by analyzing observed two-mode three-way asymmetric proximities, and it is based on that of the predecessor.

The algorithm of the predecessor is extended by adding the revision needed to obtain the orthogonal rotation matrix. A badness of fit measure of m_{jki} to the monotone transformed s_{jki}, called stress, is defined by

$$S = \sqrt{\frac{1}{N} \sum_{i=1}^{N} \left[\sum_{\substack{j=1 \\ j \neq k}}^{n} \sum_{k=1}^{n} (m_{jki} - \hat{m}_{jki})^2 / \sum_{\substack{j=1 \\ j \neq k}}^{n} \sum_{k=1}^{n} (m_{jki} - \bar{m}_i)^2 \right]}, \tag{4}$$

where $\hat{m}_{jki}$ is the monotone transformed s_{jki} called disparity, and $\bar{m}_i$ is the mean of m_{jki} for source i. The objective here is to derive $x_{jt}, r_j, w_{it}, u_s^*$ and the orthogonal rotation matrix which minimize S in a Euclidean space of a given dimensionality.

In a space of a given dimensionality p, initial configurations and values are constructed. The identity matrix $\mathbf{I}$ is used as the initial transformation matrix $\mathbf{T}$. The present iterative algorithm consists of the following five steps: (a) updating r_j and x_{jt}, (b) normalizing m_{jki} and x_{jt}, (c) updating $\mathbf{T}$ and u_s^*, (d) normalizing m_{jki} and x_{jt}, (e) updating w_{it} for existing $x_{jt}, r_j, \mathbf{T}$ and u_s^*. At each iteration. S is calculated to check for convergence. Figure 5 shows the flow of the algorithm at a given dimensional space, and corresponds to the right hand part of Figure 4 of Okada and Imaizumi (1997).

In step (c). it is not practical to derive the transformation matrix $\mathbf{T}$ and the asymmetry weight u_s^* which minimize S because of the complexity of the computation. A simpler procedure described below is used. Let $\bar{z}_{jk} = \sum_i^N [d_{jki}(r_j - r_k)/(d_{jki} - \hat{m}_{jki})]$. From Equation (2)

$$\bar{z}_{jk}^2 = (\mathbf{x}_j^* - \mathbf{x}_k^*)' \mathbf{U}^{*-1} (\mathbf{x}_j^* - \mathbf{x}_k^*), \tag{5}$$

where $\bar{z}_{jk} = \sum_i^N [d_{jki}(r_j - r_k)/(d_{jki} - \hat{m}_{jki})], \mathbf{x}_j = [x_{j1}, \cdots, x_{jp}], \mathbf{x}_j^* = \mathbf{x}_j \mathbf{T}$, and $\mathbf{U}^*$ is a diagonal matrix whose sth diagonal element is u_s^*. And $\mathbf{T}$ which

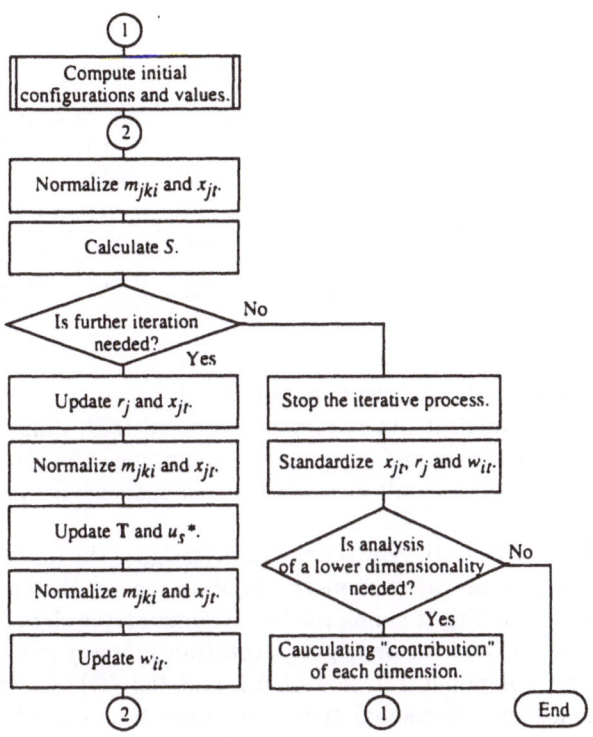

Figure 5: Flow of the algorithm at a given dimensional space.

minimizes the losss function

$$loss function = \left[\sum_{j=1}^{n} \sum_{k=1}^{n} (\bar{z}_{jk}^2 - (\mathbf{x}_j^* - \mathbf{x}_k^*)' \mathbf{U}^{*-1} (\mathbf{x}_j^* - \mathbf{x}_k^*)) \right].$$ (6)

is computed. Equation (6) indicates the procedure of deriving $\mathbf{T}$ ($\mathbf{x}_j^*$ can be obtained once $\mathbf{T}$ is derived) and u_s^* can be formulated as a linear regression problem. In the present procedure, we use only positive $\bar{z}_{jk}$. Let the estimated regression coefficient for the independent variable $x_j \times x_k$ be a_{jk}, and constructing a $p \times p$ matrix $\mathbf{A}$ whose (j, k) element is a_{jk}. Then $\mathbf{T}$ and u_s^* can be derived by decomposing $\mathbf{A}$ as

$$\mathbf{A} = \mathbf{T}' \mathbf{U}^{*-1} \mathbf{T}.$$ (7)

4. Discussion

A model and the associated algorithm of two-mode three-way asymmetric MDS, which allow different orientations of dimensions for symmetric and

Table 1: Results of the simulation.

| No. of | Rotation | S | | R^2 | |
| | | No. of objects | | No. of objects | |
Sources	(Degree)	10	20	10	20
5	0	0.232	0.220	0.891	0.913
	30	0.243	0.224	0.881	0.894
10	0	0.184	0.179	0.920	0.929
	30	0.193	0.181	0.919	0.925

asymmetric components of proximity relationships among a set of objects, are presented. Although it did not come to the surface, differently oriented dimensions for symmetric and asymmetric components might cause a difficulty in representing the result geometrically in spaces with more than two-dimensions.

A small Monte Carlo study was done to investigate the performance of the present two-mode three-way asymmetric MDS to recover the known synthetic structure (a common object configuration, a symmetry weight configuration, asymmetry weights, and an orthogonal rotation matrix) having symmetric and asymmetric components ratios of 0.7 and 0.3 (Okada and Imaizumi, 1997). Random two-dimensional structures were constructed for two levels of the number of objects ($n = 10, 20$), two levels of the number of sources ($N = 5, 10$), and two levels of the rotation from the orientation of dimensions of the symmetry weight configuration to that for asymmetry weights (0, 30 degrees counterclockwise). A synthetic data set was generated by calculating m_{jki} from each of the eight structures ($2 \times 2 \times 2$) and by adding an error term to each m_{jki}. The error term is the random normal deviate with mean 0 and the standard deviation of 0.2 times that of m_{jki}. Five replications were done for each structure. Thus we have 40 synthetic data sets. Each data set was analyzed by the present two-mode three-way asymmetric MDS in two-dimensional space by using rational initial configurations and values, and then was analyzed by using 100 different random initial common object configurations (Okada and Imaizumi, 1997). For each data set, the result showing the smallest S among these 101 results was chosen as the solution. Table 1 shows S and R^2 which is the square of the product moment correlation coefficient between m_{jki} of the synthetic structure and of recovered m_{jki}, in the two-dimensional space. Each figure shows the mean of five replications. The S value is not impressively small, but shows the moderately good fitness (S defined by Equation (3) is based on the stress formula two (Kruskal and Carroll, 1969)). The R^2 shows the synthetic structure has been recovered fairly well. Extensive Monte Carlo studies will be needed to examine more fully the performance of the present two-mode three-way asymmetric MDS.

While the present model represents differences among sources in symmetric proximity relationships by the salience of the symmetric proximity relationships along each dimension for each source, it cannot represent differences among sources in asymmetric proximity relationships (cf. Okada and Imaizumi, 1997). The present model represents the salience of the asymmetric component of proximity relationships along each of the dimensions, which are differently oriented from those of the symmetry weight configuration. The orientation of the dimensions of the symmetry weight configuration is determined by differences among sources in symmetric proximity relationships, and the orientation of the dimensions for the asymmetry weights is determined by differences among dimensions in asymmetric proximity relationships.

Asymmetry weight u_{it}, which is represented by the same dimensions with the symmetry weights, was multiplicatively decomposed as the product of the asymmetry weight for source y_i and the asymmetry weight for dimension u_t (Okada and Imaizumi, 2000c); $u_{it} = y_i u_t$. A series of models of two-mode three-way asymmetric MDS can be generated by multiplicatively decomposing symmetry and asymmetry weights.

Four levels of decompositions are possible. In the case of the asymmetry weight u_{it}, they are; (a) u_{it} (without decomposition), (b) $u_{it} = y_i u_t$, (c) $u_{it} = y_i \cdot 1$, and (d) $u_{it} = 1 \cdot u_t$. In (a), each source has its own salience along each dimension of the asymmetric component of proximity relationships among objects. In (b), each source has its own salience of the asymmetric component, and each dimension has its own salience of the asymmetric component. In (c), each source has its own salience of the asymmetric component, and all dimensions have the same salience of the asymmetric component. In (d), all sources have the same salience of the asymmetric component, and each dimension has its own salience of the asymmetric component. Similar decompositions are possible for asymmetry weight u_{is}^*, which is represented by the dimensions having the different orientation from that for the symmetry weights. Similar four decompositions are possible for symmetry weight w_{it} as well. Combining the four levels of the symmetry weight and the four levels of asymmetry weight u_{it} yields 16 models. Another set of 16 models is generated by combining the four levels of the symmetry weight and the four levels of asymmetry weight u_{is}^*.

The models, which have been previously introduced, are produced by combining one of the four decompositions of symmetry weight and one of the four decompositions of asymmetry weight. The model of Okada and Imaizumi (1997) is generated by combining (c) of the symmetry weight and (a) of the asymmetry weight u_{it}. The model of Okada and Imaizumi (2000a) is obtained by combining (a) of the symmetry weight and (a) of the asymmetry weight u_{it}. The model of Okada and Imaizumi (2000c) is generated by combining (c) of the symmetry weight and (b) of the asymmetry weight u_{it}. The present model, which has been introduced in Okada and Imaizumi

(2000b), is generated by combining (a) of the symmetry weight and (d) of the asymmetry weight u_{is}^*. These models assume that dimensions for the symmetry weight as well as for the asymmetry weight are orthogonal. Models which have oblique dimensions are possible.

Acknowledgment

The authors would like to express their gratitude to two anonymous referees for their constructive and helpful reviews.

References:

Carroll, J. D. and Chang. J. J. (1970). Analysis of individual differences in multidimensional scaling via an N-way generalization of 'Eckart-Young' decomposition. *Psychometrika*, **35**, 283-319.

Chino, N., Grorud, A., and Yoshino, R. (1996). *A complex analysis for two-mode three-way asymmetric relational data*. Proc. of the 5th Conference of the International Federation of Classification Societies at Kobe, Japan (vol. 2), 83-86.

DeSarbo, W. S., Johnson, M. D., Manrai, A. K., Manrai, L. A., and Edwards, E. A. (1992). TSCALE: A new multidimensional scaling procedure based on Tversky's contrast model. *Psychometrika*, **57**, 43-69.

Kruskal, J. B. and Carroll, J. D. (1969). Geometric models and badness-of-fit measures. *In Multivariate Analysis*, Krishnaiah, P. K. (ed.), 639-671. Academic Press, New York.

Okada, A. and Imaizumi, T. (1997). Asymmetric multidimensional scaling of two-mode three-way proximities. *Journal of Classification*, **14**, 195-224.

Okada, A. and Imaizumi, T. (2000a). *A generalization of two-mode three-way asymmetric multidimensional scaling* . Proc. of the 24th Annual Conference of the German Classification Society, 115.

Okada, A. and Imaizumi, T. (2000b). *Multidimensional scaling with different orientations of symmetric and asymmetric dimensions*. Proc. of the International Conference on Measurement and Multivariate Analysis at Banff, Canada (vol. 1), 124-126.

Okada, A. and Imaizumi, T. (2000c). Two-mode three-way asymmetric multidimensional scaling with constraints on asymmetry. *In Classification and Information Processing at the Turn of the Millennium*, Decker, R. et al. (eds.), 52-59. Springer-Verlag, Berlin.

Zielman, B. (1991). *Three-way scaling of asymmetric proximities*. Research Report RR91-01, Department of Data Theory, University of Leiden, Leiden.

Zielman, B. and Heiser, W. J. (1993). Analysis of asymmetry by a slide-vector. *Psychometrika*, **58**, 101-114.

Complex Space Models for the Analysis of Asymmetry

Naohito Chino
Aichi Gakuin University
12 Araike, Iwasaki, Nisshin-city, Aichi, Japan 470-0195

Summary: Two kinds of complex space models are discussed for the analysis of asymmetry. One is the H*E*rmitian *F*orm *A*symmetric multidimensional *S*caling for *I*nterval *D*ata (EFASID), which is a version of *H*ermitian *F*orm *M*odel (HFM) for the analysis of one-mode two-way asymmetric relational data proposed by Chino and Shiraiwa (1993). It was first proposed by Chino (1999). Some results from simulations of EFASID are reported. The other is a possible complex difference system model for the analysis of two-mode three-way asymmetric relational data. Implications of such a complex difference system model are discussed.

1. Introduction

Asymmetric MDS has been advanced in the last quarter of the last century since Young (1975) proposed his ASYMSCAL. A major school or a group proposed augmented distance models in which the observed similarity or dissimilarity is supposed to be represented by a certain *traditional symmetric metric* such as Minkowski's *r*-metric plus a prescribed asymmetric term. Typical such models are the distance-density model (Krumhansl, 1978), the Okada-Imaizumi model (Okada & Imaizumi, 1987) and its two-mode three-way version (Okada & Imaizumi, 1997), the Saito-Takeda model (Saito & Takeda, 1990), the Saito model (Saito, 1991), and the Weeks-Bentler model (Weeks & Bentler, 1982).

Another group proposed some non-distance models which do not assume any traditional metric directly. Typical models of this type are an ASYMSCAL proposed by Chino (Chino, 1977, 1978), GIPSCAL (Chino, 1990), the Gower Diagram (or sometimes called *C*anonical *A*nalysis of *SK*ew-symmetry (CASK)) (Gower, 1977; Constantine & Gower, 1978), the complex coding (or sometimes called the *H*ermitian *C*anonical *M*odel (HCM)) (Escoufier & Grorud, 1980), DEDICOM (Harshman, 1978; Harshman, Green, Wind, & Lundy, 1982), generalized GIPSCAL (Kiers & Takane, 1994).

By contrast, Sato (1988) proposed a real *asymmetric metric* model which utilizes a general asymmetric Minkowski metric.

In spite of the variability of these models stated above, they all embed objects in a *real space*. Recently, however, Chino and Shiraiwa (1993) examined a geometrical structure of the eigenvalue problem of the Hermitian matrix H constructed uniquely from any asymmetric relational data matrix S whose elements are measured in principle at a ratio scale level, and proved that objects are embedded in *complex spaces* depending on the nature of the eigenvalues of H. Here, by "in principle" we mean that HFM can deal with data measured at an interval scale level if we administer the double center-ing transformation to the data. For, this transformation in HFM cancel the additive constant contained in the interval data. As a result we can regard it as if it were a ratio data. But, to be strict, we must estimate the additive constant in such a case by an appropriate method like EFASID discussed in Section 2.

According to the theorem by Chino and Shiraiwa (1993), one can embed objects in a *finite-dimensional complex* (f.d.c.) *Hilbert space* if H is positive or negative semi-definite, and in an *indefinite metric* (complex) *space* if H is indefinite.

Since they fit a *Hermitian form* to each element of H, they call their method the *Hermitian Form Model* (HFM).

HFM is written as

$$H = X\Omega_s X^t + i\, X\Omega_{sk} X^t,$$

where $X = (U_r, U_c)$, and U_r and U_c are the real part and imaginary part of U_1, respectively. U_1 is such that $H = U_1 \Lambda U_1^*$, where U_1^* is the conjugate transpose of U_1. Here, Λ is a diagonal matrix of order $p \leq N$ (the number of objects), whose diagonals are *nonzero*-eigenvalues of H. Ω_s and Ω_{sk} are special symmetric and skew-symmetric matrices of order $2p$, and are written as

$$\Omega_s = \begin{pmatrix} \Lambda, & O_3 \\ O_3, & \Lambda \end{pmatrix}, \quad \Omega_{sk} = \begin{pmatrix} O_3, & -\Lambda \\ \Lambda, & O_3 \end{pmatrix}. \tag{1}$$

The real counterpart of HFM is written as $S = X\Omega_s X^t + X\Omega_{sk} X^t$, or in scalar form as

$$s_{jk} = \sum_{l=1}^{p} \lambda_l s_{jk}^{(l)}, \quad s_{jk}^{(l)} = (\rho_{jl}\rho_{kl} + \sigma_{jl}\sigma_{kl}) - (\rho_{jl}\sigma_{kl} - \sigma_{jl}\rho_{kl}). \tag{2}$$

where ρ_{jl} and σ_{jl} are, respectively, the real part and the imaginary part of the coordinate of object j on dimension l (that is, a complex plane) in a multidimensional complex space.

From the view point of the approximation theory, HFM has some desirable properties. Let us rewrite

$$H = \sum_{l=1}^{N} \lambda_l G_l, G_l = u_l u_l^*,$$

instead of $H = U_1 \Lambda U_1^*$. This decomposition is sometimes called the spectral decomposition, and is a special case of the *generalized Fourier expansion*. Then, it is easy to show that

$$< G_i, G_j > = 0, \quad < H, G_l > = \lambda_l,$$

where $< \bullet, \bullet >$ denotes a Frobenius inner product. Moreover, we have

$$\|H\|_E^2 = \sum_{l=1}^{N} \lambda_l^2,$$

where $\| \bullet \|_E^2$ denotes a Frobenius (Euclidean, or Hilbert- Schmidt) norm.

From the above consideration, we can conclude that HFM is an economical way of approximating asymmetric relational data in a (complex) Hilbert space.

However, there is a shortfall in HFM. For, as stated earlier, HFM assumes that the data are measured in principle at a ratio scale level. In the next section, we shall briefly refer to a method called EFASID which overcomes this shortfall (Chino, 1999), and give some results from simulations pertaining to it.

Chino-Shiraiwa's theorem provides us with a mathematical foundation in developing another approach to the analysis of asymmetry in marked contrast to MDS. Let us suppose that we are given a set of asymmetric longitudinal relational data matrices. If we analyze such a set of data via a *dynamical systems approach*, we must first define the *state space* in which dynamical behavior of objects are described. According to their theorem, we may choose an f.d.c.Hilbert space or indefinite metric space as the state space. In the final section we shall glance at such an approach. EFASID may provide a configuration of objects at the initial point in time.

2. EFASID

In the social and behavioral sciences, it is often the case that the data are measured at an interval scale level or a nominal scale level. Here, we assume that each element of the observed matrix S is measured at an interval level. Then, for fallible data with an additive constant α, the real counterpart of HFM can be written as

$$s_{jk} = \sum_{l=1}^{p} \lambda_l \left\{ (\rho_{jl}\rho_{kl} + \sigma_{jl}\sigma_{kl}) - (\rho_{jl}\sigma_{kl} - \sigma_{jl}\rho_{kl}) \right\} + \alpha + \varepsilon_{jk}, \qquad (3)$$

where ε_{jk} is an error term. In this case, the total number of unknown parameters, except for the error term, amounts to $(2N+1)p + 1(=n)$. Later, we shall denote them by $\theta_1, \theta_2, \cdots, \theta_n$, in order to number them consecutively, or denote them by $\boldsymbol{\theta} = (\theta_1, \theta_2, \cdots, \theta_n)^t$.

There exist *unitary constraints* in the coordinate parameters of HFM. The total number of these constraints is $p^2(= m)$. We shall denote them by $h_1(\boldsymbol{\theta}), h_2((\boldsymbol{\theta}), \cdots, h_m(\boldsymbol{\theta})$ in order to number them consecutively, or by

$$h(\boldsymbol{\theta}) = (h_1(\boldsymbol{\theta}), h_2(\boldsymbol{\theta}), \cdots, h_m(\boldsymbol{\theta}))^t. \qquad (4)$$

Table 1: Original asymmetric data/Chino (1978)

rater\ratee	1	2	3	4	5	6
1		1	−1	−1	0	1/2
2	−1		1	−1	−$\sqrt{2}$	−1/2
3	−1	−1		1	0	−1/2
4	1	−1	−1		$\sqrt{2}$	1/2
5	$\sqrt{2}$	0	−$\sqrt{2}$	0		$\sqrt{2}/2$
6	1/2	1/2	−1/2	−1/2	0	

One way to estimate parameters in (3) is to minimize the following function $\phi(\boldsymbol{\theta})$ under the unitary constraints:

$$\phi(\boldsymbol{\theta}) = \sum_{\substack{j \neq k}}^{N} \sum_{}^{N} c_{jk} \, \varepsilon_{jk}^2 = \sum_{\substack{j \neq k}}^{N} \sum_{}^{N} c_{jk} \left(s_{jk} - \alpha - \sum_{l=1}^{p} \lambda_l \, s_{jk}^{(l)} \right)^2, \qquad (5)$$

where c_{jk} is 1 if s_{jk} is observed, and 0 otherwise.

To do this job, we shall define the following Lagrangian function

$$\psi(\boldsymbol{\theta}) = \phi(\boldsymbol{\theta}) - \boldsymbol{\mu}^t \, \boldsymbol{h}(\boldsymbol{\theta}),$$

where $\boldsymbol{\mu}$ is composed of the Lagrange multipliers, and is written as $\boldsymbol{\mu} = (\mu_1, \mu_2, \cdots, \mu_m)^t$.

In order to obtain an extremum of $\phi(\boldsymbol{\theta})$ subject to equality constraints $\boldsymbol{h}(\boldsymbol{\theta})$. we may solve the following system of equations

$$\begin{pmatrix} \boldsymbol{\theta}_{r+1} \\ \boldsymbol{\mu}_{r+1} \end{pmatrix} = \begin{pmatrix} \boldsymbol{\theta}_r \\ \boldsymbol{\mu}_r \end{pmatrix} - \delta_r \, \boldsymbol{G}_r^+ \, \boldsymbol{g}_r, \qquad (6)$$

where $\boldsymbol{G}_r$ and $\boldsymbol{g}_r$ of order $m + n$, respectively, are the Hessian matrix and the *gradient vector* of ψ. They take special forms in the context of Lagrange's multiplier method. $\boldsymbol{G}_r^+$ is the Moore-Penrose inverse of $\boldsymbol{G}_r$.

As initial estimates of coordinates of objects, we use solutions of HFM to the doubly centered similarity matrix. This double centering transformation transforms the matrix as if it were the one measured at a ratio scale level.

3. Applications

We shall now show the results from applications of EFASID to artificial data which were cited in Chino (1978). Table 1 shows the artificial data, *without diagonals*.

Figure 1 is the artificial configuration of objects which generates the above data via a two-dimensional model of an ASYMSCAL by Chino (1978), which is considered as a special case of GIPSCAL as well as HFM.

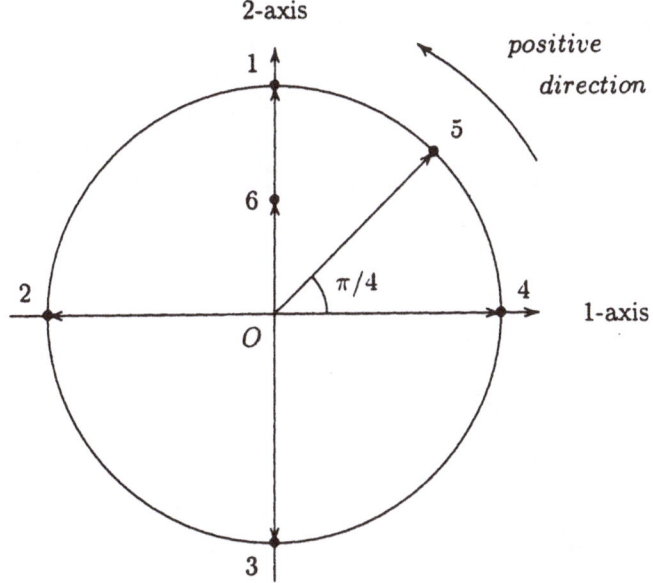

Figure 1: Original artificial configuration of objects

Table 2 shows summary results of EFASID for the artificial data cited above. We added an additive constant to the original data shown in Table 1 in each of the simulation studies, and the estimated additive constant via EFASID for it is shown in the column titled "est.add.c.". In this table, "double.c." is the abbreviation of double centering transformation of the input data.

It is apparent from the summary table that double centering transformation plays an important role in the algorithm of EFASID.

4. Possible Complex Difference Equation Models

Given a set of longitudinal relational data matrices, one may construct various types of dynamical system models if one is interested in examining changes in coordinates of objects in a multidimensional complex space over time, or predicting changes in similarities or dissimilarities among objects over time. For example, one possible model is a multidimensional complex

Table 2: Summary results of EFASID for the artificial data

simulation	add.c.	double.c.	est.add.c.	n.iter.	conv.c.
1	0.0	n	0.0018	2	0.0001
2	5.0	n	−0.2008	2	0.0001
3	5.0	y	5.3209	2	0.0001
4	55.0	y	55.0599	4	0.0001
5	55.0	y	55.0599	4	0.00001

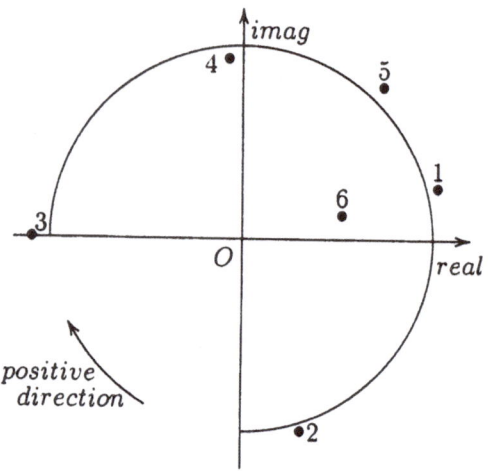

Figure 2: estimated configuration by Simulation 4

difference equation model,

$$z_{j.t+1} = z_{jt} + \sum_{k \neq j}^{N} D_{jkt} \left(z_{kt} - z_{jt} \right),$$

where

$$D_{jkt} = diag \left\{ w_{jkt}^{(1)}, \cdots, w_{jkt}^{(p)} \right\},$$

and

$$w_{jkt}^{(l)} = a_t^{(l)} r_{jt}^{(l)} r_{kt}^{(l)} \sin \left(\theta_{kt}^{(l)} - \theta_{jt}^{(l)} \right).$$

Here, $z_{j,t+1}$ and z_{jt} denote the coordinate of member j in a p-dimensional complex space at time $t + 1$ and t, respectively, while, $w_{jkt}^{(l)}$ indicates the weight of member j toward member k on dimension l at time t. Moreover,

$r_{jt}^{(l)}$ is the modulus of object j on dimension l, i.e., the lth dimensional complex plane. Finally, $\theta_{jt}^{(l)}$ and $\theta_{kt}^{(l)}$, respectively, denote the *counterclockwise* oriented angles of member j and member k on lth dimensional complex plane at time t. If the positive direction of the lth plane is estimated via HFM or EFASID as being counterclockwise, the constant $a_t^{(l)}$ might be positive, and otherwise negative.

This model asserts that members will always move toward other members with whom they feel affinities. This simple principle leads us to the above equation, especially to the above weight. The reason for this is that the following equation holds for $s_{jk}^{(l)}$ in the second equation defined previously in (2) when a new subscript t is introduced:

$$s_{jk,t}^{(l)} - s_{kj,t}^{(l)} = 2r_{jt}^{(l)} r_{kt}^{(l)} \sin(\theta_{kl}^{(t)} - \theta_{jl}^{(t)}), \tag{7}$$

noticing that

$$\rho_{jl}^{(t)} \rho_{kl}^{(t)} + \sigma_{jl}^{(t)} \sigma_{kl}^{(t)} = r_{jt}^{(l)} r_{kt}^{(l)} \cos(\theta_{kt}^{(l)} - \theta_{jt}^{(l)}), \tag{8}$$

and

$$\rho_{jl}^{(t)} \sigma_{kl}^{(t)} - \sigma_{jl}^{(t)} \rho_{kl}^{(t)} = r_{jt}^{(l)} r_{kt}^{(l)} \sin(\theta_{kt}^{(l)} - \theta_{jt}^{(l)}). \tag{9}$$

In any case, the above model is not new except for the state space specified in this paper, and similar models can be found elsewhere. The important and original point of our model is that the state space of our model is a *complex space*. As is well known, even a simple discrete dynamical system in a complex unidimensional plane exhibits extremely interesting behaviors, like fractals originated by Mandelbrot (1977).

Therefore, it will be necessary and appropriate to examine fundamental mathematical natures of the above model, in addtion to testing its fitness to empirical data. Recently, Chino (2000) has extended the above model to a more general one.

References:

Chino, N. (1977). N-ko no taishokan no hitaisyo na kannkei wo zusikikasuru tameno ichi giho. [A graphical technique for representing the asymmetric relationships between N objects]. *Proceedings of the 5th annual meeting of the Behaviormetric Society of Japan* (pp.146-149),

Chino, N. (1978). A graphical technique for representing the asymmetric relationships between N objects. *Behaviormetrika*, 5, 23-40.

Chino, N. (1990). A generalized inner product model for the analysis of asymmetry. *Behaviormetrika*, **27**, 25-46.

Chino, N. (1999). A Hermitian form asymmetric MDS for interval data. *Proceedings of the 27th annual meeting of The Behaviormetric Society of Japan*. 325-328.

Kurashiki, Japan.

Chino, N. (2000). Complex difference system models of social interaction. - Preliminary considerations and a simulation study. *Bulletin of The Faculty of Letters of Aichi Gakuin University*, **30**, 43-53.

Chino, N., & Shiraiwa, K. (1993). Geometrical structures of some non-distance models for asymmetric MDS. *Behaviormetrika*, **20**, 35-47.

Constantine, A. G. & Gower, J. C. (1978). Graphical representation of asymmetric matrices. *Applied Statistics*, **27**, 297-304.

Escoufier, Y., & Grorud, A. (1980). Analyse factorielle des matrices carrees non symetriques. In E. Diday et al. (Eds.) *Data Analysis and Informatics* (pp.263-276). Amsterdam: North Holland.

Gower, J. C. (1977). The analysis of asymmetry and orthogonality. In J.R. Barra, F. Brodeau, G. Romer, & B. van Cutsem (Eds.), *Recent Developments in Statistics* (pp.109-123). Amsterdam: North-Holland.

Harshman, R. A. (1978). Models for analysis of asymmetrical relationships among N objects or stimuli. *Paper presented at the First Joint Meeting of the Psychometric Society and The Society for Mathematical Psychology*, Hamilton, Canada.

Harshman, R. A., Green, P. E., Wind, Y., & Lundy, M. E. (1982). A model for the analysis of asymmetric data in marketing research. *Marketing Science*, **1**, 205-242.

Kiers, H. A. L., & Takane, Y. (1994). A generalization of GIPSCAL for the analysis of asymmetric data. *Journal of Classification*, **11**, 79-99.

Krumhansl, C. L. (1978). Concerning the applicability of geometric models to similarity data: The interrelationship between similarity and spatial density. *Psychological Review*, **85**, 445-463.

Mandelbrot, B. B. (1977). *Fractals: Form, chance, and dimension*. San Francisco: Freeman.

Okada, A., & Imaizumi, T. (1987). Nonmetric multidimensional scaling of asymmetric proximities. *Behaviormetrika*, **21**, 81-96.

Okada, A., & Imaizumi, T. (1997). Asymmetric multidimensional scaling of two-mode three-way proximities. *Journal of Classification.*, **14**, 195-224.

Saito, T. (1991). Analysis of asymmetric proximity matrix by a model of distance and additive terms. *Behaviormetrika*, **29**, 45-60.

Saito, T., & Takeda, S. (1990). Multidimensional scaling of asymmetric proximity: model and method. *Behaviormetrika*, **28**, 49-80.

Sato, Y. (1988). An analysis of sociometric data by MDS in Minkowski space. In K. Matsusita (Ed.), *Statistical Theory and Data Analysis II* (pp.385-396). Amsterdam: North-Holland.

Young, F. W. (1975). An asymmetric Euclidean model for multi-process asymmetric data. *Paper presented at U.S.-Japan Seminar on MDS*, San Diego, U.S.A.

Weeks, D. G., & Bentler, P. M. (1982). Restricted multidimensional scaling models for asymmetric proximities. *Psychometrika*, **47**, 201-208.

Structural Equation Modeling by Extended Redundancy Analysis[1]

Heungsun Hwang and Yoshio Takane[1]
Department of Psychology
McGill University

1205 Dr. Penfield Avenue, Montreal, Quebec, Canada H3A 1B1

Summary: A new approach to structural equation modeling, so-called extended redundancy analysis (ERA), is proposed. In ERA, latent variables are obtained as linear combinations of observed variables, and model parameters are estimated by minimizing a single least squares criterion. As such, it can avoid limitations of covariance structure analysis (e.g., stringent distributional assumptions, improper solutions, and factor score indeterminacy) in addition to those of partial least squares (e.g., the lack of a global optimization). Moreover, data transformation is readily incorporated in the method for analysis of categorical variables. An example is given for illustration.

1. Introduction

Two different approaches have been proposed for structural equation modeling (Anderson & Gerbing, 1988; Fornell & Bookstein, 1982). One analyzes covariance matrices as exemplified by covariance structure analysis (Jöreskog, 1970), while the other analyzes data matrices as exemplified by partial least squares (PLS, Wold, 1982). Typically covariance structure analysis estimates model parameters by the maximum likelihood method under the assumption of multivariate normality of variables. Yet, such a distributional assumption is often violated. A more serious problem is improper solutions (e.g., negative variance estimates), which occur with high frequency in practice. Also, factor scores or latent variable scores are indeterminate. An asymptotically distribution-free (ADF) estimator (Browne, 1984) can be used to fit non-normal data . The ADF estimation, however, is accurate only with very large samples and is still not free from improper solutions and factor score indeterminacy.

In PLS, on the other hand, latent variables are obtained as exact linear composites of observed variables and model parameters are estimated by the fixed-point algorithm (Wold, 1965). As such, PLS does not need any restrictive distributional assumptions. Moreover, PLS does not suffer from improper solutions and indeterminate factor scores. PLS, however, does not solve a global optimization problem for parameter estimation. The lack of a global optimization feature makes it difficult to evaluate an overall model fit.

---

[1]Both authors have contributed equally to the paper, and the authorship reflects the alphabetical order of the two authors.

Also, it is not likely that the obtained PLS solutions are optimal in any well defined sense (Coolen & de Leeuw, 1987).

In this paper, we propose a new method that avoids the major drawbacks of the conventional methods. It may be called extended redundancy analysis (ERA). In ERA, latent variables are estimated as linear combinations of observed variables, so that there are no improper solutions and non-unique factor scores. Also, it employs a global least squares (LS) criterion to estimate model parameters. Thus, it offers an overall model fit without recourse to the normality assumption.

2. Extended Redundancy Analysis

Let $\mathbf{Z}^{(1)}$ denote an n by p matrix consisting of observed endogenous variables. Let $\mathbf{Z}^{(2)}$ denote an n by q matrix consisting of observed exogenous variables. When an observed variable is exogenous as well as endogenous, it is included in both $\mathbf{Z}^{(1)}$ and $\mathbf{Z}^{(2)}$. Assume that the columns of the matrices are mean centered and scaled to unit variance. Then, the model for extended redundancy analysis is given by

$$\mathbf{Z}^{(1)} = \mathbf{Z}^{(2)}\mathbf{WA}' + \mathbf{E} = \mathbf{FA}' + \mathbf{E}, \qquad (2.1)$$

with

$$\operatorname{rank}(\mathbf{WA}') \leq \min(q, p),$$

where $\mathbf{W}$ is a matrix of weights, $\mathbf{A}'$ is a matrix of loadings, $\mathbf{E}$ is a matrix of residuals, and $\mathbf{F}$ ($= \mathbf{Z}^{(2)}\mathbf{W}$) is a matrix of component scores with identification restrictions $\operatorname{diag}(\mathbf{F}'\mathbf{F}) = \mathbf{I}$. In (2.1), $\mathbf{W}$ and/or $\mathbf{A}'$ are structured according to the model to be fitted. Model (2.1) reduces to the redundancy analysis model (van den Wollenberg, 1977) when no variables are shared by both $\mathbf{Z}^{(1)}$ and $\mathbf{Z}^{(2)}$, and no constraints other than $\operatorname{rank}(\mathbf{WA}')$ are imposed on $\mathbf{W}$ and $\mathbf{A}'$.

For simple illustration, suppose that there are three sets of variables, for example, $\mathbf{Z}_1 = [\mathbf{z}_1, \mathbf{z}_2]$, $\mathbf{Z}_2 = [\mathbf{z}_3, \mathbf{z}_4]$, and $\mathbf{Z}_3 = [\mathbf{z}_5, \mathbf{z}_6]$. Further suppose that there are relationships among the three sets of variables, as displayed in Figure 1. Figure 1 shows that two latent variables, one obtained from $\mathbf{Z}_1$ (i.e., $\mathbf{f}_1$), and the other from $\mathbf{Z}_2$ (i.e., $\mathbf{f}_2$), are combined to affect $\mathbf{Z}_3$. This may be expressed as

$$\begin{aligned} \mathbf{Z}_3 &= [\mathbf{Z}_1 \vdots \mathbf{Z}_2] \begin{bmatrix} w_1 & 0 \\ w_2 & 0 \\ 0 & w_3 \\ 0 & w_4 \end{bmatrix} \begin{bmatrix} a_1 & a_2 \\ a_3 & a_4 \end{bmatrix} + \mathbf{E}, \\ &= \mathbf{Z}^{(2)}\mathbf{WA}' + \mathbf{E} = \mathbf{FA}' + \mathbf{E}, \end{aligned} \qquad (2.2)$$

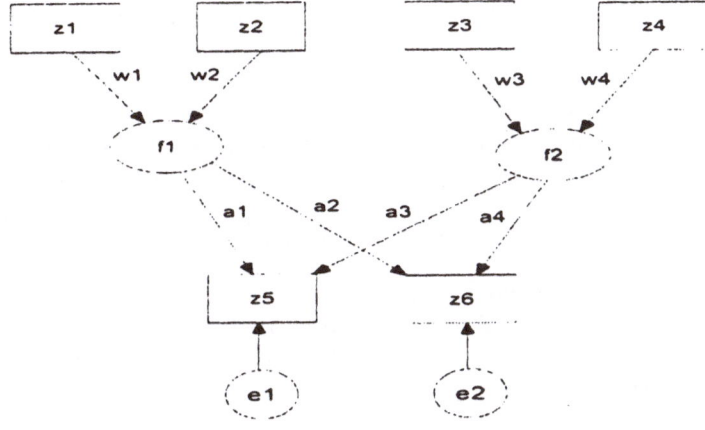

Figure 1: A two latent variable model among three sets of variables.

where $\mathbf{E} = [\mathbf{e}_1, \mathbf{e}_2]$, $\mathbf{W} = \begin{bmatrix} w_1 & w_2 & 0 & 0 \\ 0 & 0 & w_3 & w_4 \end{bmatrix}'$, $\mathbf{A}' = \begin{bmatrix} a_1 & a_2 \\ a_3 & a_4 \end{bmatrix}$, and

$\mathbf{F} = \mathbf{Z}^{(2)}\mathbf{W} = [\mathbf{f}_1 \vdots \mathbf{f}_2]$.

To estimate parameters, we seek to minimize the following LS criterion:

$$f = \mathrm{SS}(\mathbf{Z}^{(1)} - \mathbf{Z}^{(2)}\mathbf{W}\mathbf{A}') = \mathrm{SS}(\mathbf{Z}^{(1)} - \mathbf{F}\mathbf{A}'), \tag{2.3}$$

with respect to $\mathbf{W}$ and $\mathbf{A}'$, subject to $\mathrm{diag}(\mathbf{F}'\mathbf{F}) = \mathbf{I}$, where $\mathrm{SS}(\mathbf{X}) = \mathrm{tr}(\mathbf{X}'\mathbf{X})$. An alternating least squares (ALS) algorithm is developed to minimize (2.3), which is a simple adaptation of Kiers and ten Berge (1989)'s algorithm.

To employ the ALS algorithm, we may rewrite (2.3) as

$$\begin{aligned} f &= \mathrm{SS}(\mathrm{vec}(\mathbf{Z}^{(1)})) - \mathrm{vec}(\mathbf{Z}^{(2)}\mathbf{W}\mathbf{A}')) & (2.4\mathrm{a}) \\ &= \mathrm{SS}(\mathrm{vec}(\mathbf{Z}^{(1)})) - (\mathbf{A} \otimes \mathbf{Z}^{(2)})\mathrm{vec}(\mathbf{W})) & (2.4\mathrm{b}) \\ &= \mathrm{SS}(\mathrm{vec}(\mathbf{Z}^{(1)})) - (\mathbf{I} \otimes \mathbf{F})\mathrm{vec}(\mathbf{A}')) & (2.4\mathrm{c}) \end{aligned}$$

where $\mathrm{vec}(\mathbf{X})$ denotes a supervector formed by stacking all columns of $\mathbf{X}$, one below another, and $\otimes$ denotes a Kronecker product. The algorithm can then be made to repeat the following steps until convergence is reached.

(Step 1) Update $\mathbf{W}$ for fixed $\mathbf{A}'$ as follows: let $\mathbf{w}$ denote the vector formed by eliminating zero elements from $\mathrm{vec}(\mathbf{W})$ in (2.4b). Let Ω denote the matrix formed by eliminating the columns of $\mathbf{A} \otimes \mathbf{Z}^{(2)}$ in (2.4b) corresponding to the zero elements in $\mathrm{vec}(\mathbf{W})$. Then, the LS estimate of $\mathbf{w}$ is obtained by

$$\tilde{\mathbf{w}} = (\Omega'\Omega)^{-1}\Omega'\mathrm{vec}(\mathbf{Z}^{(1)}). \tag{2.5}$$

The updated $\mathbf{W}$ is reconstructed from $\tilde{\mathbf{w}}$, and $\mathbf{F} = \mathbf{Z}^{(2)}\mathbf{W}$ is normalized so that $\mathrm{diag}(\mathbf{F}'\mathbf{F}) = \mathbf{I}$.

(**Step 2**) Update $\mathbf{A}'$ for fixed $\mathbf{W}$ as follows: let $\mathbf{a}$ denote the vector formed by eliminating zero elements from $\mathrm{vec}(\mathbf{A}')$ in (2.4c). Let Γ denote the matrix formed by eliminating the columns of $\mathbf{I} \otimes \mathbf{F}$ in (2.4c) corresponding to the zero elements in $\mathrm{vec}(\mathbf{A}')$. Then, the LS estimate of $\mathbf{a}$ is obtained by

$$\tilde{\mathbf{a}} = (\Gamma'\Gamma)^{-1}\Gamma'\mathrm{vec}(\mathbf{Z}^{(1)}). \tag{2.6}$$

The updated $\mathbf{A}'$ is recovered from $\tilde{\mathbf{a}}$.

In the method, the total fit of a hypothesized model to data is measured by the total variance of the observed endogenous variables explained by the exogenous variables. This is given by

$$\mathrm{Fit} = 1 - \frac{\mathrm{SS}(\mathbf{Z}^{(1)} - \mathbf{Z}^{(2)}\mathbf{W}\mathbf{A}')}{\mathrm{SS}(\mathbf{Z}^{(1)})}. \tag{2.7}$$

This fit index ranges from 0 to 1. The larger is the fit value, the more variance of the endogenous variables is explained by the exogenous variables. The standard errors of parameter estimates can be estimated by the bootstrap method (Efron, 1982). The bootstrapped standard errors can be used to assess the reliability of the parameter estimates. The critical ratios can be used to examine the significance of the parameter estimates (e.g., a parameter estimate having a critical ratio greater than two in absolute value is considered significant at a .05 significance level).

3. Analysis of Categorical Variables by Data Transformation

ERA can readily analyze categorical variables through a certain type of data transformation, often called optimal scaling (e.g., Young, 1981). In optimal scaling, the data are parametrized as $\mathbf{S}^{(1)}$ and $\mathbf{S}^{(2)}$, which are estimated, subject to constraints imposed by the measurement characteristics of $\mathbf{Z}^{(1)}$ and $\mathbf{Z}^{(2)}$. We divide all parameters into two subsets: the model parameters and the data parameters. We then optimize a global fitting criterion by alternately updating one subset with the other fixed. Note that $\mathbf{S}^{(1)}$ and $\mathbf{S}^{(2)}$ may contain variables with different measurement characteristics. This means that a variable may not be directly comparable with other variables, so that each variable in $\mathbf{S}^{(1)}$ and $\mathbf{S}^{(2)}$ should be separately updated.

The ALS procedure with the data transformation feature proceeds as follows. Let $\mathbf{z}_i$ denote a variable in either $\mathbf{Z}^{(1)}$ or $\mathbf{Z}^{(2)}$, so that $i = 1, \cdots, p + q$. Let $\mathbf{s}_i$ denote a variable in either $\mathbf{S}^{(1)}$ or $\mathbf{S}^{(2)}$. Then, we seek to minimize

$$f = \mathrm{SS}(\mathbf{S}^{(1)} - \mathbf{S}^{(2)}\mathbf{W}\mathbf{A}') = \mathrm{SS}(\mathbf{S}^{(1)} - \mathbf{S}^{(2)}\mathbf{B}), \tag{3.1}$$

with the conditions that $\mathrm{diag}(\mathbf{W}'\mathbf{S}^{(2)'}\mathbf{S}^{(2)}\mathbf{W}) = \mathbf{I}$, $\mathbf{s}_i'\mathbf{s}_i = 1$, and $\mathbf{s}_i = \xi(\mathbf{z}_i)$, where $\mathbf{B} = \mathbf{W}\mathbf{A}'$, and ξ refers to a transformation of the observations in $\mathbf{z}_i$, which is a function of their measurement characteristics. To minimize (3.1), two main phases are alternated. One phase is the model estimation phase, in which the model parameters are estimated. The other is the data transformation phase that estimates the data parameters. The model estimation phase is analogous to the estimation procedure in Section 2. We thus focus on the data transformation phase here. The data transformation phase mainly consists of two steps. In the first step, the model prediction of $\mathbf{s}_i$ is obtained in such a way that it minimizes (3.1). In the next step, $\mathbf{s}_i$ is transformed in such a way that it maximizes the relationship between $\mathbf{s}_i$ and the model prediction under certain measurement restrictions.

The first step of the data transformation phase is given as follows. Let $\mathbf{s}_g^{(1)}$ and $\mathbf{s}_h^{(2)}$ denote the g-th and h-th variables in $\mathbf{S}^{(1)}$ and $\mathbf{S}^{(2)}$, respectively ($g = 1, \cdots, p$; $h = 1, \cdots, q$). Let $\tilde{\mathbf{s}}_i$ denote the model prediction of $\mathbf{s}_i$. Then (3.1) may be rewritten as

$$f = \sum_{i=1}^{p+q} \mathrm{SS}(\mathbf{s}_i\boldsymbol{\eta}' - (\boldsymbol{\Delta} - \boldsymbol{\Psi})). \tag{3.2}$$

In (3.2), $\boldsymbol{\eta}'$, $\boldsymbol{\Delta}$, and $\boldsymbol{\Psi}$ are defined as follows: Suppose that if $\mathbf{s}_i$ is shared by $\mathbf{S}^{(1)}$ and $\mathbf{S}^{(2)}$, it is placed in the g-th column and the h-th column of $\mathbf{S}^{(1)}$ and $\mathbf{S}^{(2)}$, respectively. Then, when the model predictions of the variables in $\mathbf{S}^{(1)}$ are updated,

$$\boldsymbol{\Delta} = \left\{ \begin{array}{ll} \mathbf{S}_{(h)}^{(2)}\mathbf{B}_{(h)} & \text{if } \mathbf{s}_i \text{ is shared} \\ \mathbf{S}^{(2)}\mathbf{B} & \text{otherwise} \end{array} \right. , \quad \boldsymbol{\Psi} = \mathbf{S}_{(g)}^{(1)}, \quad \boldsymbol{\eta}' = \left\{ \begin{array}{ll} \mathbf{e}_g' - \mathbf{b}_h' & \text{if } \mathbf{s}_i \text{ is shared} \\ \mathbf{e}_g' & \text{otherwise} \end{array} \right. .$$

When the model predictions of non-common variables in $\mathbf{S}^{(2)}$ are updated,

$$\boldsymbol{\Delta} = \mathbf{S}_{(h)}^{(2)}\mathbf{B}_{(h)}, \quad \boldsymbol{\Psi} = \mathbf{S}^{(1)}, \quad \text{and } \boldsymbol{\eta}' = \mathbf{b}_h'.$$

In the above, matrix $\mathbf{S}_{(h)}^{(2)}\mathbf{B}_{(h)}$ is a product of $\mathbf{S}^{(2)}$ whose h-th column is the n-component vector of zeros and $\mathbf{B}$ whose h-th row is the p-component vector of zeros. Matrix $\mathbf{S}_{(g)}^{(1)}$ equals to $\mathbf{S}^{(2)}$ whose g-th column is an n-component vector of zeros. $\mathbf{e}_g'$ denotes a p-component row vector whose elements are all zeros except the g-th element being unity. Vector $\mathbf{b}_h'$ corresponds with the h-th row of $\mathbf{B}$. Then, $\tilde{\mathbf{s}}_i$ is obtained by

$$\tilde{\mathbf{s}}_i = \boldsymbol{\Lambda}\boldsymbol{\eta}(\boldsymbol{\eta}'\boldsymbol{\eta})^{-1}, \tag{3.3}$$

where $\boldsymbol{\Lambda} = \boldsymbol{\Delta} - \boldsymbol{\Psi}$.

In the next step, s_i is transformed in such a way that it is close to $\bar{s}_i$ as much as possible under the appropriate measurement restrictions. In many cases, s_i is updated by minimizing a LS fitting criterion (e.g., the (normalized) residuals between s_i and $\bar{s}_i$). This comes down to regressing $\bar{s}_i$ onto the space of z_i, which represents the measurement restrictions. The LS estimate of s_i can be generally expressed as follows

$$s_i = \Upsilon_i(\Upsilon_i'\Upsilon_i)^{-1}\Upsilon_i'\bar{s}_i. \tag{3.4}$$

In (3.4), Υ_i is determined by the measurement restrictions imposed on the transformation. For example, for nominal variables, Υ_i is an indicator matrix, whose element stands for category membership, and is known in advance. For ordinal variables, Υ_i indicates which categories must be blocked to satisfy the ordinal restriction, and is iteratively constructed by Kruskal's (1964) least squares monotonic transformation algorithm. The updated s_i is then normalized to satisfy $s_i's_i = 1$. In (3.2), we see that updating a variable is dependent on other variables. To ensure convergence, we must immediately replace the previously estimated variable by the newly estimated and normalized variable. Moreover, when s_i is included in both $\mathbf{S}^{(1)}$ and $\mathbf{S}^{(2)}$, the rescaled and normalized s_i should be substituted for the corresponding columns in both $\mathbf{S}^{(1)}$ and $\mathbf{S}^{(2)}$.

4. Example

The present example is part of the so-called basic health indicator data collected by the World Health Organization. They are available through the internet (http://www.who.int). It consists of 6 variables measured in different countries: (1) infant mortality rate (IMR), defined as the number of deaths per 1000 live births between birth and exact age one year in 1998. (2) maternal mortality ratio (MMR), defined as the number of maternal deaths per 100000 live births in 1990, (3) real gross domestic product (GDP) per capita adjusted for purchasing power parity is expressed in 1985 US dollars, (4) the average number of years of education given for females aged 25 years and above (FEUD) (5) the percentage of children immunized against measles in 1997 (Measles), and (6) total health expenditures as a percentage of GDP in 1995 (Healthexp). The sample size is 51, which amounts to the number of countries for which the data are available.

Two latent variables were assumed for the last four observed variables. One latent variable called 'social and economic (SE) factor' was defined as a linear combination of GDP and FEUD, and the other called 'health services (HS) factor' as that of Measles and Healthexp. The two latent variables were in turn deemed to influence two observed endogenous variables, IMR and MMR. For this model, $\mathbf{W}$ and $\mathbf{A}'$ were identical to those in (2.2). By using ERA, the model was fitted to the data. Results are provided in Figure 2.

The goodness of fit of the model was .65, indicating that about 65% of the variance of the endogenous variables were accounted for by the two latent

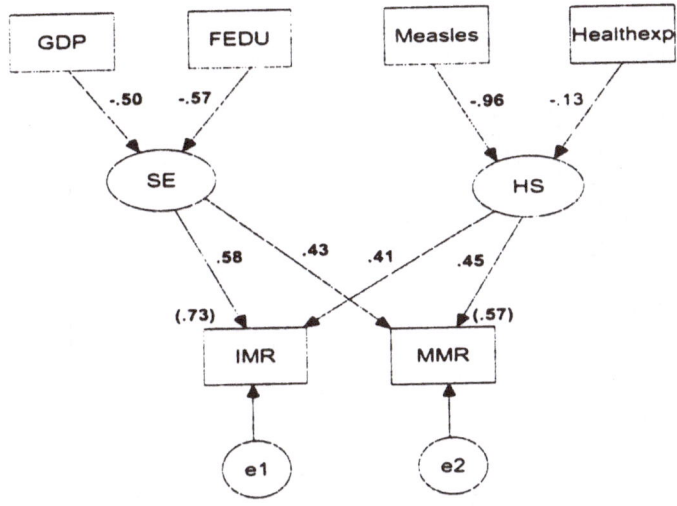

Figure 2: Results of fitting the two latent variable model for the WHO data.

variable model. The fit turned out to be significant in terms of its critical ratio obtained from the bootstrap method with 100 bootstrap samples, indicating that the fitted model was significantly different from the model which assumed $\mathbf{B} = \mathbf{0}$. The squared multiple correlations of IMR and MMR were .73 and .57, respectively. They also turned out significant according to their bootstrapped critical ratios. In Figure 2, boldfaced parameter estimates indicate that they turned out to be significant in terms of their critical ratios. The component weights associated with SE were all significant and negative. This indicates that SE was characterized as social and economic underdevelopment. Similarly, the component weights of Mealses and Healthexp were negative, indicating that HS was likely to represent a low level of health services. However, only one variable, Measles, was significantly associated with HS. Both latent variables were found to have a significant and positive effect on IMR and MMR. This indicates that social and economic underdevelopment and the low level of health services are likely to increase infant mortality rate and maternal mortality ratio. The correlation between the two latent variables was .47.

To exemplify data transformations, two observed endogenous variables, that is, IMR and MMR, were monotonically transformed. Kruskal's (1964) primary LS monotonic transformation was applied to them. This indicated that observation categories were order-preserved but tied observations might

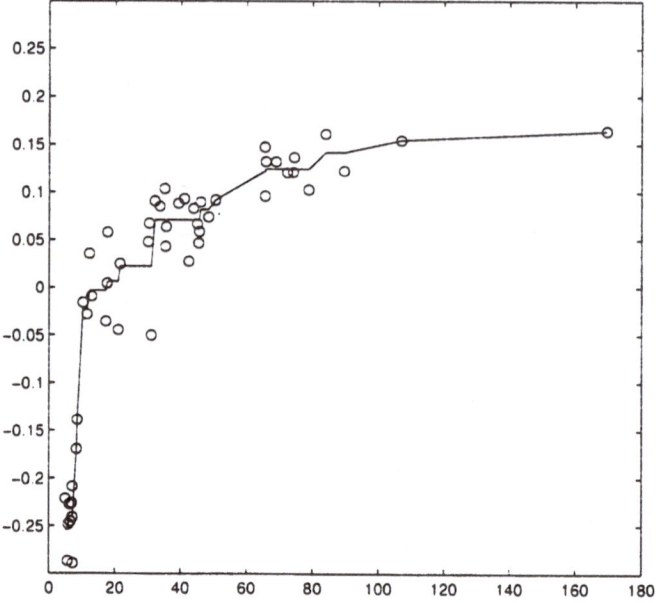

Figure 3: The LS monotonic transformation of IMR

become untied. The LS monotonic transformations of the variables are shown in Figures 3 and 4. The left-hand and right-hand figures represent the LS monotonic transformations of IMR and MMR, respectively. In both figures, the original observations (horizontal) are plotted against the transformed scores (vertical). We find that the monotonic transformations are quite steep although they contain some ties. Due to the transformation, the fit of the model was dramatically improved (.96), while providing similar interpretations of parameter estimates as those obtained when the variables were treated as numerical.

5. Concluding Remarks

A few attempts have been made to extend redundancy analysis to three sets of variables (e.g., Takane, Kiers, & de Leeuw, 1995; Velu, 1991). Yet, they are limited to model and fit a particular type of relationship among three sets of variables. Our method, on the other hand, is quite comprehensive in extending redundancy analysis, and it enables us to specify and fit various structural equation models. Although they are not presented here to conserve space, the basic ERA model can be readily extended to fit more complex relationships among variables, including direct effects of observed variables and higher-order latent variables. Moreover, it is able to perform multi-

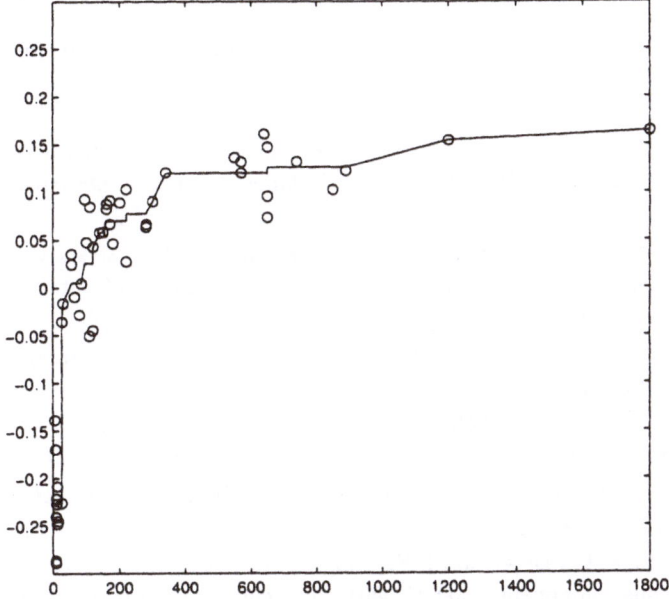

Figure 4: The LS monotonic transformation of MMR.

sample comparisons (Takane & Hwang, 2000).

The data transformation may be considered as one of the principal assets of our method. This makes the data more in line with the model, and goodness of fit may be improved. This also allows us to examine relationships among various types of data measured at different levels. This kind of data transformation is feasible because our method directly analyzes the data matrices rather than the covariance or correlation matrix. However, in PLS, which also analyzes the data matrices, this particular way of data transformation is not feasible since it requires a well-defined global criterion that is consistently optimized by updating the transformed variables.

A number of relevant topics may be considered to further enhance the capability of the method (Hwang & Takane, 2000). For instance, robust estimation may be in order since the proposed method may not be robust against outliers as far as it is based on solving a simple (unweighted) least squares criterion. Missing observations can raise a serious problem, which frequently appear in large data sets. The assumption of normality is not essential for the method due to the least squares fitting. If it is assumed, nonetheless, we can extend the current estimation method in such a way that it provides efficient estimators and allows to perform statistical significance tests without recourse to resampling methods. Future studies are needed to deal with such topics in the proposed method.

References:

Anderson, J. C., & Gerbing, D. W. (1988). Structural equation modeling in practice: A review and recommended two-step approach. *Psychological Bulletin*, **103**, 411-423.

Browne, M. W. (1984). Asymptotically distribution free methods for the analysis of covariance structures. *British Journal of Mathematical and Statistical Psychology*, **37**, 62-83.

Coolen, H., & de Leeuw, J. (1987). Least squares path analysis with optimal scaling, Paper presented at the fifth international symposium of data analysis and informatics. Versailles, France.

Efron, B. (1982). *The Jackknife, the Bootstrap and Other Resampling Plans.* Philadelphia: SIAM.

Hwang, H., & Takane, Y. (2000). Structural equation modeling by extended redundancy analysis: Extended features. Manuscript submitted for publication.

Jöreskog, K. G. (1970). A general method for analysis of covariance structures. *Biometrika*, **57**, 409-426.

Kiers, H. A. L., & ten Berge, J. M. F. (1989). Alternating least squares algorithms for simultaneous components analysis with equal component weight matrices in two or more populations. *Psychometrika*, **54**, 467-473.

Kruskal, J. B. (1964). Multidimensional scaling by optimizing goodness of fit to a nonmetric hypothesis. *Psychometrika*, **29**, 1-27.

Takane, Y., & Hwang, H. (2000). Structural equation modeling by extended redundancy analysis: Basic features. Manuscript submitted for publication.

Takane, Y., Kiers, H., & de Leeuw, J. (1995). Component analysis with different sets of constraints on different dimensions. *Psychometrika*, **60**, 259-280.

van den Wollenberg, A. L. (1977). Redundancy analysis: An alternative for canonical analysis. *Psychometrika*, **42**, 207-219.

Velu, R. P. (1991). Reduced rank models with two sets of regressors. *Applied Statistics*, **40**, 159-170.

Wold. H. (1965). A fixed-point theorem with econometric background, I-II. *Arkiv for Matematik*, **6**, 209-240.

Wold, H. (1982). Soft modeling: The basic design and some extensions. In K. G. Jöreskog and H. Wold (Eds.), *Systems Under Indirect Observations: Causality, Structure, Prediction, Vol. 2* (pp. 1-54). Amsterdam: North-Holland.

Young, F. W. (1981). Quantitative analysis of qualitative data. *Psychometrika*, **46**, 357-388.

Redundancy Index in Canonical Correlation Analysis with Linear Constraints

Akio Suzukawa and Nobuhiro Taneichi

Obihiro University of Agriculture and Veterinary Medicine
Inada, Obihiro, 080-8555, Japan

Summary: The redundancy index proposed by Stewart and Love (1968) is an index to measure the degree to which one set of variables can predict another set of variables, and is associated with canonical correlation analysis. Yanai and Takane (1992) developed canonical correlation analysis with linear constraints (CCALC). In this paper we define a redundancy index in CCALC, which is based on the reformulation of CCALC by Suzukawa (1997). The index is a general measure to summarize redundancy between two sets of variables in the sense that various dependency measures can be obtained by choosing constraints suitably. The asymptotic distribution of the index is derived under normality.

1. Introduction

Canonical correlation analysis (CCA) developed by Hotelling (1936) is a useful technique to simplify the correlation structure between two sets of variates. The first pair of canonical variates between two sets of variates $X : p \times 1$ and $Y : q \times 1$ is a pair of linear combinations $\xi_1 = \alpha_1' X$ and $\eta_1 = \beta_1' Y$ having a maximum of correlation subject to the condition that $\mathrm{Var}(\xi_1) = \mathrm{Var}(\eta_1) = 1$. And the second pair of canonical variates between X and Y is a pair of linear combinations $\xi_2 = \alpha_2' X$ and $\eta_2 = \beta_2' Y$ having a maximum of correlation subject to the condition that $\mathrm{Var}(\xi_2) = \mathrm{Var}(\eta_2) = 1$ and ξ_2 and η_2 are uncorrelated with both ξ_1 and η_1, and so on.

When data are analyzed by CCA, the i-th pair of canonical variates (ξ_i, η_i) is interpreted by investigators based on the coefficient vectors α_i and β_i. They are often interested in the correlation structure between X and Y summarized by linear combinations whose coefficient vectors satisfy a certain condition. Yanai and Takane (1992) developed a method to find a pair of maximally correlated linear combinations under certain linear constraints, and this analysis is called canonical correlation analysis with linear constraints (CCALC).

The redundancy index developed by Stewart and Love (1968) is a measure to investigate interrelationships between two sets of variables, and it is important in the interpretation of canonical analysis. Although several articles have been devoted to the study of properties of the index (Nicewander and Wood (1974,1975), Miller (1975), Gleason (1976), Cramer and Nicewander (1979)), little is known about its sampling distribution. In this paper we define a redundancy index in canonical correlation analysis with linear constraints, and derive its asymptotic distribution under normality.

Throughout this paper, we assume that $(p + q)$-variate random vector $[X', Y']'$ has a nonsingular covariance matrix Σ, and the Σ is partitioned as

$$\Sigma = \left[\begin{array}{cc} \Sigma_{xx} & \Sigma_{xy} \\ \Sigma_{yx} & \Sigma_{yy} \end{array} \right],$$

where $\Sigma_{xx} : p \times p$.

2. CCALC

Let F and H be $p \times p_2$ and $q \times q_2$ matrices with rank$(F) = p_2$ and rank$(H) = q_2$, respectively. The i-th pair of canonical variates with (F, H) is the pair of linear combinations, $\xi_i^* = \alpha_i^{*'} X$ and $\eta_i^* = \beta_i^{*'} Y$, each of unit variance and uncorrelated with the first $i - 1$ pairs and satisfying the linear constraints

$$F' \alpha_i^* = 0 \quad \text{and} \quad H' \beta_i^* = 0.$$

The correlation is the i-th canonical correlation with (F, H).

Canonical correlations and the pairs of canonical variates with linear constraints are formulated by the following theorem.

Theorem 1 [Yanai and Takane (1992)] *Let $p_1 = p - p_2$, $q_1 = q - q_2$ and $m^* = \min(p_1, q_1)$, and for $1 \leq i \leq m^*$, let ρ_i^{*2} be the i-th largest root of the equation*

$$|(I_p - C_x)\Sigma_{xy}\Sigma_{yy}^{-1}(I_q - C_y)\Sigma_{yx} - \rho^{*2}\Sigma_{xx}| = 0 \qquad (1)$$

and α_i^* and β_i^* be solutions of the equations

$$(I_p - C_x)\Sigma_{xy}\Sigma_{yy}^{-1}(I_q - C_y)\Sigma_{yx}\alpha_i^* = \rho_i^{*2}\Sigma_{xx}\alpha_i^*, \quad \alpha_i^{*'}\Sigma_{xx}\alpha_i^* = 1,$$
$$(I_q - C_y)\Sigma_{yx}\Sigma_{xx}^{-1}(I_p - C_x)\Sigma_{xy}\beta_i^* = \rho_i^{*2}\Sigma_{yy}\beta_i^*, \quad \beta_i^{*'}\Sigma_{yy}\beta_i^* = 1,$$

respectively, where

$$C_x = F(F'\Sigma_{xx}^{-1}F)^{-1}F'\Sigma_{xx}^{-1} \text{ and } C_y = H(H'\Sigma_{yy}^{-1}H)^{-1}H'\Sigma_{yy}^{-1}.$$

Then the i-th canonical correlation and the i-th pair of canonical variates with (F, H) are given by $\rho_i^* = \sqrt{\rho_i^{*2}}$ and $(\alpha_i^{*'} X, \beta_i^{*'} Y)$, respectively.

Suzukawa (1997) reformulated canonical correlations and canonical variates with linear constraints, which are based on orthogonal transformations of X and Y. The reformulation is more convenient than Theorem 1 when considering statistical inference problems.

First write

$$F' = Q_F[O, I_{p_2}]P_F' \quad \text{and} \quad H' = Q_H[O, I_{q_2}]P_H',$$

where Q_F and Q_H are $p_2 \times p_2$ and $q_2 \times q_2$ nonsingular matrices respectively, and P_F and P_H are $p \times p$ and $q \times q$ orthogonal matrices, respectively. Let

$$\Sigma^* = \begin{bmatrix} P'_F & O \\ O & P'_H \end{bmatrix} \Sigma \begin{bmatrix} P_F & O \\ O & P_H \end{bmatrix},$$

and partition Σ^* as

$$\Sigma^* = \begin{bmatrix} \Sigma^*_{11} & \Sigma^*_{12} & \Sigma^*_{13} & \Sigma^*_{14} \\ \Sigma^*_{21} & \Sigma^*_{22} & \Sigma^*_{23} & \Sigma^*_{24} \\ \Sigma^*_{31} & \Sigma^*_{32} & \Sigma^*_{33} & \Sigma^*_{34} \\ \Sigma^*_{41} & \Sigma^*_{42} & \Sigma^*_{43} & \Sigma^*_{44} \end{bmatrix},$$

where Σ^*_{11}, Σ^*_{22} and Σ^*_{33} are $p_1 \times p_1$, $p_2 \times p_2$ and $q_1 \times q_1$ matrices, respectively.

Theorem 2 [Suzukawa (1997)] *The i-th largest root of the equation (1) is given by the i-th largest root of a equation*

$$|\Sigma^*_{13}\Sigma^{*-1}_{33}\Sigma^*_{31} - \rho^{*2}\Sigma^*_{11}| = 0,$$

and α^*_i and β^*_i defined in Theorem 1 are given by

$$\alpha^*_i = P_F \begin{bmatrix} \tilde{\alpha}^*_{1i} \\ 0 \end{bmatrix} \quad and \quad \beta^*_i = P_H \begin{bmatrix} \tilde{\beta}^*_{3i} \\ 0 \end{bmatrix},$$

where $\tilde{\alpha}^*_{1i}$ and $\tilde{\beta}^*_{3i}$ are solutions of equations

$$\Sigma^*_{13}\Sigma^{*-1}_{33}\Sigma^*_{31}\tilde{\alpha}^*_{1i} = \rho^{*2}_i\Sigma^*_{11}\tilde{\alpha}^*_{1i}, \quad \tilde{\alpha}^{*'}_{1i}\Sigma^*_{11}\tilde{\alpha}^*_{1i} = 1,$$

$$\Sigma^*_{31}\Sigma^{*-1}_{11}\Sigma^*_{13}\tilde{\beta}^*_{3i} = \rho^{*2}_i\Sigma^*_{33}\tilde{\beta}^*_{3i}, \quad \tilde{\beta}^{*'}_{3i}\Sigma^*_{33}\tilde{\beta}^*_{3i} = 1,$$

respectively.

Let $\tilde{X} = P'_F X$ and $\tilde{Y} = P'_H Y$, and partition $\tilde{X}$ and $\tilde{Y}$ as $\tilde{X} = [\tilde{X}'_1, \tilde{X}'_2]'$ and $\tilde{Y} = [\tilde{Y}'_3, \tilde{Y}'_4]'$, where $\tilde{X}_1 : p_1 \times 1$ and $\tilde{Y}_1 : q_1 \times 1$. From Theorem 2 , we can see that the canonical correlations with (F, H) are equal to the usual (without any constraints) canonical correlations between $\tilde{X}_1$ and $\tilde{Y}_3$. Thus in order to calculate the canonical correlations and variates with linear constraints, we can apply the usual CCA computer program for the two sets of variates $\tilde{X}_1$ and $\tilde{Y}_3$.

Based on the reformulation, Suzukawa (1997) discussed inference problems to evaluate effects of imposing linear constraints on canonical variates and to determine the number of useful canonical variates in CCALC.

3. Redundancy Index in CCALC

Stewart and Love (1968) have proposed the redundancy index in CCA, for use in the interpretation of canonical analysis. The index is defined as

$$R^2_{Y \cdot X} = \sum_{i=1}^{q} \rho_i^2 C^2(\eta_i, Y),$$

where ρ_i is the i-th canonical correlation between X and Y, $C^2(\eta_i, Y) = r'_{\eta_i Y} r_{\eta_i Y}/q$, and $r_{\eta_i Y}$ is a q-dimensional vector of correlation coefficients between the i-th canonical variate η_i and the original variates Y.

Let P be the correlation matrix of $[X', Y']'$ and partition it corresponding to the covariance matrix Σ. Miller (1975) showed that the index can be represented as

$$R^2_{Y \cdot X} = \frac{1}{q} \mathrm{tr}(P_{yx} P_{xx}^{-1} P_{xy}).$$

Thus we can see that the redundancy index is an average of multiple correlation coefficients between components of Y and X.

In this paper, we define redundancy index in CCALC. By regarding CCALC between X and Y as CCA between the transformed variates $\tilde{X}_1$ and $\tilde{Y}_3$, we define the redundancy index with linear constraints (F, H) as

$$R^2_{Y \cdot X}(F, H) = \sum_{i=1}^{q_1} \rho_i^{*2} C^{*2}(\bar{\eta}_{3i}, \tilde{Y}_3) = \frac{1}{q_1} \mathrm{tr}(P_{31}^* P_{11}^{*-1} P_{13}^*),$$

where $\bar{\eta}_{3i} = \tilde{\beta}_{3i}^{*'} \tilde{Y}_3$, $C^{*2}(\bar{\eta}_{3i}, \tilde{Y}_3) = r'_{\bar{\eta}_{3i} \tilde{Y}_3} r_{\bar{\eta}_{3i} \tilde{Y}_3}/q_1$, $r'_{\bar{\eta}_{3i} \tilde{Y}_3}$ is a q_1-dimensional vector of correlation coefficients between $\bar{\eta}_{3i}$ and $\tilde{Y}_3$, and P^* is the correlation matrix corresponding to Σ^*.

The constrained redundancy index is a general measure to summarize redundancy between X and Y in the sense that various dependency measures can be obtained by choosing constraint matrices F and H suitably. The redundancy index $R^2_{Y \cdot X}$ without constraints can be represented as $R^2_{Y \cdot X}(O, O)$ by choosing constraint matrices as zero matrices.

If we put $F' = [0, I_{p-1}]$ and $H' = [0, I_{q-1}]$, then $\tilde{X}_1$ and $\tilde{Y}_3$ are 1-dimensional and these are the first elements X_1 and Y_1 of the original variates X and Y, respectively. Thus in this case the index $R^2_{Y \cdot X}(F, H)$ is nothing but the squared simple correlation coefficient between X_1 and Y_1.

Let $F' = O$ and $H' = [0, I_{q-1}]$, then $\tilde{X}_1$ is the original variables X, and $\tilde{Y}_3$ is Y_1 which is the first component of the original variates Y. Thus in this case the index $R^2_{Y \cdot X}(F, H)$ is the squared multiple correlation between X and Y_1.

When p and q are even and the constraint matrices are chosen as

$$F' = [I_{p/2}, -I_{p/2}] \quad \text{and} \quad H' = [I_{q/2}, -I_{q/2}]$$

then we can decompose them as

$$
\begin{aligned}
F' &= Q_F[O, I_{p/2}]P'_F \\
&= \begin{bmatrix} \sqrt{2}I_{p/2} & O \\ O & \sqrt{2}I_{p/2} \end{bmatrix} [O, I_{p/2}] \begin{bmatrix} 1/\sqrt{2}I_{p/2} & 1/\sqrt{2}I_{p/2} \\ 1/\sqrt{2}I_{p/2} & -1/\sqrt{2}I_{p/2} \end{bmatrix},
\end{aligned}
$$

$$
\begin{aligned}
H' &= Q_H[O, I_{q/2}]P'_H \\
&= \begin{bmatrix} \sqrt{2}I_{q/2} & O \\ O & \sqrt{2}I_{q/2} \end{bmatrix} [O, I_{q/2}] \begin{bmatrix} 1/\sqrt{2}I_{q/2} & 1/\sqrt{2}I_{q/2} \\ 1/\sqrt{2}I_{q/2} & -1/\sqrt{2}I_{q/2} \end{bmatrix}.
\end{aligned}
$$

Partition the original variates as $X = [X'_1, X'_2]'$ and $Y = [Y'_3, Y'_4]'$, where $X_1 : p/2 \times 1$ and $Y_3 : q/2 \times 1$. Then the transformed variates $\tilde{X}_1$ and $\tilde{Y}_3$ are $(X_1 + X_2)/\sqrt{2}$ and $(Y_3 + Y_4)/\sqrt{2}$, respectively. Thus in this case the constrained redundancy index $R^2_{Y \cdot X}(F, H)$ is a redundancy measure of the sum $Y_3 + Y_4$ to the sum $X_1 + X_2$. Let us consider a simple example, in which the same mathematics and science tests are given twice, once in the first year and the other in the second year. The scores for the mathematics tests are indicated as X_1 and X_2, and those for the science test as Y_3 and Y_4. Then it would be important to evaluate the redundancy of the science tests $Y_3 + Y_4$ to that of the mathematics tests $X_1 + X_2$.

The sample redundancy index $\hat{R}^2_{Y \cdot X}(F, H)$ with linear constraints (F, H) is defined by replacing Σ by a sample covariance matrix S based on N random samples. Since the sample covariance matrix S is a consistent estimator of Σ, it is obvious that the sample redundancy index $\hat{R}^2_{Y \cdot X}(F, H)$ is a consistent estimator of the population redundancy index $R^2_{Y \cdot X}(F, H)$. To make inference about the population redundancy index, the asymptotic distribution of the sample redundancy index is important.

We assume that $[X', Y']'$ is distributed according to $(p+q)$-variate normal distribution $N_{p+q}(\mu, \Sigma)$, where μ and Σ are unknown. First we consider the asymptotic distribution of the sample redundancy index $\hat{R}^2_{Y \cdot X} = \hat{R}^2_{Y \cdot X}(O, O)$ without constraints. Following the general theory for a function of the sample covariance matrix discussed by Siotani, Hayakawa and Fujikoshi (1985), we obtain a stochastic approximation as

$$\sqrt{n}(\hat{R}^2_{Y \cdot X} - R^2_{Y \cdot X}) = \text{tr}(AU) + O_p(n^{-1/2}), \tag{2}$$

where $U = \sqrt{n}(S - \Sigma)$, $n = N - 1$ and A is a $(p+q) \times (p+q)$ matrix defined by

$$
A = \frac{\partial \hat{R}^2_{Y \cdot X}}{\partial S}\bigg|_{S=\Sigma} = \begin{bmatrix} \dfrac{\partial \hat{R}^2_{Y \cdot X}}{\partial S_{xx}} & \dfrac{1}{2}\dfrac{\partial \hat{R}^2_{Y \cdot X}}{\partial S_{xy}} \\ \dfrac{1}{2}\dfrac{\partial \hat{R}^2_{Y \cdot X}}{\partial S_{yx}} & \dfrac{\partial \hat{R}^2_{Y \cdot X}}{\partial S_{yy}} \end{bmatrix}_{S=\Sigma}
$$

Put $S_{yy} = (s_{yy,ij})$ and $\hat{D}_y = \text{diag}(s_{yy,11}, \ldots, s_{yy,qq})$, then we can write it as

$$\frac{1}{q}\hat{R}^2_{Y \cdot X} = \frac{1}{q}\text{tr}(\hat{D}_y^{-1} S_{yx} S_{xx}^{-1} S_{xy}).$$

Thus the matrix derivatives in A are calculated as

$$\frac{\partial \widehat{R}_{Y \cdot X}^2}{\partial S_{xx}} = \frac{1}{q} \frac{\partial}{\partial S_{xx}} \text{tr}(S_{xx}^{-1} S_{xy} \widehat{D}_y^{-1} S_{yx}) = -\frac{1}{q} S_{xx}^{-1} S_{xy} \widehat{D}_y^{-1} S_{yx} S_{xx}^{-1},$$

$$\frac{\partial \widehat{R}_{Y \cdot X}^2}{\partial S_{xy}} = \frac{1}{q} \frac{\partial}{\partial S_{xy}} \text{tr}(S'_{xy} S_{xx}^{-1} S_{xy} \widehat{D}_y^{-1}) = \frac{2}{q} S_{xx}^{-1} S_{xy} \widehat{D}_y^{-1},$$

$$\frac{\partial \widehat{R}_{Y \cdot X}^2}{\partial S_{yx}} = \frac{1}{q} \frac{\partial}{\partial S_{yx}} \text{tr}(S_{yx} S_{xx}^{-1} S'_{yx} \widehat{D}_y^{-1}) = \frac{2}{q} \widehat{D}_y^{-1} S_{yx} S_{xx}^{-1},$$

$$\frac{\partial \widehat{R}_{Y \cdot X}^2}{\partial S_{yy}} = \frac{1}{q} \frac{\partial}{\partial S_{yy}} \text{tr}(\widehat{D}_y^{-1} S_{yx} S_{xx}^{-1} S_{xy}) = -\frac{1}{q} \widehat{D}_y^{-1} \widehat{D}_B \widehat{D}_y^{-1}.$$

where $\widehat{D}_B = \text{diag}(\hat{b}_{11}, \ldots, \hat{b}_{qq})$ and $\hat{b}_{ii}$ is the (i, i)-element of $S_{yx} S_{xx}^{-1} S_{xy}$. So we have

$$A = \frac{1}{q} \begin{bmatrix} -\Sigma_{xx}^{-1} \Sigma_{xy} D_y^{-1} \Sigma_{yx} \Sigma_{xx}^{-1} & \Sigma_{xx}^{-1} \Sigma_{xy} D_y^{-1} \\ D_y^{-1} \Sigma_{yx} \Sigma_{xx}^{-1} & -D_y^{-1} D_B D_y^{-1} \end{bmatrix},$$

where $D_y = \text{diag}(\sigma_{yy,11}, \ldots, \sigma_{yy,qq})$, $\Sigma_{yy} = (\sigma_{yy,ij})$, $D_B = \text{diag}(b_{11}, \ldots, b_{qq})$ and b_{ii} is the (i, i)-element of $B = \Sigma_{yx} \Sigma_{xx}^{-1} \Sigma_{xy}$.

From (2), the asymptotic distribution of $\sqrt{n}(\widehat{R}_{Y \cdot X}^2 - R_{Y \cdot X}^2)$ is same with $\text{tr}(AU)$. It is well-known that the asymptotic distribution of the symmetric matrix U is $(p+q)(p+q+1)/2$-variate normal with mean zero and covariance matrix with element $\text{cov}(u_{ij}, u_{kl}) = \sigma_{ik}\sigma_{jl} + \sigma_{il}\sigma_{jk}$ (see Theorem 2.7.2 of Siotani, Hayakawa and Fujikoshi (1985)). So the asymptotic distribution of $\text{tr}(AU)$ is normal with mean zero and variance

$$\begin{aligned} \tau^2 &= 2\text{tr}(A\Sigma)^2 \\ &= \frac{2}{q^2} \left[2\text{tr}\left\{ D_y^{-1} B D_y^{-1} \Sigma_{yy \cdot x} D_y^{-1} (D_y - D_B) \right\} \right. \\ &\quad \left. + \text{tr}\left\{ D_y^{-1} (D_B D_y^{-1} \Sigma_{yy} - B) \right\}^2 \right], \end{aligned}$$

where $\Sigma_{yy \cdot x} = \Sigma_{yy} - B$.

Let $P_{yy \cdot x}$ be a partial correlation matrix of Y for fixed X, which is defined by

$$P_{yy \cdot x} = (D_y - D_B)^{-\frac{1}{2}} \Sigma_{yy \cdot x} (D_y - D_B)^{-\frac{1}{2}},$$

and put

$$D_{Y \cdot X}^2 = \text{diag}(\rho_{Y_1 \cdot X}^2, \ldots, \rho_{Y_q \cdot X}^2),$$

where $\rho_{Y_i \cdot X}$ is the multiple correlation between X and Y_i which is the i-th element of Y. Then the asymptotic variance τ^2 can be expressed as

$$\tau^2 = \frac{2}{q^2} \text{tr} \left[\left\{ (I_q - D_{Y \cdot X}^2) P_{yy} \right\}^2 + (2D_{Y \cdot X}^2 - I_q) \left\{ (I_q - D_{Y \cdot X}^2) P_{yy \cdot x} \right\}^2 \right]. \tag{3}$$

We summarize the above result in the following

Theorem 3 *Assume that $[X', Y']'$ is distributed according to $N_{p+q}(\mu, \Sigma)$. Then the asymptotic distribution of $\sqrt{n}(\widehat{R}_{Y \cdot X}^2 - R_{Y \cdot X}^2)$ is normal with mean zero and variance τ^2 defined by (3).*

The asymptotic distribution of the sample redundancy index depends on not only $D_{Y \cdot X}^2$ but also P_{yy} and $P_{yy \cdot x}$. Even if $P_{yy} = P_{yy \cdot x} = I_q$, which means that components of Y are independent and conditionally independent for fixed X, the asymptotic variance is

$$\frac{4}{q^2} \operatorname{tr} \left\{ D_{Y \cdot X}^2 (I_q - D_{Y \cdot X}^2)^2 \right\},$$

and it is not a function of only the population redundancy index.

Next we consider the asymptotic distribution of the constrained redundancy index. Since the index $\widehat{R}_{Y \cdot X}(F, H)$ is a no-constrained sample redundancy index between $\tilde{Y}_3$ and $\tilde{X}_1$, we can obtain its asymptotic distribution immediately from Theorem 3 by replacing Σ by $\begin{bmatrix} \Sigma_{11}^* & \Sigma_{13}^* \\ \Sigma_{31}^* & \Sigma_{33}^* \end{bmatrix}$.

Theorem 4 *Under the same condition with Theorem 3, the asymptotic distribution of $\sqrt{n}\{\widehat{R}_{Y \cdot X}^2(F, H) - R_{Y \cdot X}^2(F, H)\}$ is normal with mean zero and variance*

$$\tau^{*2} = \frac{2}{q_1^2} \operatorname{tr} \left[\left\{ (I_{q_1} - D_{3 \cdot 1}^{*2}) P_{33}^* \right\}^2 + (2D_{3 \cdot 1}^{*2} - I_{q_1}) \left\{ (I_{q_1} - D_{3 \cdot 1}^{*2}) P_{33 \cdot 1}^* \right\}^2 \right],$$

where $D_{3 \cdot 1}^{*2} = diag(\rho_{\tilde{Y}_{31} \cdot \tilde{X}_1}^{*2}, \ldots, \rho_{\tilde{Y}_{3q_1} \cdot \tilde{X}_1}^{*2})$, $\rho_{\tilde{Y}_{3i} \cdot \tilde{X}_1}^*$ is the population multiple correlation between $\tilde{X}_1$ and $\tilde{Y}_{3i}$ which is the i-th element of $\tilde{Y}_3$, P_{33}^* is a correlation matrix of $\tilde{Y}_3$, and $P_{33 \cdot 1}^*$ is a partial correlation matrix of $\tilde{Y}_3$ when $\tilde{X}_1$ is fixed.

If we put $F' = O$ and $H' = [0, I_{q-1}]$, then the constrained sample redundancy index $\widehat{R}_{Y \cdot X}^2(F, H)$ is nothing but the squared sample multiple correlation between Y_1 and X, where Y_1 is the first element of Y. In this case, the asymptotic variance τ^{*2} can be simplified as

$$\begin{aligned} \tau^{*2} &= \frac{2}{1^2} \left\{ (1 - \rho_{Y_1 \cdot X}^2)^2 + (2\rho_{Y_1 \cdot X}^2 - 1)(1 - \rho_{Y_1 \cdot X}^2)^2 \right\} \\ &= 4\rho_{Y_1 \cdot X}^2 (1 - \rho_{Y_1 \cdot X}^2)^2. \end{aligned}$$

This is a well-known formula for the asymptotic variance of the squared sample multiple correlation.

132

Acknowledgements

We are grateful to the editor and referees for their helpful comments and suggestions.

References:

Cramer, E. M. and Nicewander, W. A. (1979). Some symmetric, invariant measures of multivariate association. *Psychometrika*, **44**, 1, 43–54.

Gleason, T. C. (1976). On redundancy in canonical analysis. *Psychological Bulletin*, **83**, 6, 1004–1006.

Hotelling, H. (1936). Relations between two sets of variates. *Biometrika*, **28**, 321–377.

Miller, S. K. (1975). In defense of the general canonical correlation index: reply to Nicewander and Wood. *Psychological Bulletin*, **82**, 2, 207–209.

Nicewander, W. A. and Wood, D. A. (1974). Comments on "a general canonical correlation index". *Psychological Bulletin*, **81**, 1, 92–94.

Nicewander, W. A. and Wood, D. A. (1975). On the mathematical bases of the general canonical correlation index: rejoinder to Miller. *Psychological Bulletin*, **82**, 2, 210–212.

Siotani, M., Hayakawa, T. and Fujikoshi, Y. (1985). *Modern Multivariate Statistical Analysis : A Graduate Course Handbook*. American Science Press, INC., Ohio.

Stewart, D. and Love, W. (1968). A general canonical correlation index. *Psychological Bulletin*, **70**, 160–163.

Suzukawa, A. (1997). Statistical inference in a canonical correlation analysis with linear constraints. *Journal of the Japan Statistical Society*, **27**, 1, 93–107.

Yanai, H. and Takane, Y. (1992). Canonical correlation analysis with linear constraints. *Linear Algebra and its Applications*, **176**, 75–89.

Regularized Kernel Discriminant Analysis with Optimally Scaled Data

Halima Bensmail[1], Hamparsum Bozdogan[1]

[1] Department of Statistics
326/336 Stokely Management Ctr.
Knoxville, TN 37996-0532

Summary: Linear discriminant analysis is a well known procedure for discrimination where the linear predictors define one set of variables and a set of dummy variables representing class membership which defines the other set. Here we propose a new method of discriminating between observations using a set of mixed (i.e., categorical and/or continuous) variables. This nonparametric discriminant procedure optimally scales the data and estimates the distribution of the object scores using multivariate kernel density estimation. We propose using Bozdogan's information-theoretic measure complexity ICOMP to select both the window width of the kernel density estimator as well as the dimension of the object scores matrix.

1. Introduction

Consider a data matrix $\mathbf{V}$ ($n \times m$) in which the rows corresponds to n objects measured on m variables. Let K be the number of groups G_k. So we have a sample of size n and the problem is to classify an observation $\mathbf{z}$ to one of the K groups. An important aspect of the problem is that the variables may be measured on nominal, ordinal or interval scales, or any mixture of these. In the field of discriminant analysis a number of methods has been proposed for continuous data using Gaussian assumption. In this case, the form of the class-conditional density function $f(\mathbf{z}|G_k)$ is known, and classification problems can be solved by comparing the ratio of the conditional densities with various thresholds. Often, however, the form of these class-conditional densities is not known so that a *non-parametric* or *distribution-free* method, as described in Wertz (1978) and Hand (1981), is desirable. Here we develop a new nonparametric approach to discrimination for the case of data with mixed scaling levels. In this approach, we transform the mixed data into numerical variables so that the use of the Euclidean metric is possible. This approach utilizes the optimal scaling method developed in Gifi (1990) and uses multidimensional kernel density estimation to approximate the underlying class-conditional density function $f(\mathbf{z}|G_k)$.

2. Transforming and Scaling

We presume the reader is familiar with homogeneity analysis, also known as multiple correspondence analysis or dual scaling (Nishisato (1980), Meulman (1986), van der Burg et al. (1988), Gifi (1990) and Heiser and Meulman (1995)). Let $(k_1, \ldots, k_j, \ldots, k_m)$ be the m-vector containing the number of

133

categories of each variable, and let p denote the dimensionality of the analysis that is chosen. Let each variable v_j $(j = 1, \ldots, m)$ be coded into an $(n \times k_j)$ indicator matrix G_j. An indicator matrix indicates which categories are scored by which objects. Rows of an indicator matrix usually refer to objects and columns to categories. Its elements consist of zeros (not scored) and ones (scored).

Homogeneity analysis determines *quantifications* of the categories of each of the variables such that homogeneity is maximized. If y_j, a k_j-vector, is the quantification of the categories of variable v_j, then $G_j y_j$ represents a single quantification or transformation of the n objects for variable v_j. Without additional conditions on the y_j, objects in the same categories get the same quantification. In homogeneity analysis, simultaneous quantifications for each variable are collected in the $k_j \times p$ matrices Y_j, called *multiple nominal quantifications*. Thus matrices $G_j Y_j$ induce p multiple quantifications of the objects for variable j.

Perfect homogeneity occurs when all multiple quantifications of the objects are the same for all variables. In this case, $X = G_1 Y_1 = \ldots = G_m Y_m$ and homogeneity analysis amounts to minimizing

$$\sigma(X; Y_1, \ldots, Y_m) = \frac{1}{m} \sum_{j=1}^{m} tr(X - G_j Y_j)^t (X - G_j Y_j) \qquad (1)$$

over the object scores X and multiple nominal quantifications Y_j under appropriate normalization conditions. It should be emphasized that the choice of normalization of X is crucial. In the distance approach X is orthogonal and not orthonormal (Meulman 1986); the shape of the configuration, the different amount of scatter in differents directions, is determined by the eigenvalues $(X^t X / n = \Omega^2)$, where Ω is the diagonal matrix of eigenvalues of $J P_0 J^t$, $J = (I - n^{-1} 11^t)$ a centering operator, and $P_0 = \frac{1}{m} \sum_{j=1}^{m} G_j (G_j^t G_j)^{-1} G_j^t$ is the average of all the orthogonal projectors on the subspace spanned by the columns of the indicator matrices G_j. Gifi (1990) used the case where $(X^t X / n = I)$.

When category quantifications are required to be points on a line, loss of homogeneity is still the average sum of squares across variables, but the loss measure becomes

$$\sigma(X; Y_1, \ldots, Y_m; a_1, \ldots, a_m) = \frac{1}{m} \sum_{j=1}^{m} tr(X - G_j y_j a_j^t)^t (X - G_j y_j a_j^t) \qquad (2)$$

which is minimized over object scores X and multiple nominal quantifications $y_j, (j = 1, \ldots, m)$, where a_j is the p-vector of weights and y_j gives the category quantifications (Gifi 1990, Chap. 4). The technique defined by (2) is called nonlinear PCA and is included as CATPCA in the SPSS package Categories (10.0).

Dimensionality

The dimensionality p is important here, because choosing a different dimension will lead to different transformations as we can no longer assume the solutions to be nested when we require category points to be on a line. Multiple nominal variables have different quantifications in p dimensions; the categories of non-multiple-nominal variables fit on a straight line. If there are no dependencies in the data, the maximum number of dimensions, when the first α variables are multiple nominal, is

$$\sum_{j=1}^{\alpha}(k_j - 1) + (m - \alpha). \tag{3}$$

In the case of non-multiple-nominal variables only, one can use the first p dimensions for which the eigenvalues $e_j(j = 1, \ldots, p)$ of the correlation matrix between the quantified variables is larger than $1/m$. If a dimension has an eigenvalue smaller than $1/m$, it explains less variance than an individual variable; such a dimension has little or no generalizability.

3. Optimally-Scaled Discriminant Analysis

If there are multiple nominal variables, no easy rule is proposed. It remains true that if an eigenvalue is smaller than the reciprocal of the maximum number of dimensions, the corresponding dimensions has little generalizability and could better be discarded.

Here we propose a new nonparametric method for discriminating between observations based on the optimal scaling of mixed variables. We suppose our observations are measured on a set of nominal, ordinal or numerical variables or any mixture of these variables collected in a data matrix $V(n, m)$ of n rows (objects) and m column (m variables). We further suppose that, for the $(m - \alpha)$ non-multiple nominal variables, $q_j = G_j Y_j$ where $q_j : I \longrightarrow R$ is a function assigning quantifications to the categories of variable v_j such that the transformed categories $q_j = q_j(v_j)$ are optimally scaled. This produces a p-dimensional matrix X of scores for all objects where X is ortho-normalized (on Ω) version of

$$\tilde{X} = \frac{1}{m}(\sum_{j=1}^{\alpha} G_j Y_j + \sum_{j=\alpha+1}^{m-\alpha} q_j a_j^t). \tag{4}$$

Scores in X are defined to have mean zero and uncorrelated dimensions the n observations. We then propose to use nonparametric discriminant procedures to estimate the unknown distribution of the object scores X, which are the (standardized) averages over quantified variables $q_j a_j^t$. The allocation rule, proposed in order to classify each observation z_i $(i = 1, \ldots, n)$ to one of the K possible sub-populations (groups), consists of computing the transformation $x_i(i = 1, \ldots, n)$, the p-dimensional object score, and then assigning the observation to G_k if the estimated kernel density satisfies

$$\hat{f}(\mathbf{x}_i|G_k) \geq \hat{f}(\mathbf{x}_i|G_j), \text{ for each } k \neq j \ (k, j = 1, \ldots, K), \tag{5}$$

where $\mathbf{x}_i$ denotes a column of X^t.

4. Model-Based Gaussian Discriminant Analysis

With transformed data, many rules of classification have been proposed but most rely on multivariate normality. When transforming a mixed data matrix, the matrix X of object scores always has a larger size than the initial matrix (see Equation 3). We will investigate whether eigenvalue decomposition regularized discriminant analysis (EDRDA — see Bensmail and Celeux, 1996) handles the case of a transformed data. We examine EDRDA since it is efficient for matrices with small size par comparison to the number of variables, it allows a variety of models of discrimination, and it uses the best model related to the geometrical characteristics of the data for constructing a rule of classification. EDRDA's approach is based on the spectral decomposition of the covariance matrix Σ_k of each cluster

$$\Sigma_k = \lambda_k D_k A_k D_k' \tag{6}$$

where $\lambda_k = |\Sigma_k|^{1/p}$, D_k: denotes the eigenvectors of Σ_k and A_k denotes the eigenvalues of Σ_k, $|A_k| = 1$. The eigenvalue λ_k determines volume of group G_k, the eigenvectors D_k determines the orientation, and A_k determines the shape. Various assumptions on the parameters λ_k, D_k and A_k lead to 8 general models of interest — from $[\lambda D A D^t]$ to $[\lambda_k D_k A_k D_k^t]$ — and 6 particular models — $[\lambda B]$, $[\lambda_k B]$, $[\lambda B_k]$, $[\lambda_k B_k]$, $[\lambda I]$ and $[\lambda_k I]$. The maximum likelihood estimators of the unknown parameters for each of these 14 models are computed from a training sample followed by an evaluation of the misclassification risk using the leave-one-out method. The best model is the one that minimizes this misclassification risk. Our main interest here is to provide a flexible data adaptive and user-friendly classification rule that also has a simple geometrical interpretation.

5. Model-Based Kernel Discriminant Analysis

While the Gaussian density is the most common assumption in discriminant analysis of continuous variables with probabilistic foundations, it will not typically produce the best classification rule when the (transformed) data are not Gaussian. In this section, we address this issue by introducing a flexible algorithm based on kernel density estimation. The intuition behind our approach is that kernel estimation will perform well in domains that violate the normality assumption, with little cost in domains where it holds.

Consider first the univariate kernel estimator of the marginal density function of the k-th population G_k defined by:

$$\hat{f}(\mathbf{x}|G_k) = \frac{1}{n_k h_k} \sum_{i=1}^{n_k} Ker\left(\frac{\mathbf{x} - \mathbf{x}_i}{h_k}\right) \tag{7}$$

Table 1: Description of the shape of the models, volume, and their MLEs.

Model: Σ_k	Shape	Volume	MLE
1.$[\lambda I]$	Spherical	Same	$\lambda = (np)^{-1}tr(W)$
2.$[\lambda_k I]$	Spherical	Different	$\lambda_k = (pn_k)^{-1}tr(W_k)$
3.$[B]$	Ellipsoidal	Same	$B = n^{-1}diag(W)$
4.$[\lambda B_k]$	$\neq$ Ellipsoidal	Same	$\lambda = n^{-1} \sum_{k=1}^{K} \|diag(W_k)\|^{1/P}$
			$B_k = \|diag(W_k)\|^{-1/P} diag(W_k)$
5.$[\lambda_k B_k]$	$\neq$ Ellipsoidal	Different	$\lambda_k = n_k^{-1}\|diag(W_k)\|^{1/P}$
			$B_k = \|diag(W_k)\|^{-1/P} diag(W_k)$
6.$[\Sigma]$	Linear kernel	Same	$n^{-1}W$.

where $\mathbf{x}$ and $\mathbf{x}_i$ are univariate, $(\mathbf{x}_i, i = 1, \ldots, n_k)$ is the sample from population G_k, $Ker(.)$ is a kernel function, usually a symmetric probability density function, and h_k is the *window width* (one for each population), which controls the degree of smoothness of the estimate. For our application, we use the multivariate Gaussian kernel

$$Ker(\mathbf{x}) = (2\pi)^{-P/2}|\Sigma|^{-1/2} \exp\{-\frac{1}{2}\mathbf{x}^t\Sigma^{-1}\mathbf{x}\} \qquad (8)$$

and our multivariate density estimator is

$$\hat{f}(\mathbf{x}|G_k) = \frac{1}{n_k} \sum_{i \in G_k} \frac{1}{(2\pi)^{P/2}h_k^P|\Sigma_k|^{1/2}} \exp\left\{-\frac{1}{2h_k^2}(\mathbf{x} - \mathbf{x}_i)^t\Sigma_k^{-1}(\mathbf{x} - \mathbf{x}_i)\right\} \qquad (9)$$

or, more compactly,

$$\hat{f}(\mathbf{x}|G_k) = \frac{1}{n_k} \sum_{i \in G_k} g(\mathbf{x}, \mu_i, h_k^2\Sigma_k) \qquad (10)$$

where $g(\mathbf{x}, \mu, h^2\Sigma)$ is a multivariate normal density with mean μ and covariance matrix $h^2\Sigma$, the index i ranges over the training points in the group G_k, and $\mu_i = \mathbf{x}_i$. This looks similar to Gaussian density estimation, except that now the estimated density is averaged over a large set of Gaussian densities, each centered at an individual data point.

In Table 1,

$$W_k = \sum_{i \in G_k} (\mathbf{x}_i - \bar{\mathbf{x}}_k)(\mathbf{x}_i - \bar{\mathbf{x}}_k)^t, \text{ and} \qquad (11)$$

$$W = \sum_{k=1}^{K} W_k, \quad \bar{\mathbf{x}}_k = \frac{\sum_{i \in G_k} \mathbf{x}_i}{\#i \in G_k}, \qquad (12)$$

where G_k is the k-th training is the k-th training sample or group and K is the number of groups. $Diag(M)$ denotes the diagonal matrix that has the same diagonal as the matrix M, and $tr(M)$ denotes the trace of M.

In Gaussian classification one can estimate μ_k and Σ_k by storing only the sum of the observed $\mathbf{x}$'s and the sum of their squares, which are the sufficient statistics for a normal distribution. With kernel-based classification, we must store every continuous attribute observed during training. The only sufficient statistics for the list of μ_i's is the list of $\mathbf{x}_i$'s itself. This means that, when computing $\hat{f}(\mathbf{x}|G_k)$ for a continuous attribute to classify an observation $\mathbf{x}$, we must perform n_k evaluations of g (i.e., one per observed value of $\mathbf{x}$ in class G_k).

Our kernel-based approach also requires that we estimate Σ_k and h_k. Since the computational burden involved in estimating Σ_k is often excessive, it is common practice to restrict it to be a diagonal matrix $diag(\sigma_1, \ldots, \sigma_p)$ (see, e.g., Hand 1981) (Model 3 in Table 1 above). Often even greater simplification is imposed by setting $\sigma_j = \sigma$ for $j = 1, \ldots, p$ (Model 1 in Table 1). Both restrictions are made to minimize the number of parameters to be estimated. We will also propose a variety of parsimonious models given in Table 1 to regularize the sample covariance matrices, Σ_k, from a linear shape (complex) to a spherical shape (simple). An estimate of the window width h_k, which controls the degree of smoothness of the estimated density, is given by

$$h_k = [\frac{1}{n_k p} \sum_{i \in G_k} d^2(\mathbf{x}_i, 2 - NN_i)]^{1/2} \tag{13}$$

where NN_i is the nearest neighbor of $\mathbf{x}_i$ (see Bryan, 1971).

6. Model Selection

We could leave the choice of a model to the user, or the investigator. For example, Model 3 of Table 1 corresponds to one of the most commonly used models. It gives relatively good results in many situations. However, such a choice may be arbitrary without an objective analytical tool for model selection. Therefore, here we develop a new and novel way of choosing both the best fitting model and the dimension of that model simultaneously. More specifically, we develop the information complexity $ICOMP$ criterion of Bozdogan (1994) based on the *maximal covariance complexity* index of van Emdem (1971) for model evaluation This approach has an established theoretical background and foundation. In this short paper, we introduce just the first formulation of $ICOMP$. This is based on the the covariance matrix properties of the parameter estimates of a model starting from their finite sampling distributions. In this case, we define $ICOMP$ as

$$ICOMP = -2logL(\hat{\theta}) + 2C_1(\widehat{\Sigma}(\hat{\theta})), \tag{14}$$

where $-2logL(\hat{\theta})$ is the -2 times the *maximized log likelihood function of the model* (*lack-of-fit, or badness of fit*), and

$$C_1(\widehat{\Sigma}(\widehat{\theta})) = \frac{p}{2}log\left[\frac{tr\widehat{\Sigma}(\widehat{\theta})}{p}\right] - \frac{1}{2}log\left|\widehat{\Sigma}(\widehat{\theta})\right| \qquad (15)$$

is the entropic complexity of the estimated covariance matrix of the model.

In *ICOMP* the first component measures the lack of fit, and the second component measures the complexity of the covariance matrix of the parameter estimates of a model.

We will use *ICOMP* to select the appropriate number of dimensions, p, and the best *window width*, h_k of the model. For our kernel-based approach, we define $C_1(\widehat{\Sigma}_k)$ by

$$C_1(\widehat{\Sigma}_k) = (1 + \hat{h}_k^2)C_1(\widehat{\Sigma}_k), \qquad (16)$$

where $\hat{h}_k$ is the estimated *window width* of the kernel function.

For more detailes on *ICOMP* and its other forms and theoretical justifications, we refer the readers to Bozdogan (2000).

7. Applications

We use two real data sets to test the performance of our approach for suggesting a rule of classification with a plausible geometric model (window width). We compare our kernel-based approach, abreviated as MKDA, with those of canonical discriminant analysis (CDA), and EDRDA. In performing our analysis, we assume that the prior probabilities of the groups are equal in each example. We calculate the cross-validated misclassification risks for various values of p and for all models described in Table 1.

7.1 Example 1: Cervical data

In this example, $n = 242$ cases of various grades of *neoplasia* were collected and diagnosed in a subsequently taken biopsy (Meulman et al. 1990). There were 50 cases with *mild displasia* G_1, 50 cases with *moderate displasia* G_2, 50 cases with *severe displasia* G_3, 50 cases with *carcinoma in situ* G_4, and 42 cases with *invasive carcinoma* G_5. There were 11 variables, 7 of which are categorical and 4 are numerical.

Under the EDRDA algorithm, we calculated the error rate of misclassification using cross-validation for each of the 14 models. Model 2 was chosen as the best fitting model (see Figure 1). The 14 models are described in Bensmail and Celeux (1996) and are summarized below.

In Table 3, $e_1,\ldots,e_p$ are the eigenvalues of the correlation between quantified variables and $m\sum_{j=1}^p e_j/m - 1$, is the total fit.

Using CDA and EDRDA, the cross-validated misclassification risks were **.34**, and **.16**, respectively. With MKDA this error fell to **.10**. It was obtained from two models: $[\lambda_k I]$ and $[H]$ which were chosen by *ICOMP*. The number of dimensions chosen by *ICOMP* with $[\lambda_k I]$ was $p = 8$ whereas for $[H]$ it was $p = 3$ and 4 (see Table 3, Figure 2).

Table 2: Fourteen models based on their covariance structures.

$M_1 : \lambda DAD^t$	$M_2 : \lambda_k DAD^t$	$M_3 : \lambda D A_k D^t$
$M_4 : \lambda_k D A_k D^t$	$M_5 : \lambda D_k A D_k^t$	$M_6 : \lambda_k D_k A D_k^t$
$M_7 : \lambda D_k A_k D_k^t$	$M_8 : \lambda_k D_k A_k D_k^t$	$M_9 : \lambda I$
$M_{10} : \lambda_k I$	$M_{11} : \lambda B$	$M_{12} : \lambda_k B$
$M_{13} : \lambda B_k$	$M_{14} : \lambda_k B_k$	

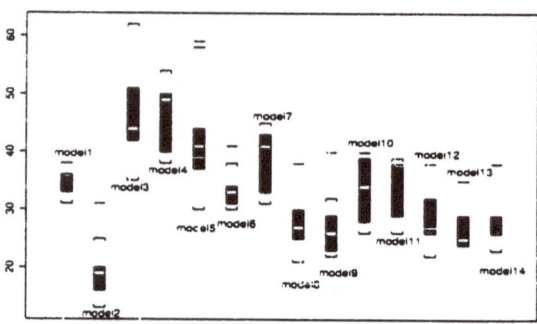

Figure 1: Box plots for the error rates of misclassification using cross-validation accross 14 models. Model 2 is chosen as the best fitting model.

Before claiming that MKDA performs better than EDRDA and CDA, we also applied the **0.632** bootstrap method with 100 bootstrap replications. The estimated misclassification risk of CDA decreases to **.20**, and for EDRDA and MKDA, it is **.17** and **.12** (see Table 4).

7.2 Example2: Cetacea

In this example we analyze Slijper's data concerning 36 differents types of cetacea (see Slijper (1973)). The whales, porpoises and dolphins in this data set were described by 15 characteristics related to their morphology, osteology or behavior (A: Morphological variables: Neck, form of the head, size of the head, beak, dorsal fin, flippers, set of teeth, longitudinal furrows on throat, blow hole, color. B: Osteological variables: Cervical vertebrae, lachrymal and jugal bones, head bones. C: Behavioral variables: Habitat and Feeding). These characteristics are considered as discriminating variables for allocating the Cetacea into 9 groups.

Using EDRDA algorithm, we calculated the error rate of misclassfication using cross-validation for the 14 models. Models 1 and 3 were chosen (see Figure 3).

Using CDA and EDRDA, the cross-validated misclassification risks were **.26** and **.13**, respectively. Using MKDA the error fell to **.11** and was obtained with three models: $[\lambda I]$, $[\lambda I]$ (spherical model), $[\lambda B]$ (ellipsoidal model with

Table 3: Example 1:Models, dimensions, and cross-validated
misclassification risks (Error-CV) for Cervical data.

No.	Model	p	Error − CV	$\frac{m\sum_{j=1}^{p} e_j}{m-1}$
1	$[\lambda I]$	11	.13	0.5
2	$[\lambda_k I]$	8	.10	0.6
3	$[\lambda B]$	6, 7	.13	0.5
4	$[\lambda B_k]$	4	.21	0.4
5	$[\lambda_k B_k]$	3, 4	.16	0.4
6	$[H]$	3, 4	.10	0.4

Table 4. Example 1: Error rates of misclassification by cross-validation
versus the bootstrapping based on the best model and dimension chosen.

Methods	Error − CV	.632 bootstrap
EDRDA	.13	.17
CDA	.34	.20
MKDA	.10	.12

the same volume) and $[H]$ (linear discriminant model). The number of dimensions chosen by *ICOMP* under $[\lambda I]$ was 7, 9 and 11, under $[\lambda_k I]$ it was 6, and under $[\lambda B]$ it was 7 and 9 , and under $[H]$ it was 4 and 6 (see Table 5, Figure 4). Using CDA, there seem to be some 5 animals in group (5) (dolphins) of which membership is not clear at all. Furthermore, our solution agrees nicely with Slijper classification and the 5 observations were classified according to the MKDA into group (6) (porpoises). The 0.632 bootstrap estimates of the misclassification risks were **.24** for CDA and **.15** for MKDA which selected $[\lambda B]$ 62 times out of 100 and selected $[\lambda I]$ 18 times of 100.

Using CDA and EDRDA, the cross-validated misclassification risks were **.26** and **.13**, respectively. Using MKDA the error fell to **.11** and was obtained with three models: $[\lambda I]$, $[\lambda I]$ (spherical model), $[\lambda B]$ (ellipsoidal model with the same volume) and $[H]$ (linear discriminant model). The number of dimensions chosen by *ICOMP* under $[\lambda I]$ was 7, 9 and 11, under $[\lambda_k I]$ it was 6, and under $[\lambda B]$ it was 7 and 9 , and under $[H]$ it was 4 and 6 (see Table 5, Figure 4). Using CDA, there seem to be some 5 animals in group (5) (dolphins) of which membership is not clear at all. Furthermore, our solution agrees nicely with Slijper classification and the 5 observations were classified according to the MKDA into group (6) (porpoises). The 0.632 bootstrap estimates of the misclassification risks were **.24** for CDA and **.15** for MKDA which selected $[\lambda B]$ 62 times out of 100 and selected $[\lambda I]$ 18 times of 100.

8. Conclusion

In the case of discriminant analysis for mixed data (continuous and cate-

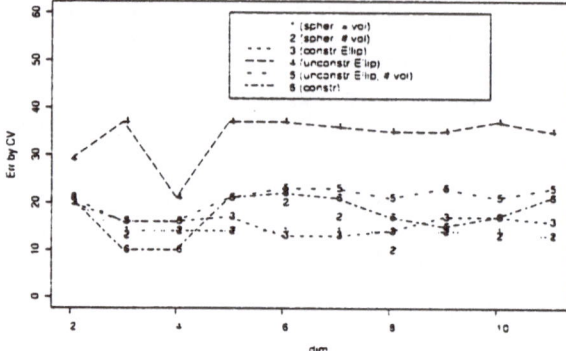

Figure 2: ICOMP based error rates by CV (Err by CV) across different dimensions of the data for six covariance structures given in Table 1.

Table 5: Example 2: Models, dimensions, and cross-validated misclassification risks (Error-CV) for Cetacea data.

No.	Model	p	Error − CV	$\frac{m\sum_{j=1}^{p} e_j}{m-1}$
1	$[\lambda I]$	7, 9, 11	.11	0.7
2	$[\lambda_k I]$	6	.11	0.6
3	$[\lambda B]$	7, 9	.11	0.6
4	$[\lambda B_k]$	3	.28	0.4
5	$[\lambda_k B_k]$	3	.26	0.4
6	$[H]$	4, 6	.11	0.6

gorical variables), in the literature not many model-based methods have been proposed. Here we proposed a novel way of combining optimal scaling and a non-parametric method for discriminating the objects using the mixed data structures by keeping the data integrity intact. First, we applied optimal scaling to transform the data in the full space. Second, we proposed a multivariate kernel discriminant analysis approach to the transformed data to estimate the posterior probabilities of group membership to allocate the observations.

A number of implementation issues arose, including which dimension to choose from the transformed data to perform the discriminant analysis and the choice of window width for multivariate kernel density estimation. We tackled these issue by proposing a variety of shaped windows, as in Bensmail and Celeux (1996), and calculated Bozdogan's (1994, 2000) *ICOMP* criterion for various models and various values of p in choosing the best fitting model structure and the dimension of the model with the minimum value of *ICOMP*.

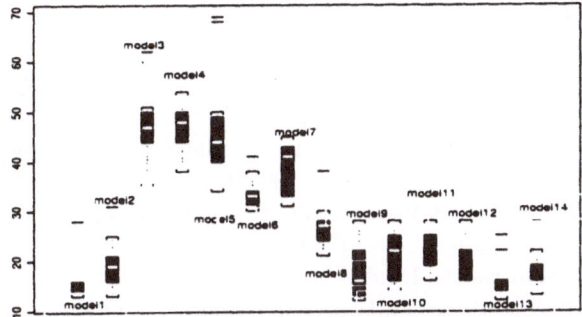

Figure 3: Box plots for the error rates of misclassification using cross-validation accross 14 models. Models 1 and 3 are chosen as the best fitting models.

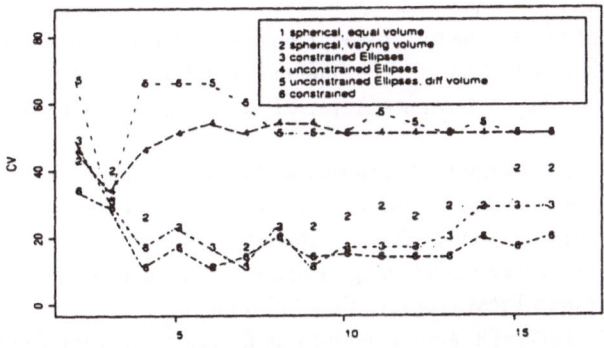

Figure 4: ICOMP based error rates by CV (Err by CV) across different dimensions of the data for six covariance structures given in Table 1.

Acknowledgements

The authors are grateful for the comments of an anonimous referee, and Dr. Peter M. Bearse for helpful comments which improved the paper further. We dedicate this paper to the long and outstanding contributions of Professor Shizuhiko Nishisato for the occasion of his retirement.

References:

Bensmail, H., and Celeux, G. (1996): Regularized Discriminant Analysis Through Eigenvalue Decomposition, Journal of the American Statistical Association, Vol. 91, No. 436, pp. 1743-1748.

Bozdogan, H. (1994): Mixture-model cluster analysis using a new informational complexity and model selection criteria. In: H. Bozdogan, ed., Multivariate Statistical Modeling, Vol. 2, Proceedings of the First US/Japan Conference on the Frontiers of Statistical Modeling: An Informational Approach, Kluwer Academic Publishers, The Netherlands, Dordrecht, 69-113.

Bozdogan, H. (2000): Akaike's Information Criterion and Recent Developments in Information Complexity. Journal of Mathematical Psychology 44, 62-91.

Devroye, L. (1983): The equivalence of weak, strong and complete convergence in L1 for kernel density estimates. The Annals of Statistics, 11, 896-904.

Gifi, A. (1990): Non linear Multivariate Analysis. Wiley Series in Probability and Mathematical Statistics, England.

Hand, D.H. (1981): Discrimination and Classification. Wiley Series in Probability and Mathematical Statistics, England.

Heiser, W.J., and Meulman, J.J. (1995): Nonlinear methods for the analysis of homogeneity and heterogeneity. In: W.J Krzanowski(Ed.). Recent Advances in Descriptive Multivariate Analysis. Clarendon Press, Oxford.

Meulman, J.J. (1986): A Distance Approach to Nonlinear Multivariate Analysis. DSWO Press, Leiden.

Meulman, J.J. et al. (1990): Prediction of Various Grades of Cervical Preneoplasia and Neoplasia on Plastic Embedded Cytobrush Samples. *Technical Report RR-90-06*, Department of Data Theory, University of Leiden.

Nishisato, S. (1980): Analysis of categorical data: dual scaling and its applcations. University of Toronto Press, Toronto Buffalo London.

Shunmugan, K. (1977): On a modified form of Parzen estimator for nonparametric pattern recognition. Pattern Recognition, 9, 167-170.

Tatsuaka, M.M. (1988): Multivariate Analysis. Techniques for Educational and Psychological research. Macmillan publishing Company, New York. Collier Macmillan Publishers, London.

Van der Burg, E. (1985): Homals classification of whales, Porpoises and Dolphins. Proceedings of the International Workshop on Data Analysis. Data Analysis in Real Life Environment: Ins and Outs of Solving Problems. J.F Marcotorchino, J. M. Proth, and J. Janssen(Eds.) Elsevier Science Publishers B. V. (North-Holland).

Van der Burg, E., De Leeuw, J., and Verdegaal, R. (1988): Homogeneity analysis with k sets of variables: an alternating least square method with optimal scaling features. Psychomerika, 53, 177-197.

Van Emden, M. H. (1971): An Analysis of Complexity. Mathematical Centre Tracts, Amsterdam.

Wertz, W. (1978): Statistical density estimation: A survey. Göttingen, Vandenhoek, and Ruprecht, Monographs in Applied Statistics and Econometrics, No. 13.

Dimension Reduction in Hierarchical Linear Models

Yoshio Takane[1] and Michael A. Hunter[2]

[1]McGill University
1205 Dr.Penfield Avenue, Montreal, Quebec, Canada H3A 1B1
[2]University of Victoria
P. O. Box 3050, Victoria, British Columbia, Canada V8W 3P5

Summary: In many disciplines of social sciences, data are often hierarchically structured. Academic performance may be measured of students who are nested in classes which are in turn nested within schools. Multi-level analysis based on the hierarchical linear model (HLM) has been effectively used to capture the hierarchical nature of such data. Most of the existing studies that employ HLM, however, use only a few predictor variables at all levels, because interpretation of parameters in HLM will become increasingly more difficult as the number of parameters increases. To alleviate the difficulty, we propose a method of reducing the dimensionality of the parameter space in HLM in a manner similar to reduced-rank regression models. We describe the two-level HLM, present a parameter estimation procedure and suggest where the rank-reduction may be applied. An example is given to illustrate the proposed method.

1. Introduction

In many fields of social sciences, data collected often have hierarchical structures. For example, academic performance may be measured of students who are nested in classes which, in turn, are nested within schools. Measurements may be repeatedly taken of an attribute from individuals grouped by the region of their domicile, and so on. Multi-level analysis based on hierarchical linear models (HLM) has been effectively used to capture the hierarchical nature of such data (Bock, 1989; Bryk & Raudenbush, 1992; Goldstein, 1987; Hox, 1995).

Most of the existing studies that employ HLM, however, use only a few (typically, one or two) predictor variables at all levels. This is primarily because interpretation of parameters in HLM becomes increasingly more difficult with the increasing number of parameters in the model. Numerical difficulties often encountered in fitting HLM with a moderate number of predictor variables may be another contributing factor to this practice. To alleviate the difficulty, we propose a method of reducing the dimensionality of the parameter space in HLM. This is done in a manner similar to reduced-rank regression models (Anderson, 1951) or equivalently redundancy analysis (Van den Wollenberg, 1977). The dimension reduction improves the interpretability of model parameters, particularly when there are a large number of parameters in the model.

2. The Method

2.1 The Model

We consider a two-level hierarchical linear model (HLM). Extensions to higher-level HLM are straightforward. Throughout this paper, we assume a fixed-effect model. While this is a bit unconventional, it may be justified on the ground that the dimension reduction is most pertinent in the exploratory mode of data analysis.

Suppose there are N_j $(j = 1, \ldots, J)$ subjects (the first-level units) nested within J groups (the second-level units). Let $\mathbf{y}_j$ denote an N_j-component vector of observations on the dependent variable for the j-th group, and let $\mathbf{X}_j$ denote an N_j by P matrix of the P first-level predictor variables for the j-th group. Although this is not an absolute requirement, we assume, for simplicity, that the number of the first-level predictor variables, P, is the same across all J groups. Let $\mathbf{w}_j$ denote an M-component vector of the second-level predictor variables for the j-th group. Then, the first-level model of the two-level HLM can be written, for a specific group j, as

$$\mathbf{y}_j = \mathbf{1}_{N_j} b_{0j} + \mathbf{X}_j \mathbf{b}_{1j} + \mathbf{e}_j, \tag{1}$$

where $\mathbf{1}_{N_j}$ is an N_j-component vector of ones, b_{0j} is the intercept parameter, $\mathbf{b}_{1j}$ is the vector of slope parameters, and $\mathbf{e}_j$ is the vector of disturbance terms. Subscripts 0 and 1 are used to distinguish between the intercept and the slope parameters for which separate second-level models are postulated:

$$b_{0j} = c_{00} + \mathbf{c}'_{01} \mathbf{w}_j + u_{0j} \quad (\text{or } b_{0j} = c_{00} + \mathbf{w}'_j \mathbf{c}_{01} + u_{0j}), \text{ and} \tag{2}$$

$$\mathbf{b}_{1j} = \mathbf{c}_{10} + \mathbf{C}'_{11} \mathbf{w}_j + \mathbf{u}_{1j} \quad (\text{or } \mathbf{b}_{1j} = \mathbf{c}_{10} + (\mathbf{I}_P \otimes \mathbf{w}'_j) \mathbf{c}_{11} + \mathbf{u}_{1j}), \tag{3}$$

where c's are the second-level regression parameters, u's are the second-level disturbance terms, $\mathbf{I}_P$ is the identity matrix of order P, $\mathbf{c}_{11} = \text{vec}(\mathbf{C}'_{11})$, and $\otimes$ indicates a Kronecker product. Substituting (2) and (3) for b_{0j} and $\mathbf{b}_{1j}$ in (1) leads to

$$\mathbf{y}_j = \mathbf{1}_{N_j} (c_{00} + \mathbf{w}'_j \mathbf{c}_{01} + u_{0j}) + \mathbf{X}_j (\mathbf{c}_{10} + (\mathbf{I}_p \otimes \mathbf{w}'_j) \mathbf{c}_{11} + \mathbf{u}_{1j}) + \mathbf{e}_j. \tag{4}$$

Let $\mathbf{y}$, $\mathbf{1}_N$, $\mathbf{u}_0$, $\mathbf{u}_1$, and $\mathbf{e}$ be super-vectors of $\mathbf{y}_j$, $\mathbf{1}_{N_j}$, u_{0j}, $\mathbf{u}_{1j}$, and $\mathbf{e}_j$ $(j = 1, \ldots, J)$, respectively. Let $\mathbf{G}$ and $\mathbf{D}_X$ denote block diagonal matrices with $\mathbf{1}_{N_j}$ and $\mathbf{X}_j$ as the j-th diagonal blocks, and let

$$\mathbf{W}'_0 = (\mathbf{w}_1, \ldots, \mathbf{w}_J), \text{ and} \tag{5}$$

$$\mathbf{W}'_1 = (\mathbf{I}_P \otimes \mathbf{w}_1, \ldots, \mathbf{I}_P \otimes \mathbf{w}_J). \tag{6}$$

Then, the model for all observations for all J groups can be written as

$$\mathbf{y} = \mathbf{1}_N c_{00} + \mathbf{G} \mathbf{W}_0 \mathbf{c}_{01} + \mathbf{G} \mathbf{u}_0 + \mathbf{X} \mathbf{c}_{10} + \mathbf{D}_X \mathbf{W}_1 \mathbf{c}_{11} + \mathbf{D}_X \mathbf{u}_1 + \mathbf{e}, \tag{7}$$

where $\mathbf{1}_N = \mathbf{G} \mathbf{1}_J$, and $\mathbf{X} = \mathbf{D}_X (\mathbf{1}_J \otimes \mathbf{I}_P)$.

The above model is not identified. To remove redundancies in the model we successively make the seven terms in the model mutually orthogonal. The model then decomposes the data vector y into seven mutually orthogonal components with each component having specific interpretation. The first term in (7) is the intercept term. The next two terms represent between-groups effects, of which the second term pertains to the portions of the between-groups effects that can be accounted for by the second-level predictor variables (w_j's), and the third term to the portions left unaccounted for by the second-level predictor variables. The remaining four terms represent within-groups effects, the first one of which (the fourth term in (7)) pertains to the main effects of the first-level predictor variables ($\mathbf{X}$), the next (the fifth term) to the within-groups interactions between the first- and the second-level predictor variables, and the sixth term represents the portions of the interaction effects between groups and the second-level predictor variables left unaccounted for by the fourth and the fifth terms. The last term represents the within-groups effects that cannot be explained by any systematic effects in the model.

2.2 Estimation

Since the model (7) is linear in parameters assumed to be of the fixed-effects type, least squares (LS) estimates of parameters (c's and u's) can be obtained in a straightforward manner. In what follows, SS_i denotes the sum of squares (SS) accounted for by the ith term in (7). We put a hat on a symbol to indicate a least squares estimate. Below, where the regular inverse cannot be taken, it may be replaced by the Moore-Penrose inverse.

We fit one term at a time sequentially. The remaining terms are orthogonalized to all previously fitted terms. Let $\bar{y} = \mathbf{1}'_N \mathbf{y}/N$, the grand mean. Then,

$$\hat{c}_{00} = \bar{y}. \tag{8}$$

$SS_1 = N\bar{y}^2$. We define $SS_t = \mathbf{y}'\mathbf{y} - SS_1$ (the total SS). To make the second term orthogonal to the first, we redefine the second term as $\mathbf{G}\tilde{\mathbf{W}}_0\mathbf{c}_{01}$, where $\tilde{\mathbf{W}}_0$ denote the columnwise centered $\mathbf{W}_0$. For later use, we also define the vector of deviation scores from the grand mean, $\mathbf{y}^* = \mathbf{y} - \mathbf{1}_N\bar{y}$. This vector represents the portions of y left unaccounted for by the grand mean. Then,

$$\hat{c}_{01} = (\tilde{\mathbf{W}}'_0\mathbf{G}'\mathbf{G}\tilde{\mathbf{W}}_0)^{-1}\tilde{\mathbf{W}}'_0\mathbf{G}'\mathbf{y}^*. \tag{9}$$

SS_2 is obtained by $SS_2 = \mathbf{y}^{*\prime}\mathbf{G}\tilde{\mathbf{W}}_0\hat{c}_{01}$. Define $\bar{\mathbf{y}}^* = (\mathbf{G}'\mathbf{G})^{-1}\mathbf{G}'\mathbf{y}^*$. This is the vector of group means in the form of deviations from the grand mean. Then,

$$\hat{\mathbf{u}}_0 = \bar{\mathbf{y}}^* - \tilde{\mathbf{W}}_0\hat{c}_{01}. \tag{10}$$

SS_3 is obtained by $SS_3 = \mathbf{y}^{*\prime}\mathbf{G}(\mathbf{G}'\mathbf{G})^{-1}\mathbf{G}'\mathbf{y}^* - SS_2$. We define $SS_b = SS_2 + SS_3$ (the between-groups SS). Let $\mathbf{y}^{(w)} = \mathbf{y}^* - \mathbf{G}\bar{\mathbf{y}}^*$. which is the vector of deviation scores from the group means and represents the portions of $\mathbf{y}^*$ (or equivalently, the portions of $\mathbf{y}$) left unaccounted for by $\mathbf{G}$. We columnwise center $\mathbf{X}_j$ within each group and denote it by $\mathbf{X}_j^*$. We then form a super-matrix of $\mathbf{X}_j^*$'s by putting them columnwise and denote it by $\mathbf{X}^*$. Then,

$$\hat{\mathbf{c}}_{10} = (\mathbf{X}^{*'}\mathbf{X}^*)^{-1}\mathbf{X}^{*'}\mathbf{y}^{(w)}. \tag{11}$$

SS_4 is given by $SS_4 = \mathbf{y}^{(w)'}\mathbf{X}^*\hat{\mathbf{c}}_{10}$. Let $\mathbf{D}_{X^*}$ denote a block diagonal matrix with $\mathbf{X}_j^*$'s as diagonal blocks. To make the fifth term orthogonal to all the previous terms, we redefine it as $\mathbf{D}_{X^*}\tilde{\mathbf{W}}_1\mathbf{c}_{11}$, where $\tilde{\mathbf{W}}_1 = \mathbf{W}_1 - (\mathbf{1}_J \otimes \mathbf{I}_P)(\mathbf{X}^{*'}\mathbf{X}^*)^{-1}\mathbf{X}^{*'}\mathbf{D}_{X^*}\mathbf{W}_1$. Then,

$$\hat{\mathbf{c}}_{11} = (\tilde{\mathbf{W}}_1'\mathbf{D}_{X^*}'\mathbf{D}_{X^*}\tilde{\mathbf{W}}_1)^{-1}\tilde{\mathbf{W}}_1'\mathbf{D}_{X^*}'\mathbf{y}^{(w)}. \tag{12}$$

SS_5 is then given by $SS_5 = \mathbf{y}^{(w)'}\mathbf{D}_{X^*}\tilde{\mathbf{W}}_1\hat{\mathbf{c}}_{11}$. Let $\mathbf{y}^{(w)*} = (\mathbf{D}_{X^*}'\mathbf{D}_{X^*})^{-1} \times \mathbf{D}_{X^*}'\mathbf{y}^{(w)}$. Then,

$$\hat{\mathbf{u}}_1 = \mathbf{y}^{(w)*} - (\mathbf{1}_J \otimes \mathbf{I}_P)\hat{\mathbf{c}}_{10} - \tilde{\mathbf{W}}_1\hat{\mathbf{c}}_{11}. \tag{13}$$

SS_6 is obtained by $\mathbf{y}^{(w)'}\mathbf{D}_{X^*}\hat{\mathbf{u}}_1$. Finally,

$$\hat{\mathbf{e}} = \mathbf{y}^{(w)} - \mathbf{D}_{X^*}\hat{\mathbf{u}}_1. \tag{14}$$

This vector represents what's left unaccounted for by all the systematic effects in the model. Let $SS_w = \mathbf{y}^{(w)'}\mathbf{y}^{(w)}$ (the within-groups SS). Then, $SS_7 = SS_w - (SS_4 + SS_5 + SS_6)$, and $SS_t = SS_b + SS_w$.

2.3 Dimension Reduction

There are several kinds of c parameters. They are c_{00} (1×1), $\mathbf{c}_{01}$ ($M \times 1$), $\mathbf{c}_{10}$ ($P \times 1$), and $\mathbf{c}_{11}$ ($PM \times 1$). None of them depend on j. The $\mathbf{c}_{11}$ represents regression coefficients for the interactions between the first- and the second-level predictor variables. Its estimate may be arranged in the form of $\mathbf{C}_{11}'$ ($P \times M$) and may be subjected to rank reduction, which is done by generalized singular value decomposition (GSVD) of $\hat{\mathbf{C}}_{11}'$ (see, for example, Takane & Shibayama, 1991). That is,

$$\hat{\mathbf{C}}_{11}' = \mathbf{TDV}', \tag{15}$$

where $\mathbf{T}$ and $\mathbf{V}$ satisfy $\mathbf{T}'\mathbf{MT} = \mathbf{I}$ and $\mathbf{V}'\mathbf{NV} = \mathbf{I}$ and $\mathbf{D}$ is diagonal and positive-definite (pd). Matrices $\mathbf{M}$ and $\mathbf{N}$ are called metric matrices, assumed pd. Appropriate metric matrices for GSVD of $\hat{\mathbf{C}}_{11}'$ are $\mathbf{M} = \mathbf{X}^{*'}\mathbf{X}^*$ on the row side and $\mathbf{N} = \tilde{\mathbf{W}}_0'\tilde{\mathbf{W}}_0$ on the column side. The GSVD of $\hat{\mathbf{C}}_{11}'$ with metrics $\mathbf{M}$ and $\mathbf{N}$ can be obtained by ordinary SVD of $\mathbf{R}_M'\hat{\mathbf{C}}_{11}'\mathbf{R}_N$, where $\mathbf{R}_M$ and $\mathbf{R}_N$ are arbitrary square root factors of $\mathbf{M}$ and $\mathbf{N}$, respectively. Let $\mathbf{R}_M'\hat{\mathbf{C}}_{11}'\mathbf{R}_N = \mathbf{T}^*\mathbf{D}^*\mathbf{V}^{*'}$ be the ordinary SVD of $\mathbf{R}_M'\hat{\mathbf{C}}_{11}'\mathbf{R}_N$. Then, the desired GSVD is obtained by $\mathbf{T} = (\mathbf{R}'_M)^{-1}\mathbf{T}^*$, $\mathbf{V} = (\mathbf{R}'_N)^{-1}\mathbf{V}^*$, and $\mathbf{D} = \mathbf{D}^*$. Matrix $\hat{\mathbf{C}}'_{11}$ may be appended by $\hat{c}_{00}$, $\hat{c}_{01}$, and $\hat{c}_{10}$ to form a super-matrix

$$\hat{\mathbf{C}} = \begin{bmatrix} \hat{c}_{00} & \hat{c}'_{01} \\ \hat{c}_{10} & \hat{\mathbf{C}}'_{11} \end{bmatrix}. \tag{16}$$

This augmented matrix may be subjected to GSVD with metrics $\mathbf{M} = [\mathbf{1}_N, \mathbf{X}^*]'[\mathbf{1}_N, \mathbf{X}^*]$ and $\mathbf{N} = [\mathbf{1}_J, \tilde{\mathbf{W}}_0]'[\mathbf{1}_J, \tilde{\mathbf{W}}_0]$. According to our experience, however, this procedure tends to facilitate the mean tendency to dominate the solution.

The u parameters are group specific. Still, one my define a P by J matrix (assuming that P remains the same across the J groups),

$$\hat{\mathbf{U}}_1 = [(\mathbf{X}_1^{*\prime}\mathbf{X}_1^*/N_1)^{1/2}\mathbf{u}_1, \ldots, (\mathbf{X}_J^{*\prime}\mathbf{X}_J^*/N_J)^{1/2}\mathbf{u}_J],$$

which may be subjected to a rank reduction by ordinary SVD. Again, $\hat{\mathbf{u}}'_0$ may be appended to $\hat{\mathbf{U}}_1$ to form a super-matrix, which may be subject to a joint rank reduction.

When N_j is constant across all J groups (say, $N_j = n$ for all j), $\hat{\mathbf{e}}$ may also be rearranged into a J by n matrix $\hat{\mathbf{E}}$, which may be subjected to a rank reduction by SVD. This may be considered a kind of error analysis, which often helps detect which crucial factors are missing in the fitted model.

3. An Illustrative Example

For illustration we use part of the data from the British Social Attitudes Panel Survey, 1983-1986 (Wiggins, Ashworth, O'Muircheartaigh, & Galbraith, 1990) on attitudes toward abortion. In a panel survey, subjects are asked to respond to the same set of questions on several occasions. This allows the stability of responses over time to be assessed, while allowing any changes in the responses to be linked to the individual characteristics of subjects.

Two hundred sixty four subjects were each interviewed four times approximately one year apart on their attitudes toward abortion. The exact format of the questions was as follows:

Here are a number of circumstances in which a woman might consider an abortion. Please say whether or not you think the law should allow an abortion in each case. Should abortion be allowed by law?

(1) The woman decides on her own she does not wish to have the child.
(2) The couple agree they do not wish to have the child.
(3) The woman is not married and does not wish to marry the man.
(4) The couple cannot afford any more children.
(5) There is a strong chance of a defect in the baby.
(6) The woman's health is seriously endangered by the pregnancy.
(7) The woman became pregnant as a result of rape.

The number of items endorsed by each subject was counted and used as a

Table 1. The breakdown of the total SS.

Source	% SS	SS	df	MS	F
SS_2	8.5%	304	8	38.0	4.5^1
SS_3	60.0%	2134	255	8.4	
SS_4	2.8%	98	9	10.9	8.4^2
SS_5	2.9%	103	72	1.4	1.1^2
SS_6	25.8%	920	711	1.3	
SS_7	0.0%	0	0		

[1] Against SS_3.
[2] Against SS_6.

measure of his/her favorableness toward abortion. The repeated measurements within subjects ($N_j = 4$) were taken as the first-level units, and the subjects ($J = 264$) were taken as the second-level units.

The second-level predictor variables used were:

1. Gender: 1) male, or 2) female.
2. Age classified into five groups: 1) up to 29 years of age, 2) up to 39, 3) up to 49, 4) up to 59, or 5) 60 and over.
3. Religion: 1) catholic, 2) protestant, 3) other, or 4) no religion.

These variables describe subject characteristics which were assumed stable throughout the four-year period of study. For our analysis, they were coded into 11 dummy variables.

The first-level predictor variables used were:

1. Year of measurements: 1) 1983, 2) 1984, 3) 1985, or 4) 1986.
2. Political party: 1) conservative, 2) labour, 3) liberal, 4) other, or 5) none.
3. Self-assessed social class: 1) middle class, 2) upper working class, or 3) lower working class.

These variables pertain to the repeated measurements within subjects and were considered time-variant. They were dummy-coded into 12 variables as in the case of the second-level predictor variables. (Both sets of predictor variables are linearly dependent, and consequently, the Moore-Penrose inverses were used to obtain LS estimates, where necessary.)

Table 1 gives the breakdown of the total SS (SS_t) explained by different terms in model (7). SS_2 represents the portions of SS_t accounted for by the main effects of the second-level predictor variables. SS_4 represents the portions accounted for by the main effect of the first-level predictor variables, and SS_5 represents the portions that can be accounted for by the interaction between the first- and the second-level predictor variables. The proportions of SS_t that can be accounted for by these effects only add up to 14.2%. However, tested against SS_3, SS_2 is statistically significant ($p < .01$), and tested against SS_6, SS_4 is significant ($p < .01$), although SS_5 is not significant ($p > .05$).

SS_3 represents the between-subjects SS left unaccounted for by the second-level predictor variables, while SS_6 represents the SS due to the subjects by first-level predictor variables interaction effects left unaccounted for by SS_4 and SS_5. These two SS's account for 85.8% of SS_t. SS_7 is equal to zero in the present case, because the number of the first-level predictor variables is larger than the number of repeated measurements per subject, and the interaction between the subjects and the first-level predictor variables captures all the within-subjects effects.

The following tables give estimates of regression coefficients for the main effect of the second-level (Table 2) and that of the first-level (Table 3) predictor variables:

Table 2. Estimate of c_{01}.

Plotting Labels	Variable Categories	Estimates $\hat{c}_{01}$
1. Gender		
w_{11}	male	.066
w_{12}	female	-.066
2. Age		
w_{21}	~29 yrs.	-.106
w_{22}	~39 yrs.	.294
w_{23}	~49 yrs.	-.206
w_{24}	~59 yrs.	.109
w_{25}	60~ yrs.	-.091
3. Religion		
w_{31}	catholic	-.778
w_{32}	protestant	.323
w_{33}	other	-.333
w_{34}	no religion	.788

Table 3. Estimate of c_{10}.

Plotting Labels	Variable Categories	Estimates $\hat{c}_{10}$
1. Year		
x_{11}	1983	.158
x_{12}	1984	-.488
x_{13}	1985	.017
x_{14}	1986	.313
2. Political Party		
x_{21}	conserv.	.006
x_{22}	labour	.167
x_{23}	liberal	-.101
x_{24}	other	.064
x_{25}	none	-.135
3. Social Class		
x_{31}	middle	-.069
x_{32}	upper w.	-.099
x_{33}	lower w.	.168

By eliminating one predictor variable at a time from the model and refitting the reduced model, we can assess the unique contribution of that variable. We found that the religion was the only variable which was statistically significant among the second-level predictor variables ($SS_{Religion} = 227.4$ with 3 df; $F(3, 255) = 9.0$). Estimated coefficients in Table 2 indicate that catholics are least favorable, and those without any religious affiliation tend to be most favorable to abortion. Somewhat unexpectedly, there was little gender difference ($SS_{Gender} = 4.3$ with 1 df; $F(4, 255) = 1.1$), and there was little variation over age ($SS_{Age} = 36.1$ with 4 df; $F(4, 255) = .5$). Among the first-level predictor variables, we found the year of measurements was the only variable significantly affecting the attitude toward abortion ($SS_{Year} = 93.2$ with 3 df; $F(3, 711) = 23.9$). Table 3 indicates that people were on average less favorable toward abortion in 1984, and more favorable in 1986, although the reason for this is not readily apparent. However, there was little systematic trend over time. Neither political alignment ($SS_{Party} = 2.6$ with 4

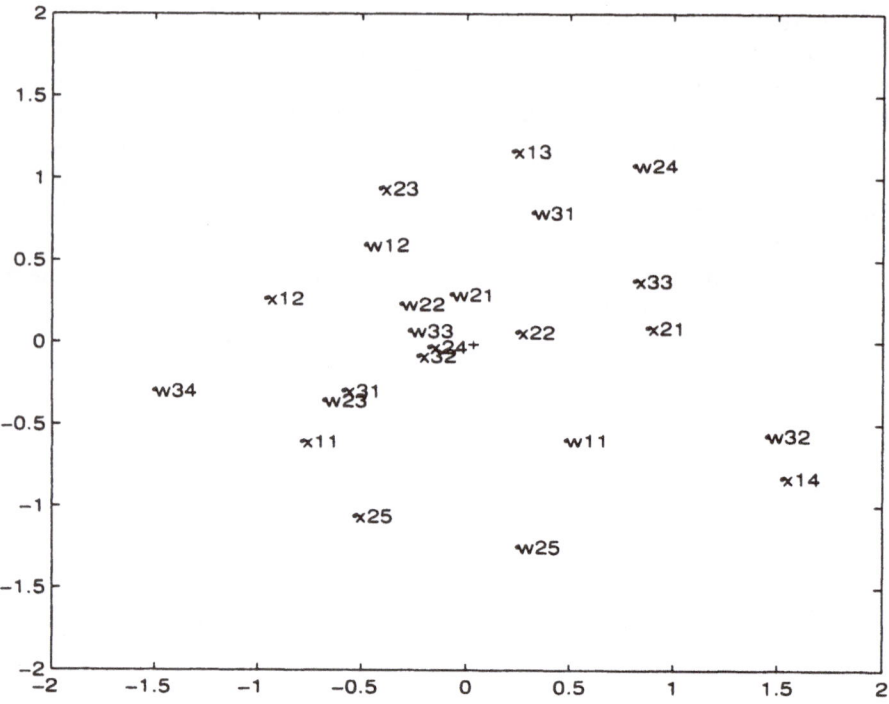

Figure 1: The two-dimensional configuration of the interaction effects.

df; F(4, 711) = .5) nor self-assessed social class (SS_{Class} = 4.5 with 2 df; F(2, 711) = 1.7) had substantial effects on attitude toward abortion.

As has been noted earlier, SS_5 representing the overall size of the interaction effects between the first- and the second-level predictor variables was not statistically significant. Still, some portions of the interaction effects may be significant. A strategy used above for testing the significance of the main effects is a bit cumbersome because there are so many of them (132), which are also partially redundant. This is where dimensionality reduction may help. We can simplify the pattern of the interaction effects by applying SVD to $\hat{C}_{11}$. Figure 1 displays the two-dimensional configuration resulting from a reduced-rank approximation to $\hat{C}_{11}$. The two dimensions account for approximately 75% of the sum of squares pertaining to the interaction effects. Plotting symbols are given in the first columns of Tables 2 and 3. (The first-level predictor variables are indicated by "x", and the second-level variables by "w", followed by the variable number and the category number.) Since what is analyzed here is the weights applied to the interaction effects between the first- and the second-level predictor variables, we look for 1) which variables come in similar directions relative to the origin (indicated by "+"), and 2) which variables come in the opposite directions. For example, w23 (protestant) and x14 (year 1986) come close to each other, indicating that protestants in the year 1986 favored abortion more than that can be expected from separately being protestants (who tend to be more favorable to abortion than average) and that the year the data were taken was 1986 (the year that people tend to be more favorable to abortion than the average year). On the other hand, w31 (catholic) and x25 (no party support) come on the opposite side. This means that catholics with no party support were even less favorable to abortion than that was expected from separately being catholics (who tend to be less favorable to abortion than average) and that the subjects had no political parties to support (who were also less favorable to abortion than average). We can make many other similar observations.

We also analyzed $\hat{U}_1$ as suggested above. However, this relates to the subject differences left unaccounted for by those already taken into account in the model, and interpretation was extremely difficult without additional information about the subjects.

4. Discussion

Reduced-rank approximations of some of the regression parameters in HLM seem useful, particularly when the dimensionality of the parameter space is very high. There are a number of things that may be done to further facilitate the use of HLM:

1. Although only the two-level HLM has been discussed in this paper, similar things can be done for higher-level HLM, although the list of terms in the model gets longer very quickly. For example, there are 15 terms in the three-level HLM. We have already derived LS estimates of parameters associated with these terms, although, due to the space limitation, they cannot be presented here.

2. Bootstrap (e.g., Efron, 1982) or other resampling techniques could be used to assess the stability of individual parameters, which may in turn be used to test their significance. Since the normality assumption is almost always in suspect in survey data, the resampling methods may also be useful to benchmark the distribution of the conventional statistics used in HLM. The number of components to be retained in the reduced-rank approximation may be determined by permutation tests in a manner similar to Takane & Hwang (in press) who developed a permutation procedure for testing the number of significant canonical correlations.

3. Additional (linear) constraints can be readily incorporated in the LS estimation procedure. This allows the statistical tests of the hypotheses represented by the constraints.

4. We exclusively discussed the univariate HLM in this paper. However, the proposed method can easily be extended to the multivariate cases. We may incorporate structures representing the hypothesized relationships among the multiple dependent variables or use the dimension reduction technique to

structure the multiple dependent variables.

5. When the u parameters are assumed to be random rather than fixed, errors are no longer statistically independent. The dependence structure among the errors may be estimated from the initial estimates of parameters (obtained under the independence assumption), which may then be used to re-estimate the parameters, and so on. This leads to an iterative estimation procedure similar to that used in Hwang & Takane (2001) in the context of structural equation modeling. The derived estimates are more efficient than those obtained under the independence assumption.

Acknowledgments

The work reported in this paper has been supported in part by NSERC individual operating grants (A6394) to the first author.

References:

Anderson, T. W. (1951). Estimating linear restrictions on regression coefficients for multivariate normal distributions. *Annals of Mathematical Statistics*, **22**, 327-351.

Bock, R. D. (1989). *Multilevel analysis of educational data.* San Diego, CA: Academic Press.

Bryk, A. S., & Raudenbush, S. W. (1992). *Hierarchical linear models.* Newbury Park, CA: Sage Publications.

Efron, B. (1982). *The Jackknife, the bootstrap and other resampling plans.* Philadelphia: SIAM.

Hwang, H., & Takane, Y.(2001). Structural equation modeling by extended redundancy analysis. This volume.

Goldstein, H. I. (1987). *Multilevel models in educational and social research.* London: Oxford University Press.

Hox, J. J. (1995). *Applied multilevel analysis.* Amsterdam: TT-Publikaties.

Takane, Y., & Hwang, H. (in press). Generalized constrained canonical correlation analysis. *Multivariate Behavioral Research.*

Takane, Y., & Shibayama, T. (1991). Principal component analysis with external information on both subjects and variables. *Psychometrika*, **56**, 97-120.

Van den Wollenberg, A. L. (1977). Redundancy analysis: an alternative for canonical correlation analysis. *Psychometrika*, **42**, 207-219.

Wiggins, R. D., Ashworth, K., O'Muircheartaigh, C. A. & Galbraith, J. I. (1990). Multilevel analysis of attitudes to abortion. *The Statistician*, **40**, 225-234.

Time Dependent Principal Component Analysis

Yasumasa Baba and Takahiro Nakamura

Institute of Statistical Mathematics
4-6-7 Minami-Azabu, Minato-ku, Tokyo 106-8569, Japan

Summary: Principal Component Analysis (PCA) is one of the useful descriptive methods for multivariate data. One aim of the methods is to construct new variables by a linear combination from original variables and illustrate the structure of variables and individuals on a new space based on new variables. In this method principal components have some meanings to summarize variables. Suppose that we obtained principal components at two time points. Then principal components obtained from data at different times will have different meanings if linear combinations of variables are different. For example, the first principal component at time 1 and that at time 2 do not always have the same meanings, as seen in the following example. The first principal component at time 1 may appear as the second or the third principal component at time 2, thus changing the meanings of the first principal components from two time points. Therefore care must be exercised for the interpretation of results from different time points.

In this paper we will propose a method to connect smoothly principal components or loadings from different time points.

1. Introduction

Principal Component Analysis (PCA) is one of the useful methods for multivariate analysis, and its theory has extensively been examined by such pioneers as Pearson (1901), Hoteling (1933), and Anderson (1963). Today it is a very popular method of multivariate data analysis, thanks to the development of statistical software. Principal component analysis has been overviewed by Jollife (1986), Flury (1995) and some others.

One of the purposes of PCA is to obtain principal components from a linear combination of variables and describe individuals or the variables on the space spanned by a few principal components. Then each principal component is recognized as a new variable which describes a new concept as a summary of all the variables. Individuals are plotted on the coordinates of such new scales and a new light may be shed on them. Therefore the meaning of each principal component is important for the description of multivariate data.

Suppose that we obtained data sets at two time points and corresponding principal components. Generally speaking, different principal components have different meanings. For example, the first principal component at time 1 may mean something different from that at time 2. Therefore we need care to interpret them.

Krzanowski (1979, 1982) proposed the idea of a formal comparison of principal components from different groups. Multiple Group Principal Component Analysis (MGPCA) by Thorpe (1983) and Common Principal Component (CPC) by Flury(1984) are well-known formal models for simultaneous principal component analysis of several groups. If we consider successive data sets from the same individuals as several groups, it may be possible to handle the current problem, that is, to interpret principal components at different times. However those currently available models are not appropriate for tracing the changes of principal components over time. In this paper we will propose a method that can trace the change of the meanings of principal components and their roles over two time points.

2. Conventional Methods

In this section we will show four methods for describing the changes of principal components between time t_1 and time t_2.

Let $X(t)$ be an $n \times p$ data matrix at time t.

$$X(t) = [x_{ij}(t)].$$

Let us define a centered data matrix as follows

$$\tilde{X}(t) = X(t) - \overline{X}(t),$$

where $\overline{X}(t)$ is composed of the mean of each variable. The discrepancy between means of t_1 and t_2 will be defined as

$$d\overline{X} = \overline{X}(t_1) - \overline{X}(t_2).$$

The covariance matrix is denoted as follows.

$$S(t_1, t_2) = \tilde{X}'(t_1)\tilde{X}(t_2)$$

Let y be the variable obtained by the transformation.

$$y = \sum_{j=1}^{p} w_j x_j$$

Here time t is not shown explicitly.

Coefficients w_j are obtained by solving eigenequation.

$$S(t, t)w = \lambda w,$$

$$w'w = 1.$$

λ is an eigenvalue and w is an eigenvector.

We will show four methods to describe changes of structure over two time points below. Hereafter we will consider $t=0$ or 1.

(1) Obtain principal components from $X(0)$, and then plot principal components from $X(0)$ and $X(1)$ on the principal axes simultaneously. The eigenequation can be written as follows.

$$S(0,0)w = \lambda w. \qquad (2.1)$$

The change of the data structure will be illustrated in terms of $t=0$.

(2) Obtain the principal components from $X(1)$, and then plot $X(0)$ and $X(1)$ on the principal axes simultaneously. The eigenequation can be written as follows.

$$S(1,1)w = \lambda w. \qquad (2.2)$$

Therefore the change will be expressed on the basis of $t=1$.

(3) Obtain the principal components from pooled data matrix F

$$F = \begin{bmatrix} X(0) \\ X(1) \end{bmatrix}.$$

The eigenequation is as follows.

$$Vw = \lambda w, \qquad (2.3)$$

where

$$V = \frac{1}{2}\{S(0,0)+S(1,1)\}+\frac{1}{4}d\overline{X}'d\overline{X}.$$

Individuals' changes can be found from those of principal components.

(4) Obtain principal components from $2p \times n$ matrix G

$$G = [X(0), X(1)]$$

The eigenequation is as follows.

$$Vw = \lambda w, \qquad (2.4)$$

where

$$V = \begin{bmatrix} S(0,0) & S(0,1) \\ S(1,0) & S(1,1) \end{bmatrix}.$$

The variables at two time points are regarded as different. Therefore we can find the change of variables in the principal plane.

3. Time Dependent PCA

The methods illustrated in section 2 did not consider time dependency, as is obvious from the fact that they did not contain t. We will propose here a method which can trace the change of the meanings and roles of principal components over time, that is, time dependent PCA. For simplicity we will consider the case of two time points.

Let us define the data matrix as follows.

$$\mathbf{X}^{*}(t) = (1-t)\mathbf{X}(0) + t\mathbf{X}(1).$$

This coincides with original matrices at points $t=0,1$. That is, $\mathbf{X}^{*}(0) = \mathbf{X}(0)$ at $t=0$ and $\mathbf{X}^{*}(1) = \mathbf{X}(1)$ at $t=1$.

The covariance matrix can be obtained as ,

$$\mathbf{V}(t) = (1-t)^{2}\mathbf{S}(0,0) + t(1-t)\{\mathbf{S}(0,1) + \mathbf{S}(1,0)\} + t^{2}\mathbf{S}(1,1).$$

The eigenequation can be written as follows.

$$\mathbf{V}(t)\mathbf{w}(t) = \lambda(t)\mathbf{w}(t). \tag{3.1}$$

with the condition that $\mathbf{w}'(t)\mathbf{w}(t) = 1$. This equation is equivalent to the equation (2.1) at $t=0$ and (2.2) at $t=1$. Therefore if t changes 0 to 1 by a small step, we can obtain a continuous change of principal components, loadings and individual scores over time.

We will pursue the change of the solution from $t = 0$ to 1 if we change t by a small amount. Let us denote eigenequations at $t = t_{0}$ and $t = t_{0} + dt$ as follows.

$$\mathbf{V}_{0}\mathbf{w}_{0} = \lambda_{0}\mathbf{w}_{0},$$

$$(\mathbf{V}_{0} + \mathbf{v})\mathbf{w} = (\lambda_{0} + \gamma)\mathbf{w},$$

$\mathbf{v}$ is the increment of the covariance matrix and γ is the increment of the eigenvalue.

Formally the relation of $\mathbf{w}_{0}$ to $\mathbf{w}$ can be deduced by the perturbation technique (Baba and Nakamura, 1997) but here we propose a numerical method with an example.

The equation (3.1) holds at any time. Therefore we can use the equation at any $t+dt$. Regarding the data matrices at two time points as two different sets of data

159

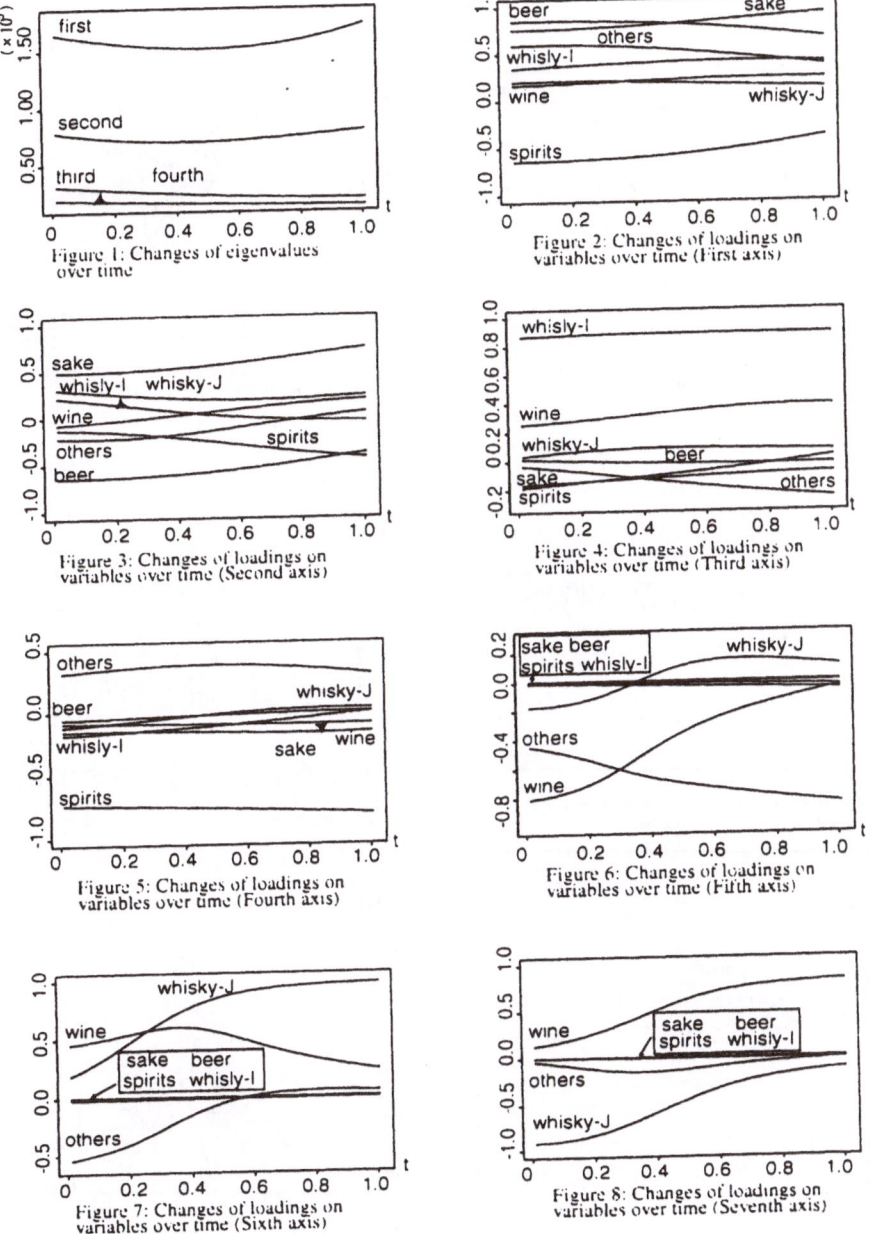

Figure 1: Changes of eigenvalues over time

Figure 2: Changes of loadings on variables over time (First axis)

Figure 3: Changes of loadings on variables over time (Second axis)

Figure 4: Changes of loadings on variables over time (Third axis)

Figure 5: Changes of loadings on variables over time (Fourth axis)

Figure 6: Changes of loadings on variables over time (Fifth axis)

Figure 7: Changes of loadings on variables over time (Sixth axis)

Figure 8: Changes of loadings on variables over time (Seventh axis)

160

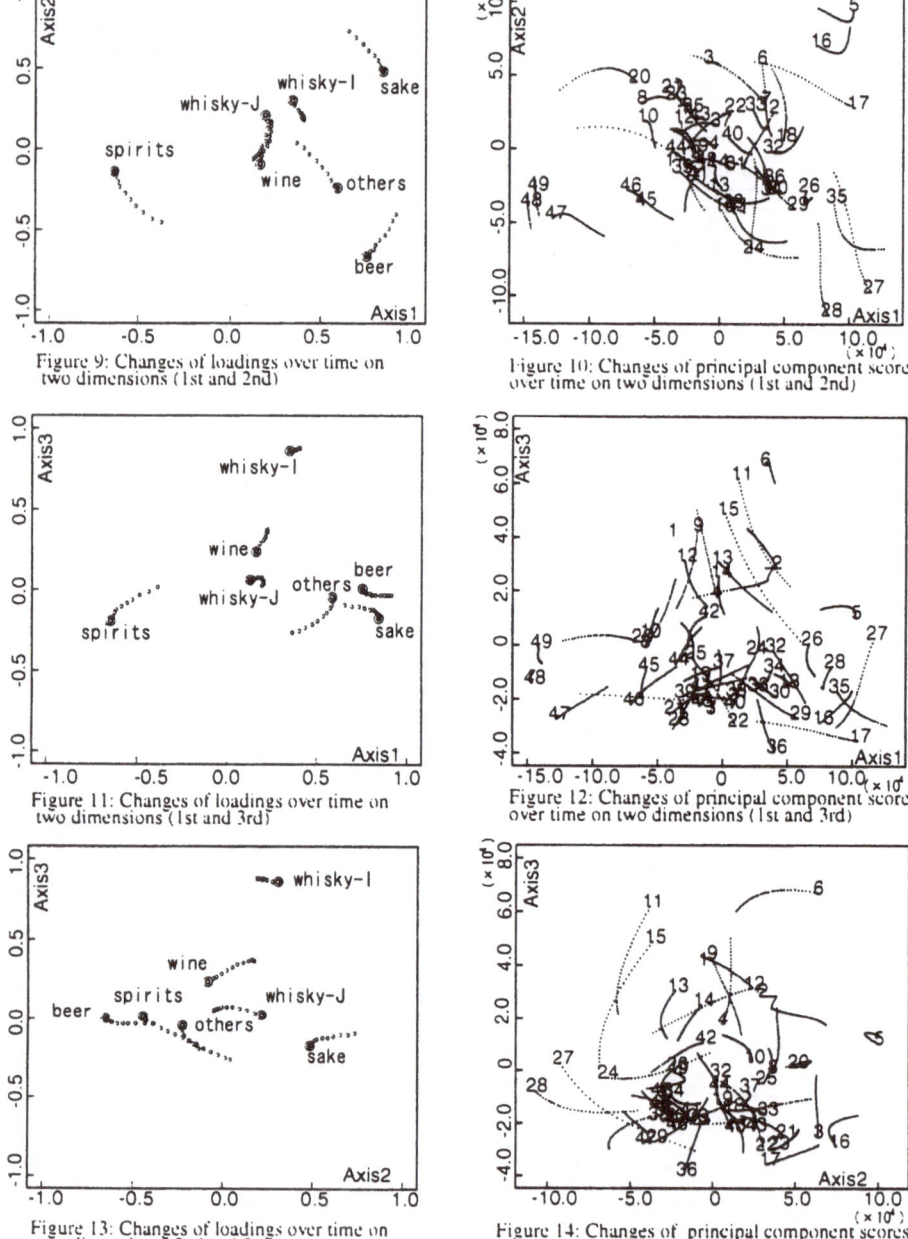

Figure 9: Changes of loadings over time on
two dimensions (1st and 2nd)

Figure 10: Changes of principal component scores
over time on two dimensions (1st and 2nd)

Figure 11: Changes of loadings over time on
two dimensions (1st and 3rd)

Figure 12: Changes of principal component scores
over time on two dimensions (1st and 3rd)

Figure 13: Changes of loadings over time on
two dimensions (2nd and 3rd)

Figure 14: Changes of principal component scores
over time on two dimensions (2nd and 3rd)

and solving the eigenequations at two time points, we obtain two sets of solutions. However it is difficult to relate those solutions from two time points especially when the meanings of principal components have likely changed.

If dt is very small, then the discrepancy between two solutions is likely to be small and we can relate them relatively easily. Let us illustrate this by an example. Our sample data set is from *Annual Report of the Family Income and Expenditure Survey* by Japanese government. We picked up yearly expenditures on alcohol beverages per family in forty nine cities of Japan in 1987 and 1992. Variables are sake, beer, wine, spirits, Japanese whisky, imported whisky, and others. We put $dt=0.01$ and connected solutions at each steps. Results are shown in figures. Numbers in Fig.10, 12, 14 are indices of forty nine cities. From figures for loadings we found that dominant variables in the first principle component are sake and beer. Examine the figures to see what have changed over time, for example, which variables are dominant at time point 1, and how about the rankings of the variables at time 0 and time 1.

4. Concluding Remarks

A method for time dependent principal component analysis was proposed in this study. For the case of two time points, it seems to work well. But the same method can easily be applied to the case where data are from more than two time points. We used a numerical method to solve and connect solutions. But, we should be aware that the sign of the eigenequation is arbitrary. This problem is removable if dt is very small because the minimum discrepancy of two sets of solutions gives us the answer to the problem. There is yet another problem, that is, if multiple solutions are involved, we may wonder how to trace changes of solutions over time. In this case, it looks as though we must consider case by case.

Acknowledgments
The authors would like to thank the referee for his helpful comments.

References:

Anderson, T. W. (1963). Asymptotic theory for principal component analysis. *Annals of Mathematical Statistics*, **34**, 122-148.

Baba, Y. and Nakamura, N. (1997). Time dependent Principal component analysis(in Japanese), *11th Annual Meeting of Japanese Computational Statistics Society*, 82-85.

Flury, B .D.(1984). Common principal components in k groups. *Journal of the American Statistical Association*, **79**, 892-898.

Flury, B. D.(1995). Developments in principal component analysis. In Krzanowski, W. J. (Ed.) *Recent Advances in Descriptive Multivariate Analysis*. Clarendon Press: Oxford, pp14-33.

Hotelling, H. (1933). Analysis of complex of statistical variables into principal components. *Journal of Educational Psychology*, **24**, 417-441, 498-520.

Jollife, I. T. (1986). *Principal Component Analysis*. Springer-Verlag, New York.

Krzanowski, W. J. (1979). Between-group comparison of principal components. *Journal of the American Statistical Association*, **74**, 703-707. (Correction note: (1981), **76**, 1022).

Pearson, K. (1901). On lines and planes of closest fit to systems of points in space. *Philosophical Magazine, Series 6*, **2**, 559-572.

Statistics Bureau Management and Coordination Agency Government of Japan(1988,1993). *Annual Report of the Family Income and Expenditure Survey* 1987, 1992.

Thorpe, R. S.(1983). A review of numerical methods for recognizing and analyzing racial differentiation. In Felsenstein, J. (Ed.) *Numerical Taxonomy*. Springer, New York., pp404-423.

Borrowing Strength from Images to Facilitate
Exploratory Structural Analysis of Binary Variables*

Robert M. Pruzek
State University of New York at Albany

This paper concerns exploratory structural analysis of relations among binary variables. Some new methods are described and illustrated, methods that appear to hold promise for general improvements in binary structural analysis. Standard common factor theory and methods, including image analysis, are used in combination with methods for data smoothing, to construct an alternative data system at the outset of analysis. Most of the new algorithms are relatively fast, easy to explain and to program, and appear to work as intended, at least for initial applications with real and simulated data. Given various advances in statistical theory and methods for prediction, as well as increasingly powerful and convenient computing facilities, there are a number of ways to extend the methods discussed here beyond the current framework.

Tucker (1983) provided a useful review of the psychometric foundations for binary factor analysis, citing especially papers by Ferguson, Guilford, Guttman and Carroll in the 1940s and 50s. Tucker noted that the "...search for factor analytic type structures in dichotomously scored items has presented nasty, exasperating problems for many years" (1983, p. 215). Waller (2000), in his manual for Microfact, a program for factor analysis of dichotomous and ordered polytomous data, reviews more recent literature on this topic, citing especially papers by Bock and Lieberman (1970), Muthén (1978), as well as Knol and Berger (1991). Other papers of note are Bock, Gibbons, and Muraki (1988), Mislevy (1986), Parry and McCardle (1991). The Knol and Berger and Waller papers are of special relevance in this context since these authors discuss questions about the relative usefulness of modern and older approaches to binary structural analysis, and because they offer practical suggestions that cut through many complications, taking account in their reviews of many of the key ideas and results that reside in the various papers connected to binary structural analysis.

It is of particular note that after numerous binary data analysis trials, Knol and Berger suggest that for many practical problems, exploratory common factor analyses of tetrachoric correlation matrices may remain the most viable approach currently available. They see this approach as especially relevant for problems based on large sets of binary variables. The Knol and Berger observation suggests that despite various advances in modern work on binary structural analysis, experience with real and simulated data suggests that more or less conventional types of exploratory common factor analysis, when initiated from what we may describe as 'reasonable starting points,' continues to hold promise for practical applications of binary data systems. Their observation is central to the current work.

---

*Presented at International Conference on Measurement and Multivariate Analysis (ICMMA), Banff, Alberta, CA May 2000

Figure 1

Phis v. tetrachorics; low.ln:midsize means, upper.ln:smallest.means

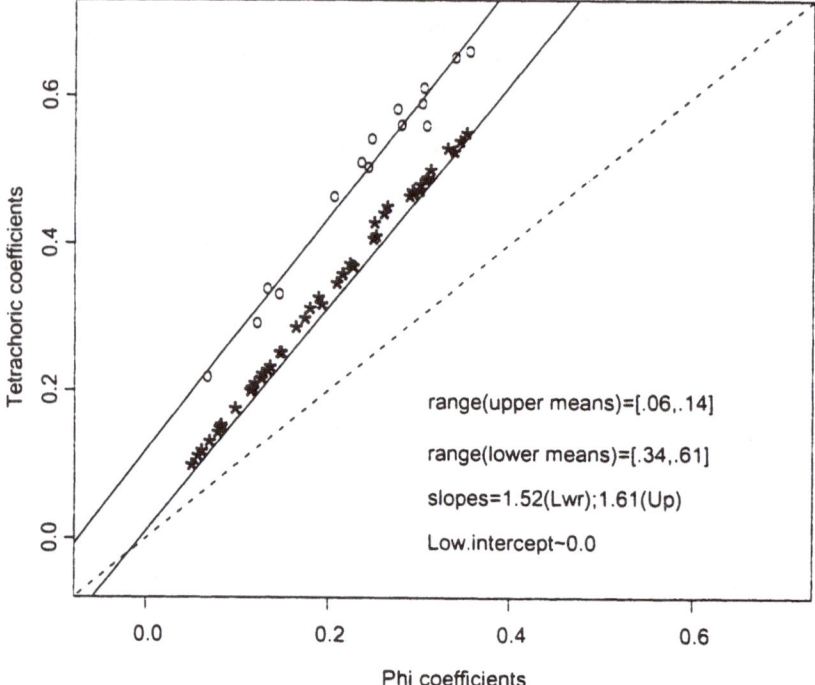

In this article it is shown that one can combine original binary scores with certain 'image' counterparts, to ameliorate basic problems that have often frustrated efforts to generate meaningful structural analyses of binary data systems. The new approach focuses on transformation and modification of the n x p [cases by variables] data matrix to devise a new data system with desirable properties.

One key advantage of the new approach is that non-Gramian covariance matrices cannot arise; a second advantage is that effects of varying means among binary variables can be controlled. The new methods can also facilitate a modern resampling approach to aid inference in binary structural analyses, although space does not permit development of this possibility. As the title notes, the general method to be developed below can be seen as borrowing strength from images to facilitate structural analyses of relations among binary variables.

To set the stage, an illustration is used to depict relationships between phi and tetrachoric coefficients, based on real data (n = 900, p =30 [cases × variables]). Figure 1 depicts the empirical associations between these two types of correlations for all pairs of two subsets of binary items. The larger set of ten items has marginal means near 0.5 (lower .30s to upper .60s), while the smaller set of six items has marginal means not far from zero (ranging from .06 to .14). (A tetrachoric-based common factor analysis of these data indicated two well-defined factors.)

Three useful points can be made from this Figure. First, if binary marginal means do not vary greatly around .5, the relationship between phi and tetrachoric correlations is especially simple; in fact the straight line fit for the lower line of points (asterisks) is almost perfect, with an intercept of virtually zero and a slope of 1.52 (in this case, and near to 1.5 for most data sets). Second, if binary variable means do differ substantially from .5, but manifest only a narrow range of values, correspondence between phi and tetrachoric coefficients continues to be nearly linear, although the straight line intercept tends to move away from zero in the latter case. (The intercept for the upper line (fit) is not given since this value is specific to the particular range of data values, and of no interest in current research.)

A third point follows by generalizing from Figure 1, to illustrate that *patterns* of phi and tetrachoric correlations correspond strongly to one another when binary marginal means are 'sufficiently homogeneous.' This of course also means that *common factor structures* derived from analyses of phi coefficient matrices may also correspond strongly to common factor structures derived from tetrachoric coefficient matrices; however, such a correspondence can break down as variation among marginal means for binary variates becomes 'sufficiently heterogeneous.'

It has long been known that results of structural analyses for these two kinds of coefficient matrices will sometimes be very similar, and sometimes differ substantially; the key point, demonstrated in Figure 1, is that variation in marginal means accounts for these differences. Indeed, the general form of this figure shows the specific way in which the relationship between the two types of correlations 'tolerates' rather wide variation among marginal binary means when means are within, say, $.5 \pm .2$. (*cf.* lower line in this figure); but it is also seen that variation among binary variable means near the extremes (say, below .3 or above .7) can lead to some scatter in the linear relationship between the two kinds of coefficients.

As noted, Figure 1 also shows why communality estimates for these two kinds of correlation matrices tend to differ markedly. This is because even when marginal means are similar, magnitudes of phi coefficients are always systematically smaller than their tetrachoric counterparts.

While these findings about relations between matrices of phi and tetrachoric are not new, we have not previously seen use of such a simple figure to show the basic relationships. One needs only to examine Figure 1, or its general form to show the correspondence between phi and tetrachoric coefficients, conditional on marginal means, and thus also to learn whether derived (common factor) structures for the two types of coefficients will be in 'strong correspondence.' Communalities derived from the two types of coefficient matrices will always differ systematically.

Image-Based Methods

The main idea of this article is the following: Given a set of binary variables for which exploratory analysis of structural relations is at issue, each manifest variable can be replaced by a weighted average of the variable itself, *i.e.*, the observed binary scores, and estimates of those scores (called images) derived from other variables in the set. It is proposed that instead of analyzing, say, phi coefficients or tetrachoric correlations, attention may better be focused initially on manifest data, especially on constructing counterparts to observed data matrices, to facilitate structural analyses. The chief aim of what follows is to learn whether it is possible, routinely, or at least for discernable classes of problems, to effectively compute weights for linear combinations so that structures derived from the new variables have desirable features.

A theoretical framework for analysis can be derived from the classic factor analytic work of Guttman (1953) and Harris (1962). Further references that serve to connect measurement issues to those of classical multivariate analysis in this context are Pruzek and Lepak (1992) and especially, Rabinowitz, Rule and Pruzek (1998). The more theoretical paper of Chen (1979) is also generally relevant.

Guttman (1953) developed a class of methods called image analysis, and showed that under certain limiting conditions image variables converge in theory to the common parts of the data, as postulated in basic common factor theory. Images, or partial images, were defined as the set of least squares regression estimates of the initial variables, based on predicting each variate from all others in the battery. Harris (1962) further elaborated on the virtues of image theory by connecting it more directly with Rao's canonical factor analysis (essentially the same as what is now called the maximum likelihood solution in common factor analysis). The elements of the Guttman-Harris algebra will be summarized; then these elements will be employed to construct smoothed data systems.

Suppose $X_{n \times p}$ represents data for n respondents on p variates; let X be column centered (by means), and further rescaled by columns, to generate n x p matrix Z so that $Z\rho Z = R$, a p x p matrix of product-moment correlations among all variates. (Correlation, not covariance metric, is used here to gain simplifications, but with no loss of generality – due to scale-freeness of this image system.)

Following Harris, including his notation, columns of Z can be rescaled using the diagonal matrix S^{-1}, where $S^{-2} = \text{diag}(R^{-1})$, to obtain $Z^* = Z S^{-1}$. The particular analysis, including Harris' rescaling, has several notable features:

1. The product $Z^* (I - S R^{-1} S) = Z^* - Z^* S R^{-1} S = M^*$, an n × p matrix of images.
2. From the svd (singular value decomposition) of Z^*, *i.e.*, $Z^* = P D_\lambda Q\rho$, $M^* = P (D_\lambda - D_\lambda^{-1}) Q\rho$. (Note that $Q\rho Q = P\rho P = I$ for any such svd.)
3. Matrix $(I - S R^{-1}S)$ contains in its off-diagonals all p-2 order partial correlations; columns (or rows, since it is symmetric) of $(I - S R^{-1}S)$ define vectors of OLS regression coefficients for predicting each variable from all others in the battery; unlike conventional standardized regression weights,

these regression coefficients necessarily fall in the range $[-1, +1]$.

4. $Z^* S R^{-1} S = Z R^{-1} S = P D_\lambda^{-1} Q\rho$, the $n \times p$ matrix of anti-images; these are OLS residual scores associated with the score estimates in M^*, also scaled in Harris's metric, so that each of anti-image variate has unit-length. It follows that $Q D_\lambda^{-2} Q\rho$ has unit diagonals. It is notable that each column of $Z R^{-1} S$ is orthogonal to the space of all $(p-1)$ other variables in the system.

5. The squared multiple correlations associated with predicting each variate from all others is given in (diagonal) matrix $D(smc) = I - S^2$, where 'smc' abbreviates squared multiple correlation.

From these algebraic relations, which assume a non-singular matrix X, but make no distributional assumptions about variables in X, it is straightforward to construct a scalar-weighted linear combination, a convex sum, of the matrices Z^* and M^*. The linear combination $Z_w^* = w Z^* + (1-w) M^*$ may be described as a *smoothed* version of the original data (after it has been rescaled).

Following appropriate substitutions, the smoothed variates matrix Z_w^* can be seen to have the simple form

$$Z_w^* = P(wD_\lambda + (1-w)[D_\lambda - D_\lambda^{-1}]) Q\rho = P(D_\lambda - (1-w) D_\lambda^{-1}) Q\rho, \qquad [1]$$

where the scalar w, $[0,1]$. If $w = 1$, $Z_w^* = Z^*$, and for $w = 0$, $Z_w^* = M^*$. It also follows that $cov(Z_w^*) = Z_w^* \rho Z_w^* = Q (D_\lambda - (1-w) D_\lambda^{-1})^2 Q\rho$, so that for the initial (standardized) scores, the covariance matrix of the smoothed counterpart $Z_w^* S = Z_w$ can be written as $cov(Z_w) = S Q (D_\lambda - (1-w) D_\lambda^{-1})^2 Q\rho S$.

Finally, it may be of interest to show that the corresponding correlation matrix for the smoothed variates can be written as $D_n S Q (D_\lambda - (1-w) D_\lambda^{-1})^2 Q\rho S D_n$ (for D_n a normalizing diagonal). Thus, for an m-factor solution it is straightforward, requiring no iteration, to construct a matrix of common factor coefficients, say $A_{w,m}$, for the matrix of correlations among smoothed variates. One such factor coefficients matrix is

$$A_{w,m} = D_n S Q_m (D_{r;m}^2 - \theta I)^{1/2}, \qquad [2]$$

where $D_{r;m}^2$ consists of m largest roots in D_r^2, for $D_r = (D_\lambda - (1-w) D_\lambda^{-1})$ and the scalar $\theta = (p-m)^{-1} \sum_{r=m+1}^{p} \lambda_r^2$, *i.e.*, θ is the mean of p-m smallest values in D_r^2. $A_{w,m}$ can be expected to reproduce off-diagonal correlations among smoothed variates to any desired degree of approximation, based on *m*, the number of factors, (Although D_r values may become negative as the scalar w is decreased, the 'best' choice for w, in numerous trials, seems to be around 0.6; thus, D_r values tend generally to be positive for realistic choices of w.)

Although the foregoing analysis does not require binary data, nor data of any specific form, the usefulness of expression [1] may be greatest when disparities among *shapes* of distributions associated with columns of X are likely to be problematic. It is well known that the maximum correlation between two variables whose marginal distributions differ substantially, such as two binary variables with widely differing means, is usually substantially less than unity; but it is easy to show that the maximum correlation between smoothed versions of two such binary variates tends to be nearer the standard upper bound of unity.

Two basic facts follow from this type of smoothing: In relation to the initial variates in X, the smoothed variates that constitute columns of Z_w^* or Z_w generally correlate more highly with one another, the more so as the scalar w is reduced from its maximum of unity, toward zero. (Indeed, as w drops below about .5, squared multiple correlations that describe predictability of each variate from all others in Z_w^* tend toward unity.) This is because each image variable is just a linear combination of all other variables in the set, and smaller values of the scalar w increase interdependencies among variables.

Second, given that magnitudes of product-moment correlations among columns of binary variables (*i.e.*, phi coefficients) are always smaller than their tetrachoric counterparts there always exists a scalar w for which the smoothed data Z_w^* yields an average correlation (or average smc) that matches its tetrachoric counterpart. In the context of a simulation, as discussed below, one can always be assured that there exists a value of w for which the smoothed version of the binary data yields an average correlation (or average smc) that corresponds exactly to the one associated with the original (pseudo-normal) variables used to construct the data.

When the last point is combined with the point made earlier, following Figure 1, that *patterns* of phi and tetrachoric correlations correspond strongly to one another when binary marginal means are 'sufficiently homogeneous,' justification can be seen for expecting both patterns and average communalities to be similar when comparing structures derived from smoothed counterparts of binary variables and structures derived from analyses of tetrachoric correlations (or, in simulation,
pseudo -normal counterparts). Results of a simulation (below) demonstrate such a correspondence.

Figure 2, with two panels *a* and *b*, depicts typical scatterplots for smoothed binary variates using two different values of scalar w, 0.8 and 0.6. These plots are for the same pair of binary variables within the same system of twelve variables seen in Table 1; but for clarity, only 300 randomly sampled points are plotted. (These plots are not based directly on columns of Z_w^*, but rather on scalings of these variables that return to the initial binary metric; such a reverse translation to binary metric is always possible in this context.) Means for the two binary variates

Figure 2

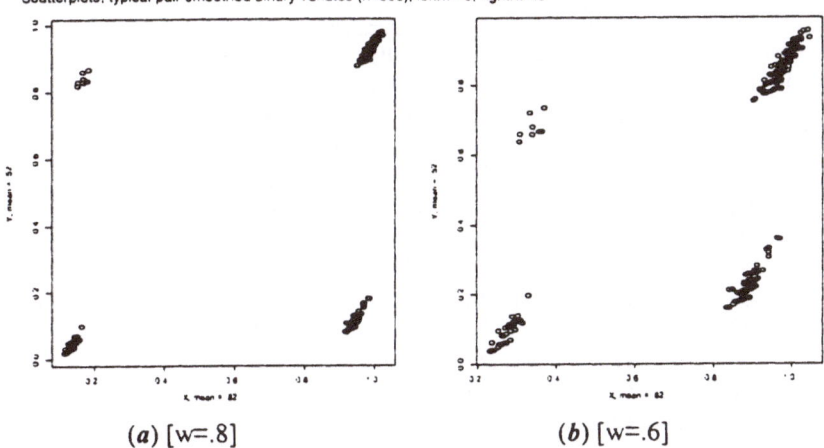

Scatterplots, typical pair smoothed binary variates (n=300); left:w= 8, right:w=.6

(*a*) [w=.8]   (*b*) [w=.6]

in Figure 2 (for the 300 case sample) are .82 and .52, respectively. The product-moment or phi correlation between the initial pair of binary variables was .36 (essentially, setting w=1), while the corresponding value for the smoother with w=.8, equals .47; with w=.6 the correlation between smoothed variates rises to .63. For smaller w's, the four point-sets or regions tend to merge or coalesce. Although these data derive from a simulation experiment, these plots are typical of real data.

Note that point scatters within each of the four regions of the plots do *not* indicate random jittering, since within each region, point configurations are generally informative. For example, considering points in the lower-left region of the right-hand panel, the points corresponding to manifest zeros for both variables at the outset, the values nearest the center of the full plot suggest relatively 'high fails,' were these variables to correspond to scores of persons who had taken a test. That is, for a points located toward the upper right part of this lower-left region, evidence exists in variables that predict these variates of relatively higher performance for these individuals on these items than for persons who also responded fail-fail to the same item pair. By contrast, extreme lower-left points in the same region tend to be corroborated as fail-fail scores for this item pair. Comparing left and right panels, which entail less and more weight to the image-variables, it is apparent that decreasing w tends to enhance correlations between two smoothed variables.

The general logic of this smoothing method is that information has been added to each binary variate, based on other (relevant) variables in the full set. That is, compared with a typical binary variate that corresponds to a column in the (standardized) matrix Z, the corresponding column of Z_w in expression [1] generally contains more information. The binary variate merely distinguishes two classes, corresponding perhaps to 'right' and 'wrong' responses to an item, but any column in Z_w contains a (weighted) average of this binary variate and all other

Table 1:
Two sets of factor coefficients, (rotated) solutions for continuous variables (left)
and smoothed binary counterparts (right) (n = 5000) [w=.7]

Initial continuous variable Solution			Smoothed binary variable solution			binary means
1	2	h^2	1	2	h^2	
.04	.71	.50	.04	.69	.47	.30
.08	.55	.30	.05	.54	.30	.69
.02	.71	.51	.01	.68	.46	.69
.06	.63	.40	.03	.66	.44	.49
.53	.27	.35	.50	.27	.32	.31
.63	.16	.42	.61	.14	.40	.50
.53	.19	.31	.51	.21	.31	.51
.50	.40	.41	.42	.42	.35	.70
.46	.39	.36	.47	.37	.36	.30
.75	.40	.72	.76	.36	.71	.30
.72	.38	.67	.72	.35	.64	.30
.65	.44	.61	.68	.38	.61	.30
SUM .24	.22	.46	.24	.21	.45	

variables that empirically predict this variate; the additional information generally
supplies further distinctions within each these two classes.

Table 1 gives an idea of the usefulness of this kind of data smoother, with
reference to Expression [1]. This Table 1 shows factor coefficients derived from
an initial matrix of continuous variables, obtained in a simulation experiment, with
$m=2$ factors and $n = 5000$ simulated respondents. The Table also displays factor
coefficients for a corresponding Z_w*, the smoothed counterpart of the initial
binary data matrix. On the left side of this Table 1 are the (maximum likelihood-
based) factor coefficients for the continuous variates, after rotation (based on
varimax). On the right side are factor coefficients for the smoothed variable
system. (Note that since computation and analysis of tetrachoric correlations is
based on the aim of trying to reproduce correlations among initial continuous
(Gaussian) variables, such #Row 'SUM' depicts variances for corresponding
common factors; and the average communality is given below each column of
communality estimates (h^2).
as those associated with the solution on the left side of Table 1, the proper target
for the smoothed binary solution is indeed the solution on the left, and not the
corresponding matrix of tetrachoric correlations; this luxury is afforded through
use of simulated data in this context.)

Details of the binary data generation process are these: First, each continuous
variate (say y, for $y \sim N(0,1)$) was arbitrarily 'cut' at a given predefined quantile to
form a binary variate. Each y quantile was chosen so that areas to the left of each
quantile corresponded to the predefined range 0.25 - 0.75; all values in this

interval were given equal probability of occurrence. This step insured that no binary variable means would be near the extremes of either zero or unity; it also helped insure that binary means would not be homogeneous. The broad goal was for the synthetic variables to have means that varied in a fashion not unlike what is often observed with real binary (item score) data. Then, starting from a 5000 x 12 matrix of standardized data, smoothed data of the form of matrix Z_w was generated, as in [1], sampling each vector y from a multivariate normal population with a prescribed covariance form: $y \sim N_p(0, \Gamma_{prescribed})$, where $\Gamma_{prescribed}$ was chosen to have two common factors (not unlike those seen in Table 1 below.)

In this solution, $w = .7$, and factor coefficients were constructed based on [2], with $m=2$. Means of the derived binary items are given in the last column of Table 1. The choice of the rather large sample size ($n = 5000$) was to help insure that stochastic variation would be minimized. The value $w = .7$ was chosen to make average communalities in the two solutions comparable.

Although study of the coefficient matrices in Table 1 suggests some minor differences in structures, it seems reasonable to suggest that interpretations of the two factors that correspond to these columns of coefficients would not change from one solution to the other, had these been real data. This is true despite modest-to-substantial variation among means. Moreover, average communalities for the smoothed data solution are nearly identical to those obtained from the initial continuous variates.

The strong similarities of the two coefficient matrices in Table 1 contrasts sharply in one respect with factor results ordinarily found when continuous variate factor solutions are compared with those derived from product-moment correlations (phi coefficients) in for binary data. Specifically, while it is usually found that binary variable solutions based on product-moment or phi coefficients yield communalities much smaller than their continuous variate (or tetrachoric) counterparts, Table 1 shows highly comparable communalities, as well as similar coefficient patterns when comparing the continuous variable solution with its smoothed binary variate counterpart.

Because the simulation methods used in generating Table 1 provided the luxury of having continuous variate (Gaussian) counterparts to the binary variates, comparison of the two matrices of factor coefficients in Table 1 avoided having to compute tetrachoric correlations, nor to worry whether corresponding tetrachoric assumptions are valid, nor whether that correlation matrix was Gramian.

Study of the two sets of factor coefficients in Table 1 entails comparison of structural results for an initial set of continuous variables (sampled from a population with known structure) with a structural analysis for binary variates, where, in effect, information had initially been 'thrown away.' That the smoothed counterparts of the binary data shows such strong similarities in coefficient patterns and communalities appears to reflect that this kind of smoothing process may be able to restore much of the information that had been lost when converting to binary form. How far such a conclusion can be generalized, however, remains unknown.

Further Considerations

The smoother Z_w* in [1] can be seen to have the simplest of forms, entailing choice of only a scalar (w) for its construction. For this reason, it is probably not realistic to think that factor structures derived from the covariance matrix of the smoothed variates in [1] will routinely or necessarily match factor structures derived from tetrachoric correlations (or to continuous variates to which the binary variables may relate). That is, the smoother in [1] makes no accommodation for what could be widely varying means among observed binary variables.

Alternative smoothers are easily defined and constructed, conceptually similar to [1], but where explicit account can be made for variation among binary means. One alternative type of smoothed data entails use of a diagonal matrix for scaling of the form

$$Z_{gw}* = Z*D_w + M*(I - D_w),$$ [3]

where weights in diagonal matrix D_w are allowed to vary, perhaps as a function of the binary means. (Of course if D_w is a scalar diagonal then [3] reduces to [1].) Harris' use of the matrix S^{-1} in rescaling Z to get $Z*$ entails giving larger weights to variables with larger smc's, smaller weight to variables with smaller smc's. In the same spirit, one may seek also to give relatively less weight in the smoother to images of binary variables with means near .5, more weight to images for those variables with means nearer the extremes. One approach of this form is to choose a diagonal weighting matrix D_w so that weights, w_j, are monotonically decreasing functions of the binary variances, *viz.*, $p_j(1-p_j)$, where p_j denotes the mean of the jth binary variable. Several alternatives of this general form have been examined, and seem to be reasonably effective. However, a fully satisfactory choice is likely to require systematic examinations over a wide range of problems and data types; no particular choice can be recommended for applications at this time.

Another approach to smoothing for binary data applications would entail use of prediction methods that depart from ordinary least squares regression. Logistic regression methods are easily implemented, and several trials have been conducted using logistic counterparts of the matrix M in [1], where each binary variable is predicted using logistic regression, based on all other variables in the set. This approach has shown itself to be viable in numerous applications, but again, more experience seems needed, over a range of problems and data types, before making recommendations about the merits or demerits of logistic models in relation to those based on standard LS regression. Classification and regression trees may also be explored as a way to obtain alternative images for use in smoothing.

Note with reference to expression [1] that even though individual smoothed variates in Z_w typically contain more information than their binary counterparts, there is no more information in the full matrix of smoothed variates than in the original binary set. This is because either matrix can be constructed from the same basis variables (as linear combinations of the variables that constitute matrix P, in expression [1]). Nevertheless, in general, as when methods such as logistic regression are used to construct fitted parts, as counterparts to those in M, it is possible that more information may reside in the full system of smoothed variates than in the original binary system. The answer to this question is currently moot. Detailed information-theoretic analysis of such possibilities would be helpful, as could further trials with real and simulated data.

That the general approach to binary structural analysis based on smoothing avoids problems with non-Gramian data, and removes certain arbitrary assumptions, may be one of its greatest practical virtues. But certain graphics, exemplified in Figures 1 and 2 above, may also have substantial value in applied data analysis since both entail highly efficient, potentially broadly effective, methods for displaying information that may be central to interpretations of binary data.

In addition, as noted, smoothing methods of this form can facilitate inferential applications. For example, it seems straightforward to resample rows of matrices such as Z_w, as in the context of bootstrapping, to assess generalizability properties of derived statistics. Results from bootstrapped common factor solutions might be used to provide standard error estimates, or confidence intervals, for several statistics of possible interest. The computational simplicity of the smoothing method associated with [1] and [2] augurs well for relatively efficient inferential studies, even for cases where the number of variables grows large, more or less regardless of the sample size. For situations where sample size is relatively small the methods of Pruzek and Lepak (1992) might be used to stabilize regression coefficients within the basic Guttman-Harris framework.

The concept of borrowing strength has once again been demonstrated to have value for analysis or interpretation of data, where in this case the context has been that of structural analysis of relations among binary variables. Further work would be desirable, to study the relative viability of alternative smoothers, such as those in expressions [1] or [3], or to learn whether methods such as logistic regression or classification trees might yield effective smoothers.

References

Bock RD, Lieberman M. (1970). Fitting a response curve model for dichotomously scored items. *Psychometrika, 35,* 179-198.

Bock, R. D., Gibbons, R., and Muraki, E. (1988) Full-information item factor analysis. *Applied Psychological Measurement, 12,* 261-280.

Chen, C. (1979) Bayesian inference for a normal dispersion matrix and its applications to stochastic multiple regression. *Journal of the Royal Statistical Society, Series B, 41,* 235 –248.

Guttman, L. (1953) Image theory for the structure of quantitative variates. *Psychometricka, 18,* 273 - 285.

Harris, C.W. (1962). Some Rao-Guttman relationships. *Psychometrika, 27,* 247 – 263.

Knol, D.L. & Berger, M.P.F. (1991). Empirical comparison between factor analysis and multidimensional item response models. *Multivariate Behavioral Research, 26*, 457-477.

Mislevy R.J. (1986) Recent developments in the factor analysis of categorical variables. *Journal of Educational Statistics, 11*, 3-31.

Muthén, B. (1978) Contributions to factor analysis of dichotomized variables. *Psychometrika, 43*, 551-560.

Parry C.D. & McArdle J. J. (1991) An applied comparison of methods for least-squares factor analysis of dichotomous variables. *Applied Psychological Measurement, 15*, 35-46.

Pruzek, R.M. & Lepak, G. (1992) Weighted structural regression: A broad class of adaptive methods for improving linear prediction. *Multivariate Behavioral Research, 27*, 95 – 129.

Rabinowitz, S.N., Rule, D. & Pruzek, R.M. (1998) Some new regression methods for predictive and construct validation. *Social Indicators Research, 45*, 201– 231.

Tucker, L. R. (1983) Searching for structure in binary data. In H. Wainer and S. Messick (Eds.), *Principals of Modern Psychological Measurement*, 215 – 235. Erlbaum and Associates, Hillsdale, N.J.

Waller, N. (2000) MicroFACT user's manual 2.0. *Assessment Systems Corp.*, St. Paul.

Comparison of Construct Mean Scores Across Populations:
A Conceptual Framework

Alain De Beuckelaer

Unilever Research Vlaardingen (URV)
& Catholic University of Brussels
URV, 120 Olivier van Noortlaan, 3130 AC Vlaardingen, The Netherlands

Summary: Applied researchers who wish to compare groups of people (e.g. different cultures) on the basis of a theoretically meaningful concept (e.g. individualism / collectivism) are faced with a number of methodological problems (e.g. measurement issues/cross-group comparability issues, … etc.). Unfortunately, the availability of 'coping strategies' is largely dependent on the nature of the underlying concept (either represented by an 'emergent' or a 'latent' construct). If the concept represents an emergent construct (i.e. as represented by formative indicators), both measurement and cross-group comparability of construct quantifications are problematical. With latent constructs (as represented by reflective indicators), however, these methodological problems are much easier to overcome. By using multigroup Confirmatory Factor Analysis (CFA) models invariance of measurement across groups can be explicitly tested, and cross-group comparisons can be made (provided sufficient evidence exists for claiming cross-group comparability of construct scores).

1. Introduction

There are many complexities that hamper the process of quantifying theoretically meaningful concepts (e.g. definition of a metric), and making cross-group comparisons on the basis of such (concept) quantifications (e.g. absence of cross-group comparability). Assuming that the researcher is fully aware of the challenges he/she is faced with, he/she can choose a strategy to 'solve' the problem at hand (or at least to adopt a pragmatic approach to it). Clearly, the choice for a solution strategy should be made by taking into account all a priori knowledge about the concept, its indicators, and reliability / validity of measurement. In the absence of such a priori knowledge, the researcher will have to (partially) base his / her choice on: (1) his/her personal preferences, and (2) any useful empirical evidence from the data (e.g. correlational patterns between indicators of a concept). As Hand & Taylor (1987) pointed out "the identification of a construct […] is more a matter of art than statistics."

The aim of this paper is to discuss some options that are available to applied researchers who wish to compare groups on the basis of a theoretically meaningful concept. When discussing the adequacy of multivariate statistical models the focus is very much on the "measurement aspects" that underpin these models. This paper reflects on work by Cole, Maxwell, Arvey & Salas (1993) who took a view of such measurement aspects. Most traditional textbooks, however, do not touch on measurement issues when discussing traditional multivariate models such as

MANOVA and Multiple Discriminant Analysis. This explains why the discussion of these models as presented in this paper may seem somewhat unfamiliar to the reader.

2. Selection of Appropriate Indicator Variables

The first challenge involves finding a common set of indicator variables (indicators) that may adequately represent the concept / construct in all groups of interest. Hox (1997) discusses some well-known methodologies (e.g. dimension / indicator analysis, semantic analysis, and facet design) which can be used in this (preliminary) phase of the research process. This paper assumes that such a common set of indicators (e.g. statements in an attitudinal questionnaire) has already been defined. It is further assumed that a scaling format with a sufficient number of scale points (e.g. a Likert-type agree/disagree-scale with at least 7 scale points) is used to score the answers to the individual questions. The models discussed in this paper should not be used when responses are coded using a binary format (e.g. a 'forced choice' between agreement and disagreement with a statement). Parametric or non-parametric Item Response models would be much more appropriate for that purpose.

3. Identifying the Nature of the Underlying Measurement Model

The second challenge is to determine the (true) causal direction between the construct and its indicators. If theory suggests that the indicators are merely observable reflections (rather than determinants) of the construct, the most plausible hypothesis is that the construct influences its indicators (and not the other way around). The construct is then called a latent construct, and the indicators referred to as reflective indicators. If, however, it is believed that the construct is the result (i.e. the combined effect) of its indicators, the construct and its indicators are referred to as an emergent construct and formative indicators, respectively.

There are many fundamental differences between both types of constructs. Reflective indicators should be highly correlated, as they measure essentially the same thing (i.e. a one-dimensional [latent] construct). With formative indicators nothing is known about the strength (i.e. high / low) nor the direction (+ / -) of their mutual correlations, unless sufficient guidance is provided by a priori theory. It is crucial to understand that the omission of a (necessary) indicator may lead to invalid measurements if the construct is emergent, as one (crucial) dimension may not be taken into account. If the construct is latent, however, no effects on the adequacy of measurement of the construct are to be expected (apart from a potential loss of reliability). Finally, from a modeling perspective, attenuation error can be taken into account if the construct is latent, but not if the construct is emergent (cf. Bollen (1984)), Bollen and Lennox (1991)).

There has been a lot of debate as to which type of construct is most relevant for the social sciences. Horn and McArdle (1992), for example, claimed that almost all concepts in the behavioral sciences represent emergent constructs. This claim is in contrast with the literature on psychological concepts (including attitudes) which suggests that latent constructs would be much more appropriate (cf. Sörbom (1981)).

If no a priori theory is available, inspection of the correlational structure between the indicators may help the researcher to find out whether the indicators could possibly reflect a (one-dimensional) latent construct. If bivariate correlations between indicator variables are low (say below 0.40), it becomes very unlikely that a one-dimensional factor model (i.e. a model representing a latent construct) will fit the data. The fact that the bivariate correlations exceed the lower bound (of 0.40) is not sufficient to conclude that a one-dimensional factor model would be an appropriate measurement model for the construct. This should be formally tested using a Confirmatory Factor-Analytic approach.

4. Defining a Metric for the Construct

The third challenge involves finding a suitable metric to indicate to what extent the construct applies to each respondent individually (i.e. the scoring or normalization of the construct).

4.1 Emergent Construct

Consider the case of an emergent construct first. Scores on the composite measure (i.e. the emergent construct) are worth interpreting only if the composite measure (i.e. a weighted sum of indicators) is meaningful from theory. The main problem is that the numerical indicator weights are 'optimal' from a statistical point of view only (e.g. providing optimal discrimination between groups of respondents, for example in MANOVA / Multiple Discriminant Analysis [MDA]), and may not be at all relevant from theory (cf. Cole et al. (1993))).

The fact that there is no theoretical basis on which a measurement model can be built makes it very hard to define emergent constructs in a reliable and valid way.

4.2 Latent Construct

The scoring (normalization) issue is relatively easier if the construct is latent. The most popular (and best) approach is making the metric of the construct equal to the metric of one of its indicators. In technical terms this implies that one particular factor loading is set equal to one (in all groups of interest). In the multigroup case a problem may arise if the (implicit) assumption of the equality of this factor loading across groups is unjustified. In that case, statistical tests based on the comparison of mean construct scores across groups may become biased. A procedure proposed by Rensvold and Cheung (1998), and Cheung and Rensvold (1999) can help to overcome this problem. Their procedure involves the following two steps:

(a) The factor-ratio test: testing all pairwise combinations of two indicators in which one indicator serves as the argument (i.e. the indicator to be tested) and the other as the referent (i.e. the indicator of which the factor loading is constrained to be one across groups)

(b) <u>The triangle heuristic</u>: applying a heuristic which helps to determine the largest possible set of indicators having identical factor loadings across the groups.

Those readers who are interested in the technical details of this procedure are encouraged to consult the original paper by Cheung & Rensvold (1999) and Rensvold & Cheung (1998).

5. Testing for Measurement Invariance Across Groups

The fourth challenge is to ensure that construct scores have the same meaning across the populations of interest. Technically speaking, the requirement is that construct scores in every single group exhibit 'measurement invariance'. As Horn & McArdle (1992) pointed out: "evidence of measurement invariance is fundamentally important to the process of building a science."

5.1 Emergent Construct

Consider the case of an emergent construct first. Multivariate techniques such as MANOVA or MDA are designed so that only one set of composite measures (i.e. 'discriminant function scores') serves as a basis for making cross-group comparisons. An implicit assumption is that these composite measures are both reliable and valid for this purpose. From a measurement perspective, this is referred to as the "homogeneity of regression" assumption (cf. Marsch and Grayson (1990)). This assumption represents in fact an extreme form of (assumed) measurement invariance. The main methodological problem is, of course, that the homogeneity of regression (measurement) assumption is not testable at all (and very unrealistic). The second problem is that the combination of the homogeneity of regression assumption and the assumption of equal variance/covariance matrices across populations make both MANOVA and MDA far too stringent multivariate techniques to be used in practice (cf. Kuehnel (1980), Stelzl and Schnabel (1992)). Some alternative multivariate techniques pose less stringent assumptions (e.g. a Partial Least Squares [PLS] approach), albeit that they also fail to provide a means of testing for measurement invariance across groups. For these reasons one has to conclude that none of the available multivariate techniques are optimal for making cross-group comparisons based on an emergent construct. This may explain why some researchers have used multivariate techniques that are appropriate for latent constructs (e.g. factor- analytic techniques), even when dealing with constructs which are likely to be emergent (cf. Cole et al. (1993)).

5.2 Latent Construct

Confirmatory Factor-Analytic (CFA) techniques, and in particular the multigroup CFA approach, are most suitable to be used with latent constructs. The main advantage of the multigroup CFA approach is that it offers a hypothesis-testing framework to evaluate the plausibility of 'competing' mathematical models. In practice, the test procedure consists of two major, consecutive phases: 1) testing for measurement invariance across groups (i.e. the measurement model), and 2)

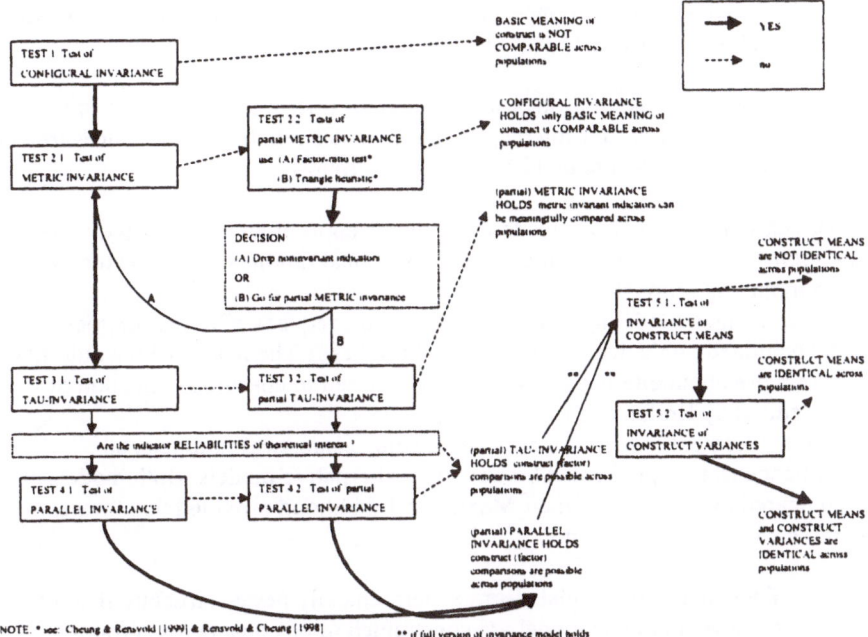

Figure 1: Hypothesis-testing framework to compare latent construct scores across different groups (populations) of interest.

testing for significant differences across populations based on a comparison of mean construct scores (and possibly also: construct variances). The hypotheses are tested using the principle of Maximum Likelihood (based on the asymptotic distributional theory). A detailed scheme of the complete hypothesis-testing process is presented in figure 1.

In phase 1 of the hypothesis-testing process different mathematical models that represent alternative definitions of measurement (i.e. factorial) invariance across populations have to be tested. The basic measurement models are:

Model 1: configural invariance model: this model assumes equal patterns of salient (nonzero) and nonsalient (zero) factor loadings across groups;

Model 2: congeneric factor model: this model is not an invariance model (i.e. it places no restrictions on the measurement parameters (across groups)), but serves as a 'baseline' model to test all subsequent invariance models using a likelihood ratio statistic;

Model 3: metric invariance model: this model assumes identical factor loadings across groups;

Model 4: tau-invariance model: this model assumes all factor loadings and indicator intercepts to be equal across groups;

Model 5: parallel invariance model: this model assumes all factor loadings, indicator intercepts, and indicator reliabilities [i.e. measurement error variances] to be identical across groups.

Basically there are two ways to evaluate the appropriateness of alternative measurement models. One of them uses a statistical rationale, whereas the other uses a modeling rationale (Little (2000)).

When using a modeling rationale the overall adequacy of a measurement model is judged on the basis of one more 'fit indices'. The reader who would like to learn more about alternative fit indices is advised to consult a collection of papers edited by Bollen and Long (1993).

If measurement models are compared using a statistical rationale, an 'equivalence test' is performed between two alternative models. Both models can be compared on the basis of their Maximum Likelihood provided that they form a 'hierarchically nested structure'.

Two alternative models form a hierarchically nested structure if some parameters in one model are constrained to be equal across groups, whereas the same parameters in the other model are freely estimated across groups.

Statistical models that do not form a hierarchically nested structure may still be statistically compared using 'goodness of fit' measures based on Information Theory (i.e. Information Criteria).[1]

As the measurement models 2-5 (see above) form a hierarchically nested structure (model 5 being the most stringent one), a Maximum Likelihood-based statistical comparison of measurement models is straightforward.

Phase 2 (making cross-group comparisons) can only start if empirical evidence supports (at least) tau-invariance across groups (i.e. tau-invariance or parallel invariance). The reason is that construct mean score comparisons across groups are only meaningful if:

(1) The indicators 'capture' changes in the underlying construct in exactly the same way in all groups (i.e. equality of factor loadings across groups); and

---

[1] Akaike (1973, 1974) introduced the first Information Criterion: Akaike's Information Criterion (AIC). Others have proposed alternative (or modified) Information Criteria (e.g. the Bayesian Information Criterion (BIC) by Schwartz (1978), and the Consistent Akaike's Information Criterion (CAIC) by Bozdogan (1987)).

(2) The metric used to measure the construct has the same 'origin' in all groups (i.e. equal indicator intercepts across groups).

Given tau-(or parallel) invariance, actual testing of the hypothesis of equal construct mean scores (and/or construct variances) is conducted by adding additional constraints to the multigroup CFA model. If empirical evidence shows that the indicators exhibit only a lower level of measurement invariance (e.g. metric invariance), construct (mean) score comparisons are likely to be biased, and thus untrustworthy. Given metric invariance, however, the groups may be legitimately compared on the basis of the individual indicator scores.

If only a subset of the indicators exhibits a certain level of measurement invariance (e.g. tau-invariance), it may be decided either to omit that indicator from the measurement model or, alternatively, consider partial forms of measurement invariance (e.g. partial tau-invariance). In the latter case the noninvariant indicators are still used to measure the latent construct, but its measurement parameters (e.g. factor loadings, intercepts) are allowed to vary across groups. Some authors (e.g. Byrne et al. (1989), Rensvold & Cheung (1998), Cheung & Rensvold (1999)) have claimed that these noninvariant indicators would not affect the cross-group comparisons to any significant extent. Provided there are enough indicators to represent the latent construct, the best option would probably be to drop the noninvariant indicators from the scale. Even though this may seem atheoretical, one must realize that the premise is that essentially all indicators measure the same thing (i.e. the latent construct), and that dropping one indicator might, therefore, be sensible.

Muthen's (1989) MIMIC (i.e. multiple-indicator multiple-cause) model may be considered an alternative to the multigroup CFA model.

References:

Akaike, H. (1973), Information theory and an extension of the maximum likelihood principle. In Second international symposium in information theory, Petrov, N & F. Csaki (Eds.), 267-281, Akademiai Kiado, Budapest.

Akaike, H. (1974), A new look at the statistical identification model, IEEE Transactions on Automatic Control, 19, 716-723.

Bozdogan, H. (1987), Model selection and Akaike's information criterion (AIC): The general theory and its analytical extensions. Psychometrika, 52, 345-370.

Bollen, K.A. (1984). Multiple indicators: internal consistency or NO necessary relationship. Quality and Quantity, 18, 377-385.

Bollen, K.A. and Lennox, R. (1991). Conventional wisdom on measurement measurement a structural equation perspective. Psychological Bulletin, 110, 2, 305-314.

Bollen, K.A. and Long, J.S. (Eds.)(1993), Testing structural equation models, Sage, Newbury Park, California.

Byrne, B.M., Shavelson, R.J., and Muthen, B. (1989). Testing for the equivalence of factor covariance and mean structures: the issue of partial measurement invariance. Psychological Bulletin, 105, 3, 456-466.

Cheung , G.W. and Rensvold, R.B. (1999). Testing factorial invariance across groups: a reconceptualisation and proposed new method. Journal of Management, 25, 1, 1-27.

Cohen, P.J., Cohen, J., Teresi, J., Marchi, M., and Velez, C.N. (1990). Problems in the measurement of latent variables in structural equations causal models. Applied Psychological Measurement, 14, 2, 183-196.

Cole, D.A., Maxwell, S.E., Arvey, R., and Salas, E. (1993). Multivariate group comparisons of variable systems: MANOVA and structural equation modeling. Psychological Bulletin, 114, 1, 174-184.

Hand, D.J. and Taylor, C.C. (1987). Multivariate analysis of variance and repeated measures: A practical approach for behavioural scientists. Chapman and Hall, New York.

Horn, J.L. and McArdle, J.J. (1992). A practical and theoretical guide to measurement invariance in aging research. Experimental Aging Research, 18, 117-144.

Hox, J.J. (1997). From theoretical concept to survey question. In Survey measurement and process quality, Lyberg, L. et al. (Eds.), 47-69, Wiley, New York.

Kuehnel, S. (1988). Testing MANOVA designs with LISREL. Sociological Methods and Research, 16, 4, 504-523.

Little, T.D. (2000). On the comparability of constructs in cross-cultural research. Journal of Cross-cultural Psychology, 31, 2, 213-219.

Marsch, H.W. and Grayson, D. (1990). Public/catholic differences in the high school and beyond data: a multi-group structural equation modelling approach to testing mean differences. Journal of Educational Statistics, 5, 199-235.

Muthen, B. (1989). Multiple-group structural modeling with non-normal continuous variables. British Journal of Mathematical and Statistical Psychology, 42, 55-62.

Rensvold, R.B. and Cheung, G.W. (1998). Testing measurement models for factorial invariance: a systematic approach. Educational and Psychological Measurement, 58, 6, 1017-1034.

Sörbom, D. (1981). Structural equation models with structured means. In Systems under indirect observation: causality structure, and prediction, Jöreskog, K.G. and Wold, D. (Eds.), 183-195, North Holland, Amsterdam.

Stelzl, I and Schnabel, K. (1992). The two-group MANOVA problem with unequal covariance matrices: a simulation study comparing Hotelling's T^2 to the LISREL approach. Methodika, 6, 54-75.

Nonlinear Generalized Canonical Correlation Analysis by Neural Network Models

Yoshio Takane and Yuriko Oshima-Takane

McGill University
1205 Dr.Penfield Avenue, Montreal, Quebec, Canada H3A 1B1

Summary: A method of K-set canonical correlation analysis capable of joint multivariate nonlinear transformations of data was proposed. The method consists of K nonlinear data transformation modules, each of which is a multi-layered feed-forward network, and one integrator module which combines information from the K transformation modules. The proposed method is useful for integrating information from K concurrent sources.

1. Introduction

We propose a method of nonlinear generalized (K-set) canonical correlation analysis (NGCANO), where $K \geq 2$. Generalized CANO (Carroll, 1967; Horst, 1961; Meredith, 1964) is an interesting technique because it subsumes a number of existing techniques for multivariate data analysis as special cases. It specializes into the usual (2-set) CANO when $K = 2$. It reduces to principal component analysis (PCA) when each of the K sets consists of a single (usually) continuous variable, and to multiple correspondence analysis (MCA) when each set consists of a matrix of dummy variables representing responses to a multiple-choice questionnaire item. Generalized CANO has been extended to allow variable-wise nonlinear transformations of input variables (Gifi, 1990), called OVERALS. In this paper we further extend it to allow for joint multivariate nonlinear transformations of input variables by artificial neural networks.

There are a number of situations in which NGCANO could be useful. We may, for example, have a problem of integrating information from different sensors, from different modalities, from different measurement tools, and so on. Several different cues are available for depth perception, e.g., binocular disparities, motion parallax, shading, textures, occluding contours, etc. Different cues are processed more or less independently up to certain levels by separate brain modules, which should be integrated into coherent images (Marr, 1982). NGCANO approximates such information integration mechanisms. Becker and Hinton (1992) developed a similar procedure for $K = 2$ based on a somewhat different fitting criterion, which they successfully applied to identify surface structures in random dot stereograms. Asoh and Takechi (1994) devised an approximate method for Becker and Hinton's method. NGCANO extends their methods to $K \geq 2$.

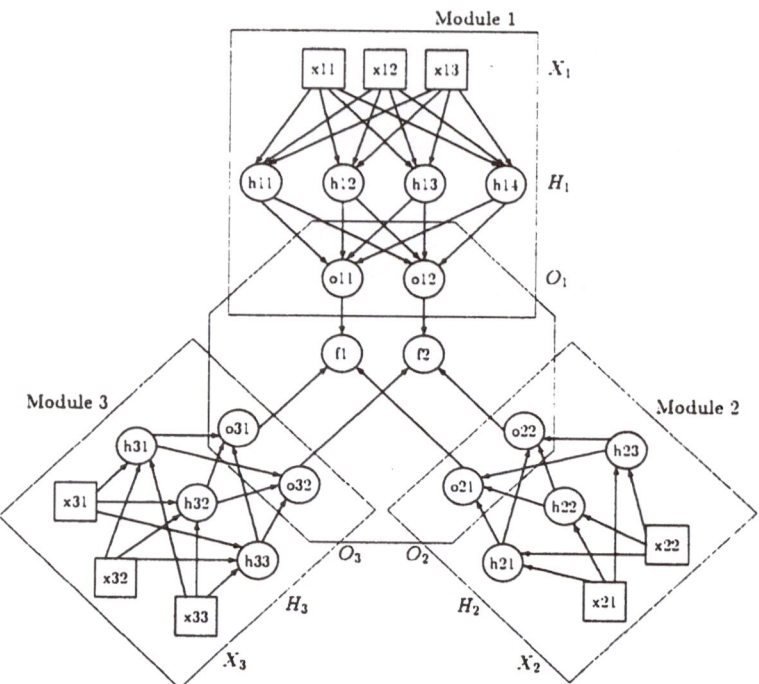

Figure 1: The basic construction of NGCANO for $K = 3$ modules with a single hidden layer in each module.

2. The Method

Figure 1 displays the basic construction of NGCANO for $K = 3$. The three modules are enclosed by large squares. Each module (corresponding to a set of input variables) consists of a multi-layered feed-forward (MLFF) network. It accepts inputs, forms linear combinations of the inputs and transforms them by sigmoid transformations to obtain hidden-layer activations, which capture nonlinear and interaction effects among the input variables. It then forms linear combinations of hidden-layer activations as outputs from the network. NGCANO attempts to make the outputs from different modules as homogeneous as possible. This information integration part is depicted inside the octagon in the figure. The outputs from all the modules are made to approximate a single common set of quantities (called common canonical variates) as closely as possible.

2.1 Optimization Criterion

Let O_k denote the matrix of outputs from module k, and let $F = [f_1, f_2]$ denote the matrix of canonical variates. Define

$$g = \sum_{k=1}^{K} g_k \qquad \text{with} \qquad g_k = \|F - O_k\|^2, \tag{1}$$

where $O_k = H_k W_k$, and $H_k = \sigma(X_k V_k)$ with σ being the sigmoid transformation. Here, X_k is the matrix of inputs, and V_k and W_k are matrices of the first and the second layer weights, respectively. We minimize g in (1) with respect to V_k, W_k ($k = 1, \ldots, K$), and F subject to:

$$F'F = I \qquad \text{and} \qquad F'1_N = 0, \qquad (2)$$

where 1_N is an N-element vector of ones (where N is the number of cases in the training sample, and 0 is a zero vector of appropriate size.

Output O_k from each module should approximate F as much as possible. Each O_k, in turn, is obtained by linear combinations of the matrix of hidden layer activations (H_k), which, in turn, are obtained by sigmoid transformations of some linear combinations of the input matrix (X_k). Constraints (2) state that F is column-wise centered and orthogonal, which are required for identification purposes.

2.2 Algorithms

There are three sets of parameters, $\{V_k, W_k, F\}$. We propose two algorithms to minimize (1). Algorithm I splits the parameter set into $\{V_k, W_k\}$ and $\{F\}$, whereas Algorithm II into $\{V_k\}$ and $\{W_k, F\}$.

Algorithm I: We alternate the following two steps until convergence is reached.

1. For $k = 1, \ldots, K$, minimize g_k with respect to V_k and W_k for fixed F by the Levenberg-Marquardt (LM) method.

2. Minimize g with respect to F subject to (2) for fixed V_k and W_k. This is done by first defining $F^* = (I_N - 1_N 1_N'/N) \sum_{k=1}^{K} O_k/K)$, where I_N is the identity matrix of order N, and then by applying the Gram-Schmidt orthogonalization method to F^* to obtain F. (Matrix $I_N - 1_N 1_N'/N$ has the effect of column-wise centering the matrix that follows.)

Step 1 in the above algorithm works just like separate MLFF networks with F as target outputs. A ready-made algorithm for MLFF networks (like the one in the Neural Network Tool Box in MATLAB) can be used for optimization in this phase.

Algorithm II: In the above algorithm, the most time consuming part is Step 1 which involves an iterative optimization method. It is best to minimize the number of parameters in this step. In Algorithm II, we alternate the following two steps.

1. For $k = 1, \ldots, K$, minimize g_k with respect to V_k for fixed F and W_k by the LM method.

2. Minimize g with respect to F and W_k for fixed V_k. This is done as follows: Define $A = (I_N - 1_N 1_N'/N)HD^{-1/2}$ where $H = [H_1, H_2, \cdots, H_K]$ and D is a block diagonal matrix with $H_k'H_k (k = 1, \ldots, K)$ as diagonal blocks. We compute the generalized singular decomposition of A with column metric matrix D to obtain F. We then calculate W_k by $W_k = (H_k'H_k)^{-1}H_k'F$.

Both algorithms are monotonically convergent. Note also that in both algorithms Step 1 can be carried out for each module separately, which significantly reduces the number of parameters updated simultaneously in each optimization problem.

3. An Illustrative Example

The data used to demonstrate the feasibility of NGCANO is part of large scale survey data collected at the Institute of Statistical Mathematics in Tokyo. The survey questions asked about traditional versus modern views on Japanese society and culture. We used six items from the survey, five of which (items 1, 2, 3, 5 and 6) had three response categories and one (item 4) had two response categories. There were 1864 subjects responding to the survey questionnaire. This is the kind of data set to which multiple correspondence analysis (MCA) is typically applied. An analytic solution exists, so we know what NGCANO is supposed to obtain.

We used the so-called analog coding instead of dummy coding (as typically done in MCA); we arbitrarily assigned numbers to the response categories and treated them as values on an input variable in each module. Each module consisted of a single (continuous) input variable, so that the situation has direct analogy to nonlinear PCA. The assigned numbers could be any distinct numbers, although the first 2 or 3 consecutive integers were used in the present example. Which integers are used to code which response categories is also essentially arbitrary. NGCANO is supposed to find the best nonlinear transformations of these prescribed numbers.

We used one less hidden unit than the number of response categories in each module and obtained a solution with two canonical variates. The derived solution was virtually indistinguishable from that obtained by MCA. Figure 2 depicts the hidden unit activations and the output activations for two output units as functions of the single input variable for items (which coincide with modules) 2, 3 and 4. The hidden unit activations were obtained by sigmoid transformations of the input variable times the weights. The activation functions are bound to be monotonic (either increasing or decreasing). The output activations were obtained by linear combinations of the hidden unit activations, and may no longer be monotonic. An important thing is that whereas in MCA we only get values at the response categories (marked by small circles), NGCANO fits a continuous nonlinear function that passes through these points. This does not mean that we can freely interpolate values between the response categories, but that in principle the input variables can take infinitely many values (in fact, a continuum of values) between prescribed values of the response categories. The output activation functions are generally not unique, but always pass through the points marked by small circles representing the response categories of the item in the module. There are infinitely many functions that pass through the three points, and different functions are typically obtained from different initial estimates of the weights

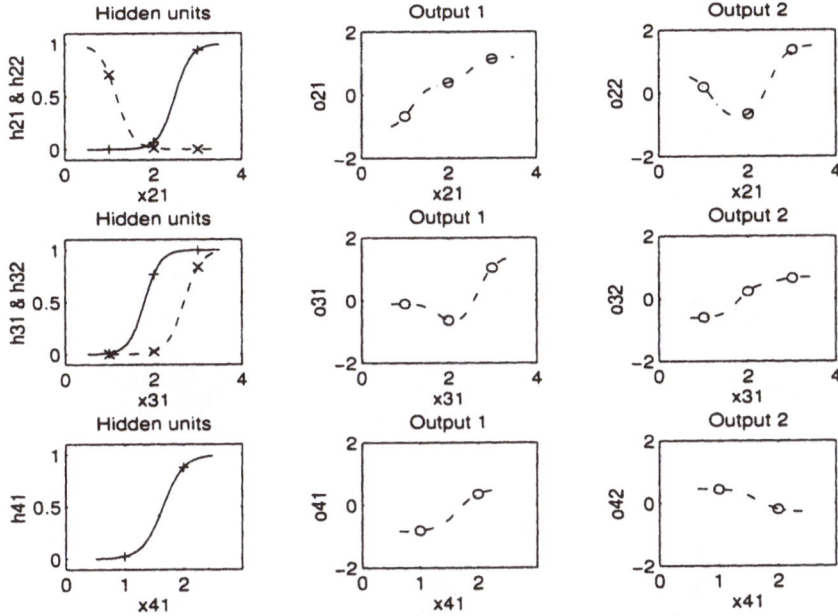

Figure 2: Hidden unit activations and best nonlinear transformations of input variables for module 2 (top), module 3 (middle), and module 4 (bottom). In each row the leftmost figure gives hidden unit activations (marked by "+" for h_1 and by "×" for h_2), and the center and the rightmost figures give output activations as functions of the input variable in the module.

in the network.

The second analysis investigates what happens if we include more than one item in one module. We included the first two items in module 1, and left the rest of the items as in the previous analysis. Again analog coding was used, and a solution with two canonical variates were obtained. There were thus nine possible input patterns (3 by 3) in module 1, which were coded as 1 1, 1 2, 1 3, 2 1, 2 2, 2 3, 3 1, 3 2, and 3 3 on the two input variables in this module. We used eight hidden units in module 1 to obtain the solution. The derived solution was essentially the same as the one obtained by the so-called interactive coding of the first two items in MCA. In the interactive coding, we create an item with nine categories by factorial combinations of the three response categories in each of the two items.

Figure 3 depicts hidden unit activations as functions of the two input variables for the eight hidden units in module 1. All of them are monotonic in a particular direction on the $x_1 - x_2$ plane. These hidden unit activations were linearly combined to obtain output activations which may no longer be monotonic in any direction on the plane. Figure 4 depicts output activations for output 1 in module 1 corresponding to the first canonical variate, which

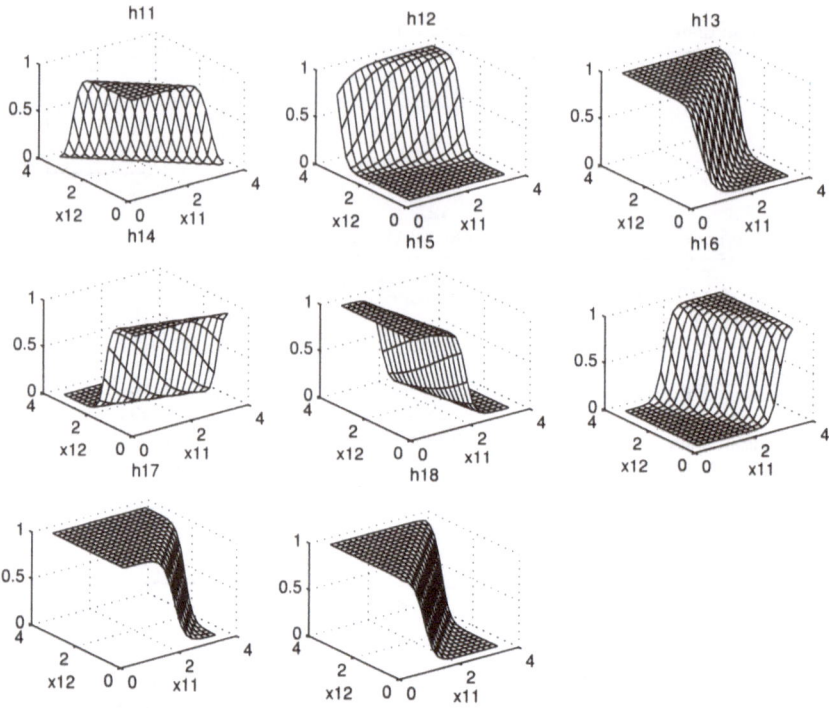

Figure 3: Hidden unit activations for the eight hidden units in module 1 as functions of the two input variables in the module, obtained by sigmoid transformations of some linear combinations of the two input variables.

is a nonlinear transformation of the two input variables in module 1. Small circles indicate the function values at the prescribed values of the response categories. Some degree of interaction effects are observed, although there does not seem to be much interactions, as indicated by near parallel lines connecting points within particular response categories in each of the two items. A similar output activation function was obtained for output 2.

4. Discussion and Future Prospects

The proposed method works in the way it is supposed to in the analysis of multiple-choice categorical data. Although results are not presented here because of space limitation, it has also been demonstrated that NGCANO works in situations where there are many more response categories in each item (to the extent that each category receives only one response). This is the situation in which nonlinear PCA is typically called for, which is another special case of NGCANO. These are very encouraging, although admittedly they are still preliminary.

The basic framework of the method presented above can be extended in

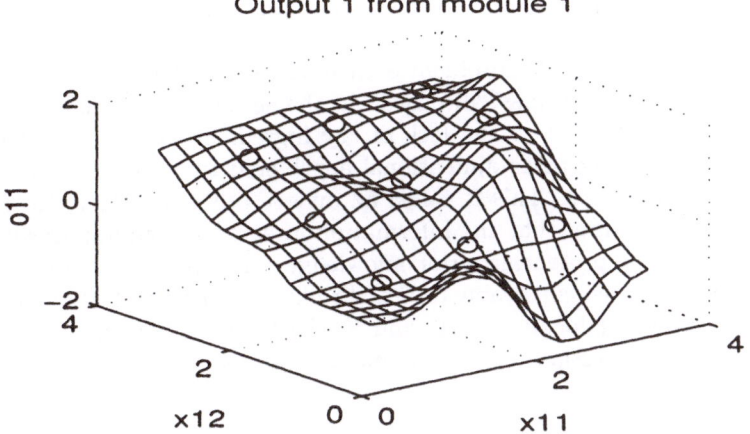

Figure 4: Output activations as functions of the two input variables in module 1 obtained by a linear combination of the hidden unit activations depicted in Figure 3.

a variety of ways. Below is a list of possible extensions, some of which have already been implemented.

1) Differential weighting of modules. Information processed by some modules is more important than others. Information from contradicting modules may duly be ignored completely. Different modules may therefore be differentially weighted to reflect their importance. The weights may be chosen *a priori* or algorithmically according to how well each module fits to the criterion values, as in robust regression analysis by iterative reweighting.

2) Different numbers of hidden layers and different types of transfer functions across modules. In the model depicted in Figure 1, all modules had only one hidden layer, and the same sigmoid transfer functions were used for all the hidden units. Information processed by different modules may be of different types and of different complexity. In such cases it would be useful to allow different numbers of hidden layers and different types of transfer functions across different modules. In the example above, one of the items had only two response categories. Since two points can always be perfectly fitted by a linear function, no nonlinear transformation was necessary for this item, while the other items required multiple nonlinear transformations. This attests the necessity of differentiating the size and the complexity of the networks for data transformation purposes.

3) Differential weighting of training patterns. Some training patterns are more important and/or more reliable than others. In a manner similar to 1) above, a differential weighting scheme may be introduced to different training patterns. This may also be useful in dealing with missing data. We give the weight of one to observed data, while that of zero to missing observations.

4) Regularizations. The problem each module is solving may be extremely complicated, sometimes "ill-posed" in the sense that no unique solutions can be obtained due to the lack of key ingredients as inputs, as in the problem of recovering three-dimensional depth structures based on two-dimensional retinal images. This is called an inverse problem (Marr, 1982). In such cases it is essential that the information integrator is equipped with regularization terms representing prior knowledge (also known as smoothing terms, penalty terms, constraint terms, shrinkage terms, etc.) about the problems to be solved (Poggio, Torre, and Koch, 1985). There may be as many such terms as necessary for each module. This can be implemented by redefining g_k in (1) as

$$g_k = \|F - O_k\|^2 + \sum_i^{n_k} \rho_{ki}\|q_{ki}(V_k)\|^2, \qquad (3)$$

where the ρ_{ki} are the penalty parameters, the q_{ki} are some functions of V_k, and n_k is the number of regularization terms for module k.

Acknowledgments

The work reported in this paper has been supported in part by NSERC individual operating grants to the authors, and in part by a team grant from Fonds pour la Formation de Chercheurs et l'Aide a la Recherche. We would like to extend our appreciation to Marina Takane who prepared Figure 1.

References:

Asoh, H., and Takechi, O. (1994). An approximation of nonlinear canonical correlation analysis by multilayer perceptrons. *ICANN 94 Proceedings of the International Conference on Artificial Neural Networks*. New York: Springer, 713–716.

Becker, S., and Hinton, G.E. (1992). Self-organizing neural network that discovers surfaces in random-dot stereograms. *Nature*, **355**, 161–163.

Carroll, J.D. (1968). A generalization of canonical correlation analysis to three or more sets of variables. *Proceedings of the 76th Annual Convention of the American Psychological Association*, 227–228.

Gifi, A. (1990). *Nonlinear multivariate analysis*. Chichester: Wiley.

Horst, P. (1961). Relations among m sets of measures. *Psychometrika*, **26**, 129–149.

Marr, D. (1982). *Vision: A computational investigation into the human representation and processing of visual information*. San Francisco: Freeman.

Meredith, W. (1964). Rotation to achieve factorial invariance. *Psychometrika*, **29**, 187–206.

Poggio, M.J., Torre, V., and Koch, C. (1985). Computational vision and regularization theory. *Nature*, **317**, 314–319.

Simultaneous Estimation of a Mean Vector Based on Mean Conjugate Priors

Takemi Yanagimoto[1] and Toshio Ohnishi[2]

[1] Institute of Statistical Mathematics
and [1,2] Graduate University for Advanced Studies

Summary: An empirical Bayes method for the simultaneous estimation of a mean vector is discussed under the mean conjugate prior. This prior, which is dual to a conjugate prior, provides us with a simple efficient estimate of the hyperparameter. Two real examples are presented.

1. Introduction

A current urgent need for developing statistical methods pertains to inference of a high-dimensional vector parameter in a model. Such a model is necessary to describe complex data sets. The empirical Bayes method provides us with practical inferential procedures for a vector parameter. The James-Stein estimator (James and Stein 1961) is a breakthrough estimator in this area, which is regarded as an empirical Bayes method (Efron and Morris 1973). A difficulty arises in constructing an empirical Bayes estimator, which is due to a restricted number of existing families of prior densities. In this paper we introduce a useful family of prior densities, mean conjugate priors. It is regarded as a dual family to that of natural conjugate priors.

2. Empirical Bayes Method

We will focus our discussion on the case where the parameter in the study is a mean vector $\mu(\in R^p)$ in the exponential family. The density function $p(x; \mu)$ is proportional to $\exp\{x'\eta - \psi(\eta)\}$ where η is the canonical parameter, which is a function of μ, $\eta = \eta(\mu)$.

Let $\pi(\mu; \delta)$ be a prior density. Then the marginal likelihood and the posterior density are written as

$$
\begin{aligned}
\pi_m(x; \delta) &= \int p(x; \mu)\pi(\mu; \delta)\mathrm{d}\mu \\
\pi_p(\mu; \delta \mid x) &= p(x; \mu)\pi(\mu; \delta) / \pi_m(x; \delta)
\end{aligned}
\tag{2.1}
$$

respectively. Then the hyperparameter δ is estimated in terms of the marginal likelihood, and the mean parameter μ is estimated by the posterior mean or the posterior mode. The method of moments or the maximum likelihood method is usually employed for estimating a hyperparameter.

An interesting example is found in the normal distribution, where an estimator close to the James-Stein estimator is obtained.

Example 1: Consider the estimation of the mean vector μ from a sample $x \sim N(\mu, I)$. Assume a prior distribution on μ as $\mu \sim N(0, (1/\delta)I)$. Then it is easily shown that the marginal distribution, $\pi_m(x; \delta)$, is $N(0, \{(1+\delta)/\delta\}I)$ and the posterior distribution of μ, $\pi_p(\mu; \delta \mid x)$, is $N((1/(1+\delta))x, (1/(1+\delta))I)$. Thus the posterior mean of μ, which is also the posterior mode, is $\{1/(1+\delta)\}x$. Equating $\|x\|^2 = E(\|x\|^2) = p(1+\delta)/\delta$ we obtain an estimate of δ as $1/\hat{\delta} = [\|x\|^2 - p]^+/p$ with $[y]^+ = \text{Max}\{y, 0\}$. This estimate is also the maximum (marginal) likelihood estimate of δ. Then the empirical Bayes estimate of μ is expressed as $\{[\|x\|^2 - p]^+/\|x\|^2\}x$. On the other hand, equating $1/\|x\|^2 = E(1/\|x\|^2) = (1+\delta)/(p-2)\delta$, we obtain another estimator of δ, $1/\hat{\delta}_{JS} = \{\|x\|^2 - (p-2)\}/(p-2)$, by allowing the estimator to take a negative value. This estimator yields the well known James-Stein estimator $\hat{\mu}_{JS} = \{1 - (p-2)/\|x\|^2\}x$ (Efron and Morris, 1973).

In the empirical Bayes method the choice of a prior distribution is the key to constructing an estimate. We will claim advantages of mean conjugate priors in subsequent sections.

3. Mean Conjugate Priors

A widely employed family of prior densities in Bayesian theory is that of natural conjugate priors. The density is expressed for an assumed mean value μ_0 as

$$\Pi_n = \{\pi(\mu; \delta) \mid \pi(\mu; \delta) \propto \exp \delta(\mu_0'\eta - \psi(\eta)), \ \delta \in \Delta \subset R^+\}. \quad (3.1)$$

The history of the family of natural conjugate priors is long. Actually, it is a naive, easy-to-handle family of prior densities. Raiffa and Schlaifer (1961) are regarded as original contributors on this family. A mathematically rigorous definition was given by Diaconis and Ylvisaker (1979). For example, the normal distribution $N(\mu_0, (1/\delta)I)$ is a family of natural conjugate priors to the normal distribution $x \sim N(\mu, (1/\tau_0)I)$, where μ_0 and τ_0 are assumed to be known. A notable property of a natural conjugate prior is that the posterior mean is expressed as a linear combination of the sample mean and the assumed mean μ_0, when a regularity condition is satisfied (Diaconis and Ylvisaker, 1979). However, note that a natural conjugate prior does not satisfy the regularity condition in many cases, for example the inverse Gaussian distribution of $p(x; \mu)$. More importantly, an interpretation on a prior distribution is often difficult, since it is defined not on the mean parameter μ but on the canonical parameter η.

To introduce a mean conjugate prior we consider here the Kullback-Leibler separator between $p(x; \theta_1)$ and $p(x; \theta_2)$ as

$$D(\theta_1, \theta_2) = E\{\log(p(x; \theta_1)/p(x; \theta_2)) \mid p(x; \theta_1)\}.$$

When $p(x; \mu)$ is proportional to $\exp\{x\eta - \psi(\eta)\}$, the separator with $\theta = \mu$ is written as $\mu_1(\eta_1 - \eta_2) - \psi(\eta_1) + \psi(\eta_2)$. Thus it is shown that

$$\Pi_n = \{\pi(\mu; \delta) \mid \pi(\mu; \delta) \propto \exp -\delta D(\mu_0, \mu)\}. \quad (3.2)$$

Table 1: Two conjugate prior distributions for familiar sampling distributions.

Sample	Natural conjugate	Mean conjugate
Normal	$\mu \sim$ Normal	$\mu \sim$ Normal
Gamma	$1/\mu \sim$ Gamma	$\mu \sim$ Gamma
Inv. Gauss.	$1/\mu^2 \sim$ Inv.Gauss.*	$\mu \sim$ Inv.Gauss.

* It does not have the favorable property regarding the posterior mean.

Therefore we find that a natural conjugate prior density is expressed through the Kullback-Leibler separator. This suggests that a family of mean conjugate prior densities is defined as

$$\Pi_m = \{\pi^m(\boldsymbol{\mu}; \delta) \mid \pi^m(\boldsymbol{\mu}; \delta) \propto \exp -\delta D(\boldsymbol{\mu}, \boldsymbol{\mu}_0)\}. \qquad (3.3)$$

Note that dual structures are found between $D(\boldsymbol{\theta}_1, \boldsymbol{\theta}_2)$ and $D(\boldsymbol{\theta}_2, \boldsymbol{\theta}_1)$, and also between the mean and the canonical parameters in the exponential family.

In this concern a mean conjugate prior is dual to a natural conjugate prior. Three examples of families of dual conjugate priors are given in Table 1. In these examples a mean conjugate prior distribution is the same with a sampling distribution. It looks easier to understand the meaning of a mean conjugate prior than that of a natural conjugate prior, since the latter pertains to the canonical parameter η rather than the mean parameter μ.

4. The Gamma Case

As an application of mean conjugate priors we discuss the empirical Bayes method for the simultaneous estimation of a mean vector in the gamma distribution. This problem has attracted some authors such as Athrea (1986) and Yanagimoto (2000). A reason comes from the fact that this problem appears in the estimation of variances in the design of experiments. The density function $p(\boldsymbol{x}; \boldsymbol{\mu})$ is assumed to be written as $\Pi p(x_i; \mu_i, \tau_0)$ with a known τ_0, that is, components x_i's are mutually independent. In the case where $p(x; \mu, \tau_0)$ is the gamma density, $Ga(\mu, \tau_0)$, the density function is expressed as

$$p(x; \mu, \tau_0) = \frac{\tau_0^{\tau_0} x^{\tau_0 - 1}}{\Gamma(\tau_0) \mu^{\tau_0}} e^{-x\tau_0/\mu}.$$

A familiar form of a natural conjugate prior in the standard textbook is that $\pi(\boldsymbol{\mu}; \delta) = \Pi \pi(\eta_i; \delta)$ with $\eta_i = 1/\mu_i$, and $\eta_i \sim Ga(1/m, \delta)$. Note that this prior is slightly modified so as to satisfy $E(\eta_i) = 1/m$. On the other hand, a mean conjugate density is defined as $\mu_i \sim Ga(m, \delta)$, which is again slightly modified so as to satisfy $E(\mu_i) = m$. This yields the model where $x_i \sim Ga(\mu_i, \tau_0)$ and $\mu_i \sim Ga(m, \delta)$, which looks easy to be understood.

Hudson (1978) introduced an estimator by extending the Stein identity. Unfortunately, his estimator takes a negative value with a positive probability. Another non-Bayesian estimator is found in Yanagimoto (2000). The estimator proposed there is rather heuristic, but simulation studies show that its performance is satisfactory.

The posterior mean under a modified mean conjugate prior is expressed in an explicit form in terms of the modified Bessel function of the third kind $K_\nu(z)$, $(1/2) \int x^{\nu-1} \exp -z(x + 1/x)/2dx$. Setting $\nu = -\tau_0 + \delta$, $u_i = \sqrt{\tau_0 x_i m/\delta}$ and $z_i = 2\sqrt{\tau_0 x_i \delta/m}$, we find that the posterior mean of μ_i is written as $\hat{\mu}_i = u_i K_{\nu+1}(z_i)/K_\nu(z_i)$. In addition, the method of moments can be applied to yielding an estimate of δ, which is the solution of the following equation.

$$p\left(\log \hat{\delta} - \psi(\hat{\delta})\right) = [\Sigma\{x_i/m - \log x_i/m - 1\} - p \log \tau_0 + p\psi(\tau_0)]^+ .$$

This equation appears also in the maximum likelihood estimation of δ in the gamma distribution. Numerical computation of this equation is easily conducted, for example, Yanagimoto (1988).

On the other hand, we can obtain that the posterior mean of η_i under a modified natural conjugate prior is $(\tau_0 + \delta)/(x_i\delta + m\tau_0)$. Thus the estimator of μ, $\hat{\mu} = (\hat{\mu}_1, \ldots, \hat{\mu}_p)$, is written as $\hat{\mu}_i = (x_i\delta + m\tau_0)/(\tau_0 + \delta)$. A difficulty arises in obtaining an estimate of δ, since the marginal likelihood is rather complicated. Although the posterior mean is written in a simple form, the amount of computation for obtaining the empirical Bayes estimator under a natural conjugate prior is greater than that under a mean conjugate prior. Further recall that the maximum likelihood estimator does not always perform well in the mixture model.

5. Application

As is emphasized in the previous section, the empirical Bayes estimator of the gamma under a mean conjugate prior does not require elaborate numerical computation. Since the estimator is expressed in a simple form, a computer program can be easily coded. Table 2 shows an example of the program using a commercial software, Mathematica. Among 12 lines of the program the first 2 lines consist of data input and setting initial values. Thus the program has essentially 10 lines.

In the following two subsections we give two examples of analyzing data sets in the literature.

5.1 Failure Time Data

The data set in Table 3 is concerned with time to breakdown of an insulating fluid, which is cited from Table 2.1 in Chapter 7 in Nelson (1982). A distribution on time is assumed to be exponential. We assume that a mean conjugate prior distribution is the gamma distribution with the mean 200 and an unknown dispersion. The empirical Bayes estimates are given in the

Table 2: Mathematica program for the empirical Bayes estimate under a mean conjugate prior of a mean vector in the multi-dimensional gamma distribution.

```
xdata = {30,33,41,87,93,98,116,258,461,1180,1350,1500}
xm = 200.; tau0 = 1.;
    nn = Length[xdata];xdata1 = xdata/xm;xlog = Log[xdata1];
    xdv = Apply[Plus, xdata1]-Apply[Plus, xlog] -nn; res = xdv/nn;
    psaa = res -( -Log[1./tau0] - PolyGamma[tau0]);
    val = Max[psaa, 0]; vinit = val + 1/3*Log[1 + 3*val];
    psif[xx_] = -Log[xx] - PolyGamma[1/xx];
    deltae = xx/.FindRoot[ psif[xx]==val, {xx, vinit}];
    deltae = 1/deltae; myu = deltae - tau0 ; myu1 = myu + 1;
    eta = Sqrt[xdata*tau0*xm/deltae];
    zv = 2*Sqrt[xdata*tau0*deltae/xm];
    est = eta*BesselK[myu1,zv]/BesselK[myu,zv]
```

Table 3: Data for time to insulating fluid breakdown, cited from Nelson (1982), with the empirical Bayes estimate under a mean conjugate prior in the parentheses.

Time	Estimate	Time	Estimate
30	(52.84)	33	(57.49)
41	(69.51)	87	(130.88)
93	(138.15)	98	(144.11)
116	(164.87)	258	(302.82)
461	(456.95)	1180	(848.29)
1350	(922.97)	1500	(985.35)

parentheses of Table 3. We learn that the estimates are largely different from the observed values, the maximum likelihood estimates. We learn also that there are large differences among the estimates, suggesting inhomogeneity of the original populations.

5.2 Dioxin Concentration

Because of a possibly serious environmental pollution problem governmental agencies regularly conduct nation-wide surveys of dioxin concentrations. A portion of results of a survey are reproduced in Table 4. It is believed that observed concentrations are subject to fluctuations due to measurement errors and also due to location differences. We assume that the measurement error follows the gamma distribution with an unknown mean and the dispersion $\tau_0 = 3$.

Fluctuations due to location differences are described by a mean conjugate

Table 4: Dioxin concentration data with the empirical Bayes estimate under a mean conjugate prior (Data source is a government report, see reference)

Location	Concentration	Estimate	Location	Concentration	Estimate
1	0.0071	(0.0094)	2	0.0095	(0.011)
3	0.038	(0.027)	4	0.057	(0.035)
5	0.030	(0.024)	6	0.037	(0.027)
7	0.0068	(0.0092)	8	0.0075	(0.0098)
9	0.019	(0.018)	10	0.062	(0.036)
11	0.018	(0.017)	12	0.010	(0.012)
13	0.016	(0.016)	14	0.0053	(0.0077)
15	0.0075	(0.0098)			

prior, which is assumed to follow the gamma distribution with the mean 0.01 and an unknown dispersion. The empirical Bayes estimates are given in the parentheses in Table 4.

It is not easy to obtain clear findings from such a practical data set. It looks, however, that the estimates obtained are reasonable. More importantly, our choice of a mean conjugate prior distribution is easier to be accepted, since it is the same with the measurement error distribution. In fact, the reciprocal gamma distribution, which appears in a natural conjugate prior distribution, is less familiar than the usual gamma distribution.

References:

Athreya, K.B. (1986). Another conjugate family for the normal distribution. *Statist. Probab. Lett.*, **4**, 61-64.

Diaconis, P. and Ylvisaker, D. (1979). Conjugate priors for exponential families. *Ann. Statist.*, **7**, 269-281.

Efron, B. and Morris, C. (1973). Stein's estimation rule and its competitors - An empirical Bayes approach. *J. Am. Statist. Assoc.*, **68**, 117-130.

Hudson, H.M. (1978). A natural identity for exponential families with applications in multiparameter estimation. *Ann. Statist.*, **6**, 473-484.

Nelson, W. (1982). *Applied Life Data Analysis*, John Wiley, New York.

Panel of Experts on Dioxin Risk Assessment, The Environment Agency of Japan (1997). *Risk Assessment of Dioxins* (in Japanese), Central Regulations Publication, Tokyo.

Raiffa, H. and Shlaifer, R. (1961). *Applied Statistical Decision Theory*, Graduate School of Business Administration, Harvard Univ. Boston.

Yanagimoto, T. (1988). The conditional maximum likelihood estimator of the shape parameter in the gamma distribution. *Metrika*, **35**, 161-175.

Yanagimoto, T. (2000). A pair of estimating equations for a mean vector. *Statist. Probab. Letters*, **50**, 97-103.

Using Several Data to Structure Efficient Estimation of Intraclass Correlation Coefficients

S. E. Ahmed and S. M. Khan

University of Regina
Regina, Saskatchewan, Canada S4S 0A2

Summary: The estimation of several intraclass correlation coefficients is considered when independent samples are drawn from multivariate normal distributions. We propose improved shrinkage estimator for the correlation parameters. It is shown analytically and computationally that the positive-part shrinkage estimator outperforms usual shrinkage estimator. The asymptotic relative performance of the estimators in the light of their bias and risk is presented both analytically and numerically. A Monte Carlo study is carried out to assess the performance of the proposed estimators for small samples.

1. Introduction

Let $\mathbf{X}_{(p \times 1)}^{(l)}$ be a random vector which has multivariate normal distribution with mean vector $\boldsymbol{\mu}_{(p \times 1)}^{(l)} = (\mu_l, \cdots, \mu_l)'$ and covariance matrix Σ_l, $l = 1, \cdots, q$.

When the covariance matrix for n_l random variables has the structure

$$\Sigma_{l_{(p \times p)}} = \sigma_l^2 [(1 - \rho_l)\mathbf{I} + \rho_l \mathbf{J}]$$

where $\mathbf{I}_{(p \times p)}$ is the identity matrix, $\mathbf{J} = \mathbf{1}\mathbf{1}'$ and $\mathbf{1}_{(p \times 1)} = (1, \cdots, 1)'$, then the set of variables $\mathbf{X}_{(p \times 1)}^{(l)}$ is said to possess intraclass correlation. The random variables have equal variances, and covariances. The number of parameters in the covariance matrix is reduced to 2. This patterned covariance matrix, which often describes the correspondence among certain biological variables such as size of living things, has the above form. The resulting correlation matrix

$$\rho_l = (1 - \rho_l)\mathbf{I} + \rho_l \mathbf{J}$$

is also the covariance matrix of the standardized variables.

The intraclass correlation coefficient may be used in many situations. We may use it for an example to measure the degree of resemblance between siblings concerning certain characteristics, such as blood pressure, stature, body weight or lung capacity. Various aspects of intraclass correlation coefficient have been studied by host of researchers. Kapata (1993) has studied hypothesis testing concerning a constant intraclass correlation for families of varying size. Donner (1986) has provided a comprehensive review for the inference procedure in one-way random effect model, which shows that the problem of estimating the correlation parameter frequently arises in many medical and bio-statistical applications. Suppose that the researcher has collected data from different (say q) locations under similar conditions. Further, the data may have been collected at various times from the same location. The interest is in estimating the intraclass correlation coefficient ρ_l, simultaneously on the basis of q random samples when it is reasonable to assume the homogeneity of population correlations. In other words, the problem is to estimate ρ_l when we have the *nonsapmle information* (NSI) in the form of the null hypothesis

$$H_o : \rho_1 = \cdots = \rho_q = \rho \quad (unkown). \tag{1.1}$$

In this communication, we examine the properties of an improved version of shrinkage estimator for estimating the common parameter ρ. A plan of this paper is as follows. The estimators are formally introduced in section 2. The properties of the estimators are investigated in sections 3 and their risk performance is assessed in section 4. A simulation study is presented in section 5 and section 6 summarizes the findings.

2. Improved Estimators

Let $\mathbf{X}_i^{(l)} = \left(\mathbf{X}_{1i}^{(l)}, \cdots, \mathbf{X}_{pi}^{(l)}\right)$, $i = 1, 2, \cdots, n_l$ be a random sample of size n_l from a p-variate normal distribution with mean vector $\boldsymbol{\mu}_l = (\mu_l, \cdots, \mu_l)'$ and covariance matrix $\boldsymbol{\Sigma}_l$, where $\boldsymbol{\Sigma}_l$ has an intraclass correlation structure.

The *unrestricted estimator (UE)* of ρ is defined as

$$r_l = \frac{\sum\limits_{i=1}^{n_l} \sum\limits_{j \neq k}^{P} (x_{ji}^{(l)} - \bar{x}_l)(x_{ki}^{(l)} - \bar{x}_l)}{(p-1) \sum\limits_{i=1}^{n_l} \sum\limits_{j=1}^{P} (x_{ji}^{(l)} - \bar{x}_l)^2}, \quad l = 1, 2, \cdots, q, \tag{2.1}$$

where $\bar{x}_l = \frac{1}{pn_l} \sum\limits_{i=1}^{n_l} \sum\limits_{j=1}^{P} x_{ji}^{(l)}$. Consider the transformation

$$\rho_l^* = \sqrt{\frac{(p-1)}{2p}} ln \left(\frac{1 + (p-1)\rho_l}{1 - \rho_l}\right), \tag{2.2}$$

where ln means *logarithm to the base e*. Following Ahmed *et al.* (2001), the UE of $\rho^* = (\rho_1^*, \rho_2^*, \cdots, \rho_q^*)'$ can be obtained by replacing unknown ρ_l with its empirical estimate. Thus, $\hat{\rho}^* = (\hat{\rho}_1^*, \hat{\rho}_2^*, \cdots, \hat{\rho}_q^*)'$ is defined as

$$\hat{\rho}_l^* = \sqrt{\frac{(p-1)}{2p}} ln \left(\frac{1 + (p-1)r_l}{1 - r_l}\right), \quad l = 1, 2, \cdots, q. \tag{2.3}$$

However, the normal approximation becomes poor as the value of p increases. Konishi (1985) proposed the following transformation for large p

$$\theta_l = \rho_l^* - \frac{(7 - 5p)}{n_l\sqrt{18p(p-1)}}.$$ (2.4)

Thus, UE of θ_l is

$$\hat{\theta}_l = \hat{\rho}_l^* - \frac{(7 - 5p)}{n_l\sqrt{18p(p-1)}},$$ (2.5)

which follows an approximately normal distribution with mean $\theta_l = \rho_l^* - \frac{(7-5p)}{n_l\sqrt{18p(p-1)}}$ and variance $\frac{1}{(n_l-2)}$ respectively. The equality of q intraclass correlations under consideration is equivalent to the test of the equality of the values of θ_l. The hypothesis $H_o : \rho_1 = \cdots = \rho_l$ is thus equivalent to

$$H_o : \theta_1 = \cdots = \theta_l.$$ (2.6)

An estimate of the common intraclass correlation parameter is

$$\tilde{\boldsymbol{\theta}} = (\tilde{\theta}, \cdots, \tilde{\theta})', \quad \tilde{\theta} = \frac{1}{n}\sum_{l=1}^{q}(n_l - 2)\hat{\theta}_l, \quad n = \sum_{l=1}^{q}(n_l - 2),$$ (2.7)

and we call it as *Pooled Estimator (PE)* (Elston (1975)).

It is well documented in the literature that $\tilde{\boldsymbol{\theta}}$ generally has a smaller dispersion than $\hat{\boldsymbol{\theta}} = (\theta_1, \cdots, \theta_q)$ near the null hypothesis. However, $\tilde{\boldsymbol{\theta}}$ becomes considerably biased and inefficient when the null hypothesis may not be tenable. Thus, the performance of the restricted procedure depends upon the correctness of the NSI. We consider a class of improved estimators namely shrinkage estimators.

The usual shrinkage estimator denoted by $\hat{\boldsymbol{\theta}}^S = (\hat{\theta}_1^S, \cdots \hat{\theta}_q^S)'$ is given as follows:

$$\hat{\boldsymbol{\theta}}^S = \tilde{\boldsymbol{\theta}} + \{1 - (q - 3)D_n^{-1}\}(\hat{\boldsymbol{\theta}} - \tilde{\boldsymbol{\theta}}), \quad q \geq 4,$$ (2.8)

where

$$D_n = n(\hat{\boldsymbol{\theta}} - \tilde{\boldsymbol{\theta}})'\boldsymbol{\Lambda}(\hat{\boldsymbol{\theta}} - \tilde{\boldsymbol{\theta}}),$$ (2.9)

with

$$\boldsymbol{\Lambda} = \text{Diag}(\lambda_l), \quad \lambda_{n_l} = \frac{(n_l - 2)}{n}.$$ (2.10)

Ahmed *et al.* (2001) demonstrated that the shrinkage estimator dominates $\hat{\boldsymbol{\theta}}$. They were able to show that $\hat{\boldsymbol{\theta}}^S$ combines the sample information and NSI in a superior way. The resulting estimator improves upon $\hat{\boldsymbol{\theta}}$ regardless the correctness of the nonsample information. However, the $\hat{\boldsymbol{\theta}}^S$ may shrink beyond the hypothesis resulting in the change of sign of $\hat{\boldsymbol{\theta}}$. We therefore, propose a superior alternative to the usual shrinkage estimator by considering its positive part. The positive-part estimator not only manages the over-shrinking problem but is also superior to $\hat{\theta}^S$ in the entire parameter space. The proposed estimator prevents changing of the sign of the usual estimator, which is an inherited problem of the shrinkage estimator. Ahmed (1992) discussed the asymptotic properties of this estimator in estimating correlation coefficients from bivariate normal populations. For a review of the shrinkage estimators, readers are referred to Kubokowa (1998) among others.

We propose a *positive-part Stein-rule estimator (PSE)* as follows

$$\hat{\theta}^{S+} = \tilde{\theta} + \{1 - (q-3)D_n^{-1}\}^+(\hat{\theta} - \tilde{\theta}), \quad q \geq 4. \tag{2.11}$$

This estimator can be re-written as

$$\hat{\theta}^{S+} = \tilde{\theta} + (1 - (q-3)D_n^{-1})I(D_n > (q-3))(\hat{\theta} - \tilde{\theta}),$$

where $I(A)$ is the indicator function defined on the set A.

We consider the *asymptotic distributional risk (ADR)* for a sequence $\{K_{(n)}\}$ of local alternatives

$$K_{(n)} : \theta = \theta_n, \quad \text{where} \quad \theta_n = \theta 1_q + \frac{\delta}{\sqrt{n}}, \tag{2.12}$$

and δ is a fixed real vector and $1_q = (1, \cdots, 1)'$.

The *asymptotic distribution function (ADF)* of $\{\sqrt{n}(\theta^* - \theta)\}$ is given by

$$G(\mathbf{y}) = \lim_{n \to \infty} P\{\sqrt{n}(\theta^* - \theta) \leq \mathbf{y}\}, \tag{2.13}$$

where θ^* is any estimator of θ for which the limit in (2.13) exists. Also, let

$$\mathbf{Q} = \int \int \cdots \int \mathbf{y}\mathbf{y}'dG(\mathbf{y}). \tag{2.14}$$

Then, the *asymptotic distributional risk (ADR)* is defined by

$$ADR(\theta^*; \theta) = \text{trace}(\mathbf{W}\mathbf{Q}), \tag{2.15}$$

where $\mathbf{W}$ is a positive definite matrix.

In the following section, we obtain the expressions for the bias and asymptotic distributional risk (ADR) of the estimators.

3. Main Results

Lemma 3.1: Under (2.12)

$$\mathbf{X}_n = \sqrt{n}(\hat{\theta}_n - \theta) \sim N_q(\delta, \Lambda^{-1}),$$

and

$$\mathbf{Y}_n = \sqrt{n}(\hat{\boldsymbol{\theta}}_n - \tilde{\boldsymbol{\theta}}_n), \sim N_q(\boldsymbol{\beta}, \boldsymbol{\Lambda}^{-1}\mathbf{H}'),$$

where

$$\boldsymbol{\beta} = \mathbf{H}\boldsymbol{\delta}, \quad \mathbf{H} = \mathbf{I}_q - \mathbf{J}\boldsymbol{\Lambda}, \quad \mathbf{J} = \mathbf{1}_q\mathbf{1}_q'.$$

The above result follows immediately after noting that $\mathbf{Y}_n = \sqrt{n}\mathbf{H}\hat{\boldsymbol{\theta}}_n$ (Ahmed et al. (2001)) .

Lemma 3.2: $\mathbf{Z}_n = \sqrt{n}(\tilde{\boldsymbol{\theta}}_n - \boldsymbol{\theta}) \sim N_q(0, \mathbf{J})$, here we assume that $\boldsymbol{\lambda}'\boldsymbol{\delta} = 0$, where $\boldsymbol{\lambda} = (\lambda_1, \cdots, \lambda_q)$.

Consequently,

$$\begin{pmatrix} \mathbf{X}_n \\ \mathbf{Y}_n \end{pmatrix} \sim N_{2q} \left\{ \begin{pmatrix} \boldsymbol{\delta} \\ \boldsymbol{\beta} \end{pmatrix}, \begin{pmatrix} \boldsymbol{\Lambda}^{-1} & \boldsymbol{\Lambda}^{-1}\mathbf{H}' \\ \mathbf{H}\boldsymbol{\Lambda}^{-1} & \boldsymbol{\Lambda}^{-1}\mathbf{H}' \end{pmatrix} \right\},$$

$$\begin{pmatrix} \mathbf{Z}_n \\ \mathbf{Y}_n \end{pmatrix} \sim N_{2q} \left\{ \begin{pmatrix} 0 \\ \boldsymbol{\beta} \end{pmatrix}, \begin{pmatrix} \mathbf{J} & 0 \\ 0 & \boldsymbol{\Lambda}^{-1}\mathbf{H}' \end{pmatrix} \right\}.$$

Ahmed *et al.* (2001) demonstrated that the test statistic $D_n = n(\hat{\boldsymbol{\theta}} - \tilde{\boldsymbol{\theta}})'\boldsymbol{\Lambda}(\hat{\boldsymbol{\theta}} - \tilde{\boldsymbol{\theta}})$ is distributed asymptotically as a noncentral chi-square distribution with $(q-1)$ degrees of freedom and noncentrality parameter $\Delta = \boldsymbol{\beta}'\boldsymbol{\Lambda}\boldsymbol{\beta}$. Here $\boldsymbol{\beta} = \mathbf{1}_q'\mathbf{1}_q - \mathbf{1}_q'\boldsymbol{\Lambda}$. Hence, under the null hypothesis for large n, D_n closely follows the central chi-square distribution with $(q-1)$ degrees of freedom. For given α, the critical value of D_n may be approximated by $\chi^2_{q-1,\alpha}$, the upper $100\alpha\%$ point of the chi-square distribution with $(q-1)$ degrees of freedom.

Let us denote a q-dimensional multivariate normal distribution function having mean vector $\mathbf{u}$ and dispersion matrix $\boldsymbol{\Sigma}$ by $\Phi_q(\mathbf{x}; \mathbf{u}, \boldsymbol{\Sigma})$ and the corresponding probability density function (p.d.f) by $\varphi_q(\mathbf{x}; \mathbf{u}, \boldsymbol{\Sigma})$. Also, let $H_q(x; \Delta)$ stands for the non-central chi-square distribution function with non-centrality parameter Δ and q degrees of freedom. Further,

$$E\left(\chi_q^{-2m}(\Delta)\right) = \int_0^\infty x^{-2m} dH_q(x; \Delta).$$

The expressions for the *asymptotic distributional bias (ADB)* of the proposed estimator is given in the following relation.

$$\begin{aligned} ADB(\hat{\boldsymbol{\theta}}^{S+}) &= \lim_{n\to\infty} E\{\sqrt{n}(\hat{\boldsymbol{\theta}}^{S+} - \boldsymbol{\theta})\} = \\ &\quad -\boldsymbol{\delta}\left[H_{q+1}(q-3; \Delta) + E\{\chi_{q+1}^{-2}(\Delta)I(\chi_{q+1}^2(\Delta)) > (q-3))\}\right]. \end{aligned}$$

$$(3.1)$$

The bias of $\hat{\boldsymbol{\theta}}^S$ is

$$ADB(\hat{\boldsymbol{\theta}}^S) = -(q-3)\boldsymbol{\delta} E(\chi_{q+1}^{-2}(\Delta)). \qquad (3.2)$$

Figure 1: Bias Analysis of the Estimators

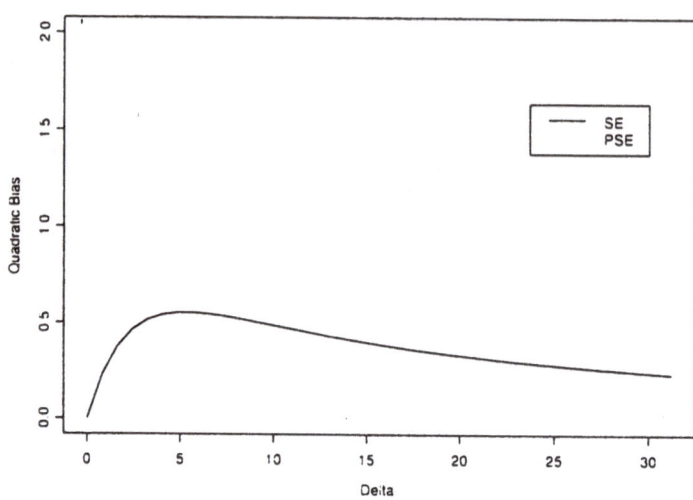

In order to obtain a clear-cut analysis of the various bias functions, first we transform these functions into a scalar (quadratic) form by defining

$$B(.) = [ADB(\boldsymbol{\theta}^*)]'\Lambda[ADB(\boldsymbol{\theta}^*)]$$

as the quadratic bias of an estimator $\boldsymbol{\theta}^*$ of parameter vector $\boldsymbol{\theta}$. Then, we have

$$B(\hat{\boldsymbol{\theta}}^{S+}) = \Delta \left[H_{q+1}(q-3; \Delta) + E\{\chi_{q+1}^{-2}(\Delta)I(\chi_{q+1}^2(\Delta)) > (q-3))\} \right]^2,$$

$$B(\hat{\boldsymbol{\theta}}^{S}) = (q-3)^2\Delta[E(\chi_{q+1}^{-2}(\Delta))]^2.$$

The quadratic bias of $\hat{\boldsymbol{\theta}}^{S}$ starts from 0 at $\Delta = 0$, then increases to a point, then decreases towards 0, since $E(\chi_{q+1}^{-2}(\Delta))$ is decreasing log-convex function of Δ. The behavior of $\hat{\boldsymbol{\theta}}^{S+}$ is similar to $\hat{\boldsymbol{\theta}}^{S}$. However, the quadratic bias curve of $\hat{\boldsymbol{\theta}}^{S-}$ remains below the curve of $\hat{\boldsymbol{\theta}}^{S}$ for all values of Δ. Figure 1 displays these features of the estimators for $q = 6$.

Theorem 3.1: Under the local alternatives we obtain the ADR functions of the estimators

$$
\begin{aligned}
ADR(\hat{\theta}^{S+}) = {} & ADR(\hat{\theta}^S) - \text{trace}(\mathbf{W}\Lambda^{-1}) \\
& E[\{1 - (q-3)\chi_{q+1}^{-2}(\Delta)\}^2 I(\chi_{q+1}^2(\Delta) \le (q-3)] + \\
& \Delta_w[E[2\{1 - (q-3)\chi_{q+1}^{-2}(\Delta)\}I(\chi_{q+1}^2(\Delta) \le (q-3)] - \\
& E[\{1 - (q-3)\chi_{q+3}^{-2}(\Delta)\}^2 I(\chi_{q+3}^2(\Delta) \le (q-3)]],
\end{aligned}
$$

$$(3.3)$$

where

$$
\begin{aligned}
ADR(\hat{\theta}^S, \hat{\theta}) = {} & ADR(\hat{\theta}) + \Delta_w(q-3)(q+1)E(\chi_{q+3}^{-4}(\Delta)) - \\
& (q-3)\text{trace}(\mathbf{W}\Lambda^{-1})\{2E(\chi_{q+1}^{-2}(\Delta)) - (q-3)E(\chi_{q+1}^{-4}(\Delta))\}.
\end{aligned}
$$

$$(3.4)$$

Here $ADR(\hat{\theta}) = \text{trace}(\mathbf{W}\Lambda^{-1})$, $\Delta_w = \beta'\mathbf{W}\beta$.

Proof. Using Lemma 3.1 and Lemma 4 of Ahmed (2001) the above relations are established.

4. Risk Comparisons

In this section, the large sample properties of the proposed estimators are discussed in the light of the quadratic loss function.

Comparing the risk performance of $\hat{\theta}^{S+}$ and $\hat{\theta}^S$, it is concluded from relations (3.3) and (3.4) that

$$
\frac{ADR(\hat{\theta}^{S+})}{ADR(\hat{\theta}^S)} \le 1, \quad \text{for all} \quad \delta,
$$

with strict inequality for some δ. The above inequality follows from the fact

$$
I\left(\chi_{q+k}^2(\Delta) \le (q-3)\right) = \begin{cases} 1 & \text{if } \chi_{q+k}^2 \le q-3, \ k = 1,3 \\ 0 & \text{otherwise.} \end{cases}
$$

Therefore each indicator function in relation (3.3) has either a value of 1 if $\chi_{q+l}^2(\Delta) \le (q-3)$ or value 0. This implies that $1 - \frac{q-3}{\chi_{q+k}^2(\Delta)} \le 0$, for $k = 1, 3$. Hence $ADR(\hat{\theta}^{S+}) - ADR(\hat{\theta}^S) \le 0$. Therefore, $\hat{\theta}^{S+}$ asymptotically dominates $\hat{\theta}^S$ under local alternatives and $\hat{\theta}^{S+}$ is superior to $\hat{\theta}^S$. Thus, $\hat{\theta}^{S+}$ is also superior to $\hat{\theta}$.

In order to facilitate numerical computation of the risk functions, we consider the particular case $\mathbf{W} = \Lambda$. In this case $\text{trace}(\mathbf{W}\Lambda^{-1}) = q$ and $\delta'\mathbf{W}\delta = \Delta$. The ADR expressions are evaluated numerically through programming.

We have plotted $ADR(\hat{\theta})$, $ADR(\hat{\theta}^S)$ and $ADR(\hat{\theta}^{S+})$ versus Δ to assess the performance of the proposed positive part shrinkage estimator in relation to the usual shrinkage estimator and unrestricted estimator.

204

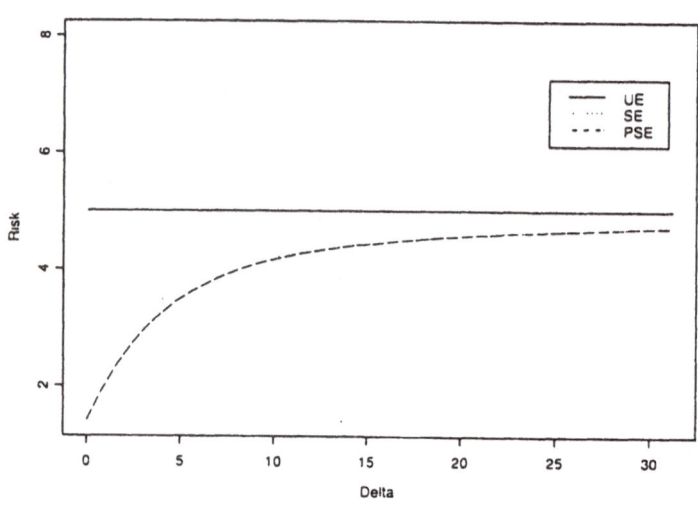

Figure 2: Risk Analysis of the Estimators

Figure 2 shows that Stein-type estimators dominate $\hat{\theta}$ for all values of Δ. We notice that both shrinkage estimators have maximum risk gain as compared to $\hat{\theta}$ when the null hypothesis is true. Proposed positive part shrinkage estimator also dominates shrinkage estimator.

5. Simulation Results

To investigate the risk behavior of the proposed estimators when the samples are small, we have considered a simulation study. Based on simulation results the percentage improvement in risk is calculated by using the formula

$$PI(\Delta^\star) = \frac{100(R_1(\Delta^\star) - R_b(\Delta^\star))}{R_1(\Delta^\star)}, \quad b = 2, 3,$$

where $\Delta^\star = \sum_{i=1}^{q}(\theta - \theta_o)^2$. The simulated data is based on the samples generated from multivariate normal populations having mean vector $\mu = 1$ and $\sigma^2 = 1$. The random numbers are generated by $IMSL$ sub-routine $RNMVN$ for given q and n_i. Based on simulated values of $\hat{\theta}, \tilde{\theta}$, the test statistic D_n is computed first and then using these quantities shrinkage estimators are calculated. First, the random numbers were generated for $\Delta^\star = 0$. Then in order to investigate the behavior of the estimators for $\Delta^\star > 0$, further samples were generated from multivariate normal populations which assumed a shift to the right by an amount $\Delta^\star$ when $\theta \neq \theta_o$. To assess the impact of the sample size on the performance of these estimators, we carried out simu-

Table 1: Simulated percentage risk improvement of PSE and SE over UE for $q = 4$, $n_i = 5, 10, 20, 30$

$n_1 = 5$			$n_1 = 10$		
Δ^*	SE	PSE	Δ^*	SE	PSE
.00	24.33	30.97	.00	26.74	32.50
.03	22.21	30.79	.03	24.94	31.79
.10	5.97	30.11	.10	24.45	29.98
.23	18.15	28.92	.23	20.02	27.27
.42	17.86	27.25	.42	15.72	23.79
.67	15.29	25.19	.67	14.24	19.91
1.01	13.39	22.72	1.01	12.57	15.89
1.44	11.67	19.93	1.44	9.62	12.17
2.01	10.55	16.90	2.01	7.17	8.68
2.76	9.27	13.77	2.76	4.48	5.60
3.79	5.84	10.59	3.79	2.75	3.06
5.28	5.33	7.41	5.28	.73	1.52

$n_1 = 20$			$n_1 = 30$		
Δ^*	SE	PSE	Δ^*	SE	PSE
.00	18.94	31.56	.00	24.66	33.16
.03	20.95	30.38	.03	23.14	31.15
.10	16.72	26.81	.10	16.22	25.79
.23	14.62	21.54	.23	13.94	18.97
.42	10.56	15.82	.42	5.80	12.31
.67	7.03	10.74	.67	5.61	7.61
1.01	5.44	6.94	1.01	4.02	4.43
1.44	3.97	4.27	1.44	2.80	2.80
2.01	2.35	2.36	2.01	1.85	1.85
2.76	1.00	1.01	2.76	1.17	1.17
3.79	.34	.34	3.79	.82	.82
5.28	.07	.07	5.28	.59	.59

lations for $n_i = 20, 30, 50$. The number of replications for these experiments is set to 500. We have checked the simulation results for large replications and results are not affected. Table 1 below provides the simulated percentage improvements in risk of $\hat{\theta}^{S+}$ and $\hat{\theta}^{S}$ over $\hat{\theta}$.

It is clear from the above table that both shrinkage estimators have maximum risk gain over $\hat{\theta}$ at $\Delta^* = 0$ and the value of the improvement is a decreasing function of Δ^*. Thus, the performance of proposed estimator for small samples is similar as in the case of large samples. Hence, the simulation results validate the asymptotic results presented in this investigation. The proposed methodology can be used even for sample as small as 5. It may be noted that relationship between the efficiency and the sample size may not be monotone due to sampling variation. Tables for the simulated percentage improvements in risk of $\hat{\theta}^{S+}$ and $\hat{\theta}^{S}$ over $\hat{\theta}$ for some other values of q are prepared but not provided here to save space. However, the large sample analytical results derived in the previous sections are well supported by the simulation results presented here.

6. Final Comments

The Stein-type estimation strategies are asymptotically superior to a strategy based on sample information only. Further, the usual Stein-type estimator is asymptotically dominated by its positive part. However, we must stress that the important issue here is not only the improvement in the sense of lowering the risk by using the positive part of the $\hat{\theta}^{S}$. By considering positive part of $\hat{\theta}^{S}$, the resulting estimator $\hat{\theta}^{S+}$ removes the funny behavior of $\hat{\theta}^{S}$ when the test statistic takes the values near zero and it does not change the sign of the estimators. We recommend that the usual shrinkage estimator should be used as a tool for developing the positive part estimator and should not be used as an estimator in its own right. The large sample analytical analysis of the proposed estimation methodology is well supported by computational/simulated results presented by graphs and table. The computations for the figures and table have been carried out with FORTRAN programs.

Acknowledgments

The authors would like to thank the referee and editors for their helpful comments. The research work was partially supported by a grant from the Natural Sciences and Engineering Research Council of Canada.

References:

Ahmed, S. E. (2001). Shrinkage estimation of regression coefficients from censored data with multiple observations. *In Empirical Bayes and Likelihood Inference*, S. E. Ahmed and N. Reid (eds.), 103-120, Springer Verlag, NewYork.

Ahmed, S. E., Gupta, A. K., Khan, S. M. and Nicol, C. J. (2001). Simultaneous estimation of several intraclass correlation coefficients. *To appear in Annals of the Institute of Statistical Mathematics*.

Ahmed, S. E.(1992). Large-sample pooling procedure for correlation. *The Statistician*, **41**, 425-438.

Donner, A. (1986). A review of inference procedure for the intra class correlation coefficient in the one-way random effect model. *International Statistical Review*, **54**, 67-82.

Elston, R.C.(1975). Correlation between correlations. *Biometrika*, **62**, 133-148.

Kapata, R. S. (1993). A test of hypothesis on familial correlations. *Biometrics*, **49**, 569-576.

Kubokawa, T. (1998). The Stein phenomenon in simultaneous estimation: A review. *Nonparametric Statistics and Related Topics* In S.E. Ahmed et al. (Eds.), 143-173, Nova Science, NewYork.

Konishi, S. (1985). Normalizing and variance stabilizing transformations for intraclass correlations. *Ann. Inst. Statist. Math.*, **37**, 87-94.

Bayesian Analysis of Structural Equation Modeling

Kazuo Shigemasu, Takahiro Hoshino, and Takuya Ohmori

University of Tokyo
3-8-1 Komaba, Megroku, Tokyo 153-8902 Japan

Summary: A Bayesian procedure to make exact distributional inferences about all structural parameters and latent variables was proposed. This procedure handles the problem associated with the fixed parameters by means of conditinalization, and uses the Gibbs sampler to derive the posterior distribution for each unknown quantitiy. A simulation study was conducted to evaluate the performance of the proposed procedure.

1. Introduction

The purpose of this paper is to propose a method which makes exact inferences about all parameters in Structural Equation Models(SEMs). Here, in this paper, Bayesian approach is employed to make it possible to derive posterior distributions for all parameters, in terms of which exact inference can be made for all parameters and latent variables(factor scores) even for small samples. Moreover, in Bayesian analysis, you can include prior knowledge as prior distributions. The prior distributions play crucial role in dealing with underidentified models or models which have rather flat likelihood.

In fact, recently several papers have appeared which use Bayesian approach for Structural Equation Models (Arminger and Muthen,1998; Scheines, Hoijtink and Boomsma.1999; Shi and Lee,1997,1998,2000).

Scheins, Hoijtink and Boomsma(1999) used the Gibbs sampler to numerically derive the posterior distributions. Because the full conditional distributions derived in their paper do not belong to the standard distributions, the authors used rejection sampling. The prior distribution used is (truncated) normal distribution for all parameters.

Arminger and Muthén(1998) analyzed the LISREL type model in which latent variables are connected by nonlinear functions. They used Gibbs Sampler to derive posterior distributions for structural parameters but used the Metropolis-Hastings algorithm rather than Gibbs sampler for the latent variables because they did not derive a form of the conditional distribution from which values of latent variables can be sampled easily.

Shi and Lee (1997) estimated factor scores in the factor analysis model when the observed variables are polytomous, censored or truncated. The structural parameters are assumed known or substituted by their estimates. Shi and Lee (2000) obtained the ML estimates of factor scores in the same situation, using the Monte Carlo EM algorithm. Shi and Lee (1998) obtained estimates for parameters as well as the latent variables in the factor analysis model based on the continuous and polytomous data. They compared the performance of Bayesian estimates using three different kinds of prior distributions. The first type of prior distributions was determined empirically using ML estimates, while the second type was determined subjectively, and the third type is noninformative prior distributions.

The proposed method in this paper uses generalized conjugate priors (Press, 1982) and the likelihood includes factor scores so that we obtain the full conditional distributions analytically which belong to standard distributions. In other words, we do not integrate out latent variables to obtain the marginal likelihood. The advantages given by introducing generalized natural conjugate priors and full likelihood are (1) the hyperparameters are meaningful so that the assessment of their values are easier without any gimmicks, (2) programs of Gibbs sampler is straightforward and numerical evaluation of the posterior distributions is easier, and (3) we can make inferences for factor scores in the integrated fashion.

Typically, some of the regression parameters such as factor loading matrix in SEM model are fixed. The full conditional distributions used in Gibbs Sampler should be conditional on these fixed elements as well as the other parameters and latent variables. The proposed method handles this conditionality in a proper way making use of the rather simple form of the conditional distributions. This can be counted as the fourth advantage of our approach.

2. The Method

2.1 The SEM Model and Assumptions

SEMs can be written in various ways, and we use the following expression emphasizing the connection with the usual exploratory factor analysis model. The data matrix $X(n \times p)$ has n subjects(rows) and p variables(columns) and follows the following equation.

$$X = F\Lambda^t + E_1. \tag{1}$$

Putting $F = [F_1, F_2]$, the causal model for factor scores can be written as follow;

$$F_1 = F_1 B_0^t + F_2 \Gamma^t + E_2 \tag{2}$$

Further, we put the following distributional assumptions for each row of F_2 (f_{2i}), each row of E_1 (ε_{1i}), and each row of E_2 (ε_{2i}). That is, $f_{2i} \sim N(0, \Phi)$, $\varepsilon_{1i} \sim N(0, \Psi)$, and $\varepsilon_{2i} \sim N(0, \Omega)$. Defining $B = (I - B_0)$, the model distribution for X and F is as follows.

$p(X, F | \Lambda, B, \Gamma, \Phi, \Psi, \Omega)$

$$\propto |\Omega|^{-\frac{n}{2}} |\Psi|^{-\frac{n}{2}} |\Phi|^{-\frac{n}{2}} \exp\left\{-\frac{1}{2}\mathrm{tr}[(F_1 B^t - F_2 \Gamma^t)\Omega^{-1}(F_1 B^t - F_2 \Gamma^t)^t\right.$$

$$\left. + F_2 \Phi^{-1} F_2^t + (X - F\Lambda^t)\Psi^{-1}(X - F\Lambda^t)^t]\right\}$$

2.2 Prior Distributions

As for Φ, Ψ, and Ω, we use the inverse Wishart distributions rather than Jeffereys type non-informative distributions in order to make the resulting estimates more stable. But at the same time, we want to "let the data speak for themselves" as much as possible. Considering this objective, the degrees of freedom can be minimal (should be greater than $2p + 2$). The hyper parameter matrices (G_Φ, G_Ψ and G_Ω) are set to be diagonal. Note that the parameters (Φ, Ψ and Ω) are not assumed to be diagonal so that the model is more general. As has been pointed out often, these covariance matrices represent the covariation among the residual vectors rather then just measurement errors. Even though there should be a reason to believe that they are diagonal apriori when we set up the model, it is better to allow for some effects of unexpected factors and unexplained relationships. Thus, the prior distributions for Φ, Ψ and Ω are

$$p(\Phi) \;\propto\; |\Phi|^{-\nu_\Phi/2} \exp\left\{ -\frac{1}{2}\mathrm{tr}\Phi^{-1}G_\Phi \right\},$$

$$p(\Psi) \;\propto\; |\Psi|^{-\nu_\Psi/2} \exp\left\{ -\frac{1}{2}\mathrm{tr}\Psi^{-1}G_\Psi \right\}, and$$

$$p(\Omega) \;\propto\; |\Omega|^{-\nu_\Omega/2} \exp\left\{ -\frac{1}{2}\mathrm{tr}\Omega^{-1}G_\Omega \right\}.$$

These prior distributions are abbreviated as $W^{-1}(\nu_\Phi, G_\Phi)$, $W^{-1}(\nu_\Psi, G_\Psi)$ and $W^{-1}(\nu_\Omega, G_\Omega)$.

As for the other parameters Λ, B and Γ, we assume the uniform distributions for convenience.

2.3 Gibbs Sampling

Gibbs Sampling is the most popular method among many Markov Chain Monte Carlo Sampling Method and a special case of Metropolis-Hastings Algorithm(Geman and Geman, 1984; Gelfand and Smith, 1990; See Chen, Shao, and Ibraim 2000 for textbook level explanation.) A reason of this popularity is that it is easier to implement the procedure and see the structure of the program. Gibbs Sampling requires the full conditional distributions from which we can generate random numbers. Suppose that the relevant parameter vector $\theta(d \times 1)$, then we need the full conditional distributions:
$p(\theta_1|\theta_2, \cdots, \theta_d), p(\theta_2|\theta_1, \theta_3, \cdots, \theta_d), \cdots, p(\theta_d|\theta_1, \cdots, \theta_{d-1})$.

We generate the random numbers regarding the other parameters are fixed at the previous cycle.

When we deal with more complicated model, it makes easier to program if we factor the parameter vector into a number of meaningful subsets.

Shigemasu and Nakamura(1993) applied Gibbs sampling to derive posterior distribution for the parameters of factor analysis model. Their key idea was to derive conditional distributions of the parameters always including factor scores. By doing so, we can use standard distributions as conditional distributions. Here we apply the same kind of technique, and the factoring of parameters and the full conditional distributions for each of a set of parameters will be given in the next subsection.

2.4 Conditional Distributions

The full conditional distributions for $\Phi, \Psi, and \Omega$ given are $W^{-1}(\nu_\Phi + n, G_\Phi + F_2^t F_2)$, $W^{-1}(\nu_\Psi + n, G_\Psi + (X - F\Lambda^t)^t(X - F\Lambda^t))$, $W^{-1}(\nu_\Omega + n, (F_1 B^t - F_2\Gamma^t)^t(F_1 B^t - F_2\Gamma^t))$, respectively. Also, the conditional distributions for F_1 and F_2 are the following matrix normal distributions. [1]

$$
\begin{aligned}
F_1 &\sim N\left\{[(X - F_2\Lambda_2^t)\Psi^{-1}\Lambda_1 + F_2\Gamma^t\Omega^{-1}(B^{-1})^t]C_1^{-1}, \ I_n \otimes C_1^{-1}, \right\} \\
F_2 &\sim N\left\{[(X - F_1\Lambda_1^t)\Psi^{-1}\Lambda_2 + F_1 B^t\Omega^{-1}\Gamma]C_2^{-1}, \ I_n \otimes C_2^{-1}, \right\}
\end{aligned}
$$

where
$$
\begin{aligned}
C_1 &= \Lambda_1^t\Psi^{-1}\Lambda_1 + (B^{-1})^t\Omega^{-1}B^{-1}, \\
C_2 &= \Phi + \Lambda_2^t\Psi^{-1}\Lambda_2 + \Gamma^t\Omega^{-1}\Gamma
\end{aligned}
$$

The derivation of posterior distributions for Λ, B and Γ needs more careful handling, because some of their elements are fixed to be zero. The parameter matrices of Λ, B and Γ have the following density functions;

$$
p(\Lambda|X, F, B, \Gamma, \Phi, \Psi, \Omega)
$$
$$
\propto \ \exp\left\{-\frac{1}{2}\mathrm{tr}\Psi^{-1}(\Lambda - X^t F(F^t F)^{-1})(F^t F)(\Lambda - X^t F(F^t F)^{-1})^t\right\},
$$

$$
p(B|X, F, \Lambda, \Gamma, \Phi, \Psi, \Omega)
$$
$$
\propto \ \exp\left\{-\frac{1}{2}\mathrm{tr}\Omega^{-1}(B - \Gamma F_2^t F_1(F_1^t F_1)^{-1})(F_1^t F_1)(B - \Gamma F_2^t F_1(F_1^t F_1)^{-1})^t\right\}
$$

and
$$
p(\Gamma|X, F, \Lambda, B, \Phi, \Psi, \Omega)
$$
$$
\propto \ \exp\left\{-\frac{1}{2}\mathrm{tr}\Omega^{-1}(\Gamma - B F_1^t F_2(F_2^t F_2)^{-1})(F_2^t F_2)(\Gamma - B F_1^t F_2(F_2^t F_2)^{-1}\right\}.
$$

---

[1] The matrix normal distribution is a conventional notation for the multivariate normal distribution. The notation used here is $N_{n,p}(B, C \otimes D)$ for the (n, p) matrix X, when it has the following density function.

$$
p(X|B, C \otimes D)
$$
$$
\propto \ |C|^{-\frac{n}{2}}|D|^{-\frac{p}{2}}\exp\left\{-\frac{1}{2}\mathrm{tr}[C^{-1}(X - B)D^{-1}(X - B)^t].\right\}
$$

In this case, $x = \mathrm{vec}(X^t)$ distributes as $N_{np}(\mathrm{vec}(B), C \otimes D)$.

These densities appear to suggest the matrix normal distributions for
$N(X^t F(F^t F)^{-1}, \Psi \otimes (F^t F)^{-1})$, $N(\Gamma F_2^t F_1(F_1^t F_1)^{-1}, \Omega \otimes (F_1^t F_1)^{-1})$
and $N(B F_1^t F_2(F_2^t F_2)^{-1}, \Omega \otimes (F_2^t F_2)^{-1})$. But what is needed is the density
function of non-zero elements of the parameter matrices. Let vec mean the
operator which transforms a matrix into a vector by stacking the columns of
the matrix. Thus, $\lambda = vec(\Lambda^t)$, $\beta = vec(B^t)$ and $\gamma = vec(\Gamma^t)$. Also, for
convenience, let $\bar{\lambda} = vec[(X^t F(F^t F)^{-1})^t]$, $\bar{\beta} = vec[(\Gamma F_2^t F_1(F_1^t F_1)^{-1})^t)]$
and $\bar{\gamma} = vec[(B F_1^t F_2(F_2^t F_2)^{-1})^t]$.

With these vector notations, the densities of relevant conditional distributions are given as

$p(\lambda | X$ and other parameters$)$

$$\propto \quad \exp\left\{-\frac{1}{2}(\lambda - \bar{\lambda})^t(\Psi \otimes (F^t F)^{-1})^{-1}(\lambda - \bar{\lambda})\right\},$$

$p(\beta | X$ and other parameters$)$

$$\propto \quad \exp\left\{-\frac{1}{2}(\beta - \bar{\beta})^t(\Omega \otimes (F_1^t F_1)^{-1})^{-1}(\beta - \bar{\beta})\right\}, and$$

$p(\gamma | X$ and other parameters$)$

$$\propto \quad \exp\left\{-\frac{1}{2}(\gamma - \bar{\gamma})^t(\Omega \otimes (F_2^t F_2)^{-1})^{-1}(\gamma - \bar{\gamma})\right\}.$$

Now let the permutation matrix K_λ which changes the location of the
elements of λ so that the non-zero elements of the vector (λ^*) at the top and
the zero elements at the bottom.

That is,

$$\begin{pmatrix} \lambda^* \\ 0 \end{pmatrix} = K_\lambda \lambda.$$

Similarly

$$\begin{pmatrix} \beta^* \\ 0 \end{pmatrix} = K_\beta \beta, \quad and \quad \begin{pmatrix} \gamma^* \\ 0 \end{pmatrix} = K_\gamma \gamma.$$

For simplicity, we write $\bar{\lambda}^* = K_\lambda \bar{\lambda}, \bar{\beta}^* = K_\beta \bar{\beta}, \bar{\gamma}^* = K_\gamma \bar{\gamma}, \Sigma_\lambda^{-1} = K_\lambda(\Psi \otimes (F^t F)^{-1})^{-1} K_\lambda^t, \Sigma_\beta^{-1} = K_\beta(\Omega \otimes (F_1^t F_1)^{-1})^{-1} K_\beta^t$, and $\Sigma_\gamma^{-1} = K_\gamma(\Omega \otimes (F_2^t F_2)^{-1})^{-1} K_\gamma^t$.

Thus, the conditional distributions for λ^* with fixed zero elements is as
follows,

$p(\lambda^* | X$ and other parameters$)$

$$\propto \quad \exp\left\{-\frac{1}{2}\left[(\lambda^* - \bar{\lambda}_1^*)^t \Sigma_{11\lambda}(\lambda^* - \bar{\lambda}_1^*) - 2\bar{\lambda}_2^{*t}\Sigma_{21\lambda}(\lambda^* - \bar{\lambda}_1^*)\right]\right\}$$

$$= \quad \exp\left\{-\frac{1}{2}(\lambda^* - \bar{\lambda}_c^*)^t \Sigma_{\lambda_c}^{-1}(\lambda^* - \bar{\lambda}_c^*)\right\}$$

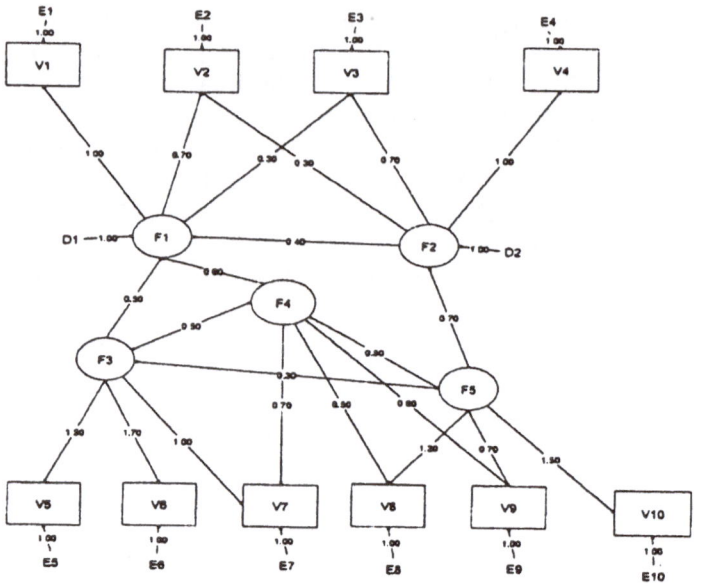

Figure 1:Structural Equation Model

where $\lambda_c^* = \bar{\lambda}_1^* - \Sigma_{12\lambda}\Sigma_{22\lambda}^{-1}\bar{\lambda}_2^*$, $\Sigma_{\lambda c} = \Sigma_{11\lambda} - \Sigma_{12\lambda}\Sigma_{22\lambda}^{-1}\Sigma_{21\lambda}$, $\Sigma_\lambda = \begin{bmatrix} \Sigma_{11\lambda} & \Sigma_{12\lambda} \\ \Sigma_{21\lambda} & \Sigma_{22\lambda} \end{bmatrix}$, $\bar{\lambda}^* = \begin{pmatrix} \bar{\lambda}_1^* \\ \bar{\lambda}_2^* \end{pmatrix}$.

The conditional distribution for λ^* is $N(\bar{\lambda}_c^*, \Sigma_{\lambda c})$. Similarly,

$$\beta^* \sim N(\bar{\beta}_c^*, \Sigma_{\beta,c}), \quad \bar{\beta}_c^* = \bar{\beta}_1^* - \Sigma_{12\beta}\Sigma_{22\beta}^{-1}\bar{\beta}_2^*, \quad \Sigma_{\beta c} = \Sigma_{11\beta} - \Sigma_{12\beta}\Sigma_{22\beta}^{-1}\Sigma_{21\beta},$$

$$\gamma^* \sim N(\bar{\gamma}_c^*, \Sigma_{\gamma,c}), \quad \bar{\gamma}_c^* = \bar{\gamma}_1^* - \Sigma_{12\gamma}\Sigma_{22\gamma}^{-1}\bar{\gamma}_2^*, \quad \Sigma_{\gamma c} = \Sigma_{11\gamma} - \Sigma_{12\gamma}\Sigma_{22\gamma}^{-1}\Sigma_{21\gamma}$$

The above formulas appear notationally formidable, but what are actually required is only the relocation of the elements of relevant vectors or matrices.

3. Simulation

3.1 Method of Numerical Experiment

The Bayesian estimates were obtained for simulated data, which were generated according to the model described in Figure 1. Two kinds of simulated data were generated and one assumed 100 subjects and the other assumed 200 subjects. Each has 100 sets of data.

Using Gibbs Sampling technique, we obtained the posterior distributions and hence the estimates of the parameters and their standard deviations. More concretely, Gibbs sampling started with the classical estimates as initial values for Λ and F, and the initial values for Ψ, Ψ and Ω were set to equal to

Table 1-a : Comparison of Bayesian Estimation and CALIS(n = 100)

Parameters			Squared Loss		Standard Deviation	
TRUE	BAYES	CALIS	BAYES	CALIS	BAYES	CALIS
λ_{21} 0.7	0.6882	0.6885	0.3951	0.4415	0.0721	0.0676
λ_{22} 0.3	0.3259	0.3177	1.1742	1.1006	0.1048	0.0941
λ_{31} 0.3	0.2967	0.3032	0.3109	0.3483	0.0683	0.0645
λ_{32} 0.7	0.7299	0.7084	0.8042	0.9025	0.1148	0.1065
λ_{53} 1.5	1.5536	1.4984	2.0739	1.6243	0.1206	0.1215
λ_{63} 1.7	1.7548	1.6954	2.0338	1.5291	0.1328	0.1343
λ_{73} 1.0	1.0370	1.0024	1.7420	1.4349	0.1749	0.1273
λ_{74} 0.7	0.7114	0.6843	1.2240	1.0643	0.1408	0.1100
λ_{84} 0.5	0.5018	0.5200	1.5764	1.9392	0.1392	0.1193
λ_{85} 1.2	1.2584	1.1819	2.4771	2.5919	0.1574	0.1402
λ_{94} 0.8	0.7975	0.7980	1.3764	1.7554	0.1512	0.1264
λ_{95} 0.7	0.7465	0.6919	1.6821	2.0226	0.1744	0.1357
λ_{105} 1.5	1.5182	1.4851	1.6717	1.5470	0.1259	0.1243
ϕ_{12} 0.5	0.5103	0.4980	0.7795	0.8041	0.1235	0.1062
ϕ_{13} 0.3	0.2871	0.2764	0.9876	1.0765	0.1019	0.0995
ϕ_{23} 0.5	0.4996	0.4868	1.0668	1.1846	0.1322	0.1153
ψ_1 0.2	0.2185	0.1948	0.3027	0.4285	0.0602	0.0595
ψ_2 0.2	0.2111	0.1999	0.1101	0.0977	0.0421	0.0385
ψ_3 0.2	0.2093	0.1979	0.1844	0.2009	0.0448	0.0419
ψ_4 0.2	0.2359	0.2007	0.4158	0.4594	0.0644	0.0651
ψ_5 0.3	0.3092	0.2942	0.3992	0.5050	0.0778	0.0755
ψ_6 0.3	0.3165	0.2926	0.7079	0.9438	0.0919	0.0906
ψ_7 0.3	0.3048	0.2836	0.3797	0.5262	0.0782	0.0707
ψ_8 0.3	0.2883	0.2820	0.4766	0.6920	0.0741	0.0721
ψ_9 0.3	0.3202	0.2919	0.4772	0.5829	0.0792	0.0734
ψ_{10} 0.3	0.3514	0.2812	1.3085	2.1130	0.1158	0.1200
γ_{11} 0.5	0.5019	0.4856	1.5441	1.5625	0.1428	0.1069
γ_{21} 0.6	0.6233	0.6161	1.8365	1.8035	0.1380	0.1146
γ_{32} 0.7	0.7033	0.6905	0.8272	0.7870	0.0915	0.0880
ω_1 0.2	0.1819	0.1680	0.3913	0.8250	0.0695	0.0720
ω_2 0.3	0.2918	0.3030	0.4629	0.5994	0.0768	0.0779
β_{21} -0.4	-0.4339	-0.4111	1.2538	1.3649	0.1300	0.1156
			32.6536	34.7587		

their prior expected values. The hyperparameters G_Φ, G_Ψ and G_Ω are set to be $0.2I \times$(degree of freedom), which is $2p + 3$. The first 500 random draws in the Gibbs run were discarded as a "burn-in" and 5000 draws were used to derive posterior distribution numerically. The Bayesian estimates were obtained as EAP (Expected A Posteori) estimates . The Bayesian estimates are compared with the traditional ML estimates using the CALIS program in SAS in terms of Mean Squared Error of Estimation.

4. Results

Table 1-a shows the Bayesian estimates and the ML estimates by CALIS for n=100 data and Table 1-b shows the results for n=200 data. The numbers shown in Table 1 are true values of simulation data, the mean of estimates of Bayes and ML estimates, Mean Squared Error of Estimation, Bayes posterior standard deviations and traditional standard errors of ML estimates. Figures 2 (a,b,c) show the posterior distribution for some selected parameters. If necessary, we can do any kind of statistical inference for relevant parameters, for example, credibility interval estimation and Bayesian posterior p-value check (Gelman, Meng and Stern, 1996). Convergence was checked by BOA (Smith, 2000), and all parameters were judged to reach convergence. For example, Gelman and Rubin's criterion (see, Gelman and Rubin, 1992) was below 1.2, which is their suggested value for convergence (See Figure 3). Figure 3 shows the shrink factor (solid line) for a parameter, and its upper

Table 1-b : Comparison of Bayesian Estimation and CALIS (n = 200)

Parameters			Squared Loss		Standard Deviation	
TRUE	BAYES	CALIS	BAYES	CALIS	BAYES	CALIS
λ_{21} 0.7	0.7057	0.7041	0.2541	0.2814	0.0511	0.0494
λ_{22} 0.3	0.3008	0.2998	0.5053	0.5319	0.0702	0.0669
λ_{31} 0.3	0.2895	0.2939	0.2222	0.2277	0.0484	0.0470
λ_{32} 0.7	0.7223	0.7099	0.5872	0.6249	0.0789	0.0764
λ_{53} 1.5	1.5089	1.4829	0.6933	0.6944	0.0822	0.0852
λ_{63} 1.7	1.7038	1.6752	0.8681	0.8681	0.0906	0.0941
λ_{73} 1.0	0.9993	0.9866	0.7279	0.6890	0.1030	0.0877
λ_{74} 0.7	0.7045	0.6888	0.7172	0.6893	0.0846	0.0777
λ_{84} 0.5	0.4929	0.5041	0.7279	0.7514	0.0921	0.0853
λ_{85} 1.2	1.2168	1.1789	1.0093	1.0977	0.1040	0.0994
λ_{94} 0.8	0.8034	0.8024	0.7577	0.7637	0.0971	0.0908
λ_{95} 0.7	0.7010	0.6799	0.8189	0.8800	0.1102	0.0965
λ_{105} 1.5	1.5077	1.4889	0.7763	0.7776	0.0865	0.0882
ϕ_{12} 0.5	0.4946	0.4946	0.4581	0.5120	0.0829	0.0744
ϕ_{13} 0.3	0.2900	0.2888	0.4883	0.5300	0.0714	0.0704
ϕ_{23} 0.5	0.5013	0.4973	0.6552	0.6583	0.0901	0.0814
ψ_1 0.2	0.2165	0.2015	0.1704	0.1782	0.0434	0.0494
ψ_2 0.2	0.2032	0.1938	0.0748	0.0839	0.0286	0.0669
ψ_3 0.2	0.2043	0.1975	0.0759	0.0803	0.0307	0.0470
ψ_4 0.2	0.2207	0.2018	0.2090	0.2221	0.0459	0.0764
ψ_5 0.3	0.3051	0.2955	0.0858	0.2648	0.0555	0.0852
ψ_6 0.3	0.3111	0.2978	0.3770	0.4002	0.0661	0.0941
ψ_7 0.3	0.3085	0.2952	0.0240	0.2166	0.0556	0.0877
ψ_8 0.3	0.3063	0.3044	0.2615	0.3146	0.0539	0.0777
ψ_9 0.3	0.3102	0.2938	0.2952	0.2996	0.0564	0.0859
ψ_{10} 0.3	0.3298	0.2888	0.6358	0.8702	0.0855	0.0994
γ_{11} 0.5	0.4799	0.4746	0.5135	0.5311	0.0841	0.0728
γ_{21} 0.6	0.6120	0.6008	0.5743	0.5619	0.0846	0.0802
γ_{32} 0.7	0.7027	0.6959	0.3871	0.3901	0.0628	0.0622
ω_1 0.2	0.1933	0.1906	0.2471	0.3064	0.0525	0.0516
ω_2 0.3	0.2932	0.2967	0.2819	0.3221	0.0546	0.0546
β_{21} -0.4	-0.4141	-0.4077	1.0407	1.3649	0.0893	0.0828
			15.5210	16.9844		

bpund (dotted line) of the 95% interval.

In terms of Mean Squared Error, Bayesian method performs a little better, but the difference is minimal. If you compare the performance of the Bayesian estimates with that of traditional estimates only in terms of the Mean Squared Error, there may not be apparent reason to employ Bayesian approach. But we again emphasize that it provides the posterior distributions for any parameters and latent variables, which can be used for any kind of statistical reasoning. Also, it may be noteworthy that Bayesian estimates of standard errors were generally larger than the approximate standard errors of ML estimates. Scheines, Hoijtink and Boomsma (1999) found the similar result. This should be investigated further, but note that estimates of the standard errors based on Fisher information matrix only provide the lower bounds by Cramer-Rao inequality. It has not been clear about their behavior for the parameters of the complex statistical models for moderate or small samples.

We believe that the proposed method is simple and straightforward compared to the preceding studies, and therefore it is easy to generalize it to the polytomous data, generalized regression type of structural relationships, and the like.

Figure 2-a : Posterior Distribution Figure 2-b : Posterior Distribution
for λ_{21} for γ_{32}

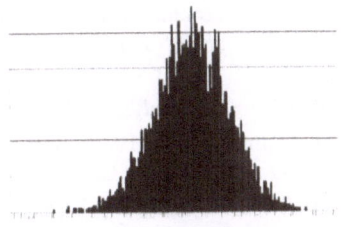

Figure 2-c : Posterior Distribution for ϕ_{13}

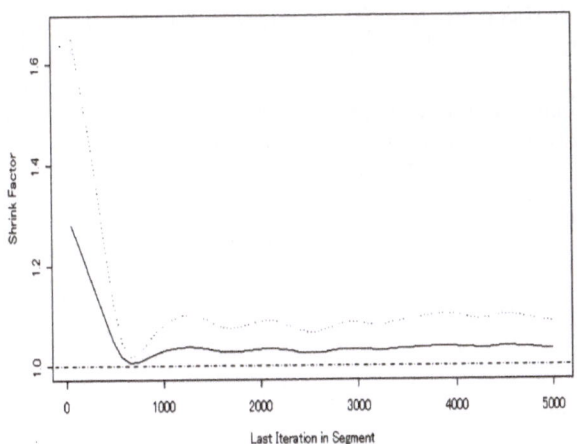

Figure 3 : Gelman & Rubin Srink Factors for λ_{23}

References:

Arminger, G., & Muthén, B.O. (1998). A Bayesian approach to nonlinear latent variable models using the Gibbs sampler and the Metropolis-Hasting algorithm. *Psychometrika, 63*, 271–300.

Chen. M-H., Shao, Q-M., & Ibrahim, J.G. (2000), *Monte Carlo methods in Bayesian computation*. New York: Springer-Verlag.

Gelfand, A.E., & Smith, A.F.M. (1990), Sampling-based approaches to calculating marginal densities, *Journal of the American Statistical Association, 85*, 398–409.

Gelman, A., & Rubin, D.B. (1992), Inference from iterative simulation using multiple sequences (with discussion). *Statistical Science, 7*, 457-511.

Gelman, A., Meng, X.-L., & Stern, H.S. (1996). Posterior predictive assessment of model fitness via realized discrepancies(with discussion). *Statistica Sinica, 6*, 733–807.

Geman, S., & Geman, D. (1984). Stochastic relaxation,Gibbs distribution and the Bayesian restoration of images. *IEEE Transactions on Pattern Analysis and Machine Intelligence, 6*, 721–741.

Press, S.J. (1982), *Applied multivariate analysis*. Florida: Krieger Publishing.

Scheines, R., Hoijtink, H., & Boomsma, A. (1999). Bayesian estimation and testing of structural equation models. *Psychometrika, 64*, 37–52.

Shi, J-Q., & Lee, S-Y. (1998). Bayesian sampling-based approach for factor analysis models with continuous and polytomous data. *British Journal of Mathematical and Statistical Psychology, 51*, 233-252.

Shi, J-Q., & Lee, S-Y., (1997). A Bayesian estimation of factor score in confirmatory factor model with polytomous, censored or truncated data. *Psychometrika, 62*, 29-50.

Shi, J-Q., & Lee, S-Y., (2000). Latent variable models with mixed continuous and polytomous data. *Journal of the Royal statistical Society, Series B, 62*, 77-87.

Shigemasu,K & Nakamura,T(1993) A Bayesian Numerical Estimation Procedure in Factor Analysis Model. *E.S.T. Research Report, 93-6*, Tokyo Institute of Technology.

Smith,B.J. (2000) *Bayesian Output Analysis Program(BOA) Version 0.5.0 User-manual (http://www.public-health.uiowa.edu/BOA)*.

Estimation of the Fifth Percentile Using a Subset of Order Statistics [1]

Kōsei Iwase[1] and Koji Kanefuji[2]

[1] Hiroshima University
1-4-1 Kagamiyama, Higashi-Hiroshima
Hiroshima 739-8527, JAPAN
[2] The Institute of Statistical Mathematics
4-6-7, Minami-Azabu, Minato-ku
Tokyo 106-8569, JAPAN

Summary: Estimation of the fifth percentile using a subset of order statistics is closely related to the water quality standards in the United States. We focus on the subsample size to estimate it. Because the possibility and consequences of skewness of the original data give a good reason to restrict the subsample size, the subsample size of 2 is recommended in this paper. We also evaluate the estimator of a Final Acute Value (FAV) analytically, based on the subsample size of 2 closest to 0.05.

1. Introduction

Calculation of the Final Acute Value (FAV) is an important part of the procedure for deriving water quality criteria for aquatic organisms. On June 1988, the U.S. Environmental Protection Agency (EPA) published "Erickson and Stephan, Calculation of the final acute value for water quality criteria for aquatic organisms". The FAV is defined to be lower than all except a small fraction of the Mean Acute Values (MAVs) that are available for the materials (COPPER, DDT, CADMIUM, DIELDRIN, etc.). The fraction was set at 0.05.

Firstly, we introduce the procedure for determining the water quality guidelines in the United States, which are described in the EPA. Next, we consider the distribution of the MAVs and simulate the estimator of the FAV to select appropriate subsample sizes. Finally, we evaluate the estimator of the FAV in the case of subsample size of 2.

2. EPA Procedure

The general instructions of the recommended procedure for FAV calculation are presented in this section. The FAV is a concentration of a material derived from an appropriate set of MAVs. The presented procedure for FAV calculation is as follows:

---

[1]This study was supported by Grant-in-Aid for Scientific Research (C)(1) 13680378 and ISM Cooperative Research Program(2001-ISM-CRP-2064).

1. Based on data set size (n), determine for ranks (r) with four cumulative probabilities $(P_r = r/(n + 1))$ closest to 0.05; for $n < 59$, this will be $r = 1$ through 4; for $60 \leq n < 79$, $r = 2$ through 5; for $80 \leq n < 99$, $r = 3$ through 6; etc.

2. From the data set, select the four MAVs with the desired ranks and calculate P_r for each of these MAVs.

3. Fit a line to $\ln(MAV)$ vs. $\sqrt{P_r}$ using the following equations for slope $(\hat{S})$ and intercept $(\hat{L})$ and calculate the $\widehat{FAV}$:

$$\hat{S} = \sqrt{\frac{\sum^4 (\ln MAV)^2 - (\sum^4 \ln MAV)^2/4}{\sum^4 P_r - (\sum^4 \sqrt{P_r})^2/4}},$$

$$\hat{L} = (\sum^4 \ln MAV - \hat{S} \sum^4 \sqrt{P_r})/4, \qquad \widehat{FAV} = \exp[\hat{S}\sqrt{0.05} + \hat{L}].$$

EPA guidelines used ten thousand generated random samples from four two-parameter distributions (Laplace, normal, uniform, and triangular distributions) to estimate the fifth percentile of these distributions for each combination of the following sample sizes and subsample sizes. Sample sizes (n) of 8, 15, and 30 and subsample sizes (k) of 4, 8, 15, and 30 were selected. Based on the simulations, EPA guidelines recommended the four points to estimate the FAV. Because there was no substantial advantage with respect to the precision, by using subsample sizes greater than 4 in EPA guidelines.

3. Distributions

Calculation of a FAV from a typically small set of MAVs requires that the set be considered a sample from a statistical population and that the FAV be considered an estimate of the fifth percentile of its population. Methods for estimating the fifth percentile of a population require at least some assumptions about the distributional characteristics of the population. For example, all data sets in the EPA documents are highly positively skewed and estimation would be benefited by the logarithmic transformation of MAVs.

Twenty data sets for freshwater species and seventeen for saltwater species were considered to be acceptable for the purposes of EPA guidelines because they contained Species Mean Acute Values (SMAVs) from at least eight families in a variety of taxonomic and functional groups. These data sets contain from 8 to 45 SMAVs for a variety of organic and inorganic materials.

In this paper, five distributions of X are introduced by the relation that $\frac{\ln(X/\mu)}{c}$ obeys the five standardized symmetrical distributions in the section 5, where $\ln(x)$ means the natural logarithm of x, $c > 0$, and $\mu > 0$. $\frac{\ln(X/\mu)}{c} \sim N(0, 1)$ corresponds to the standardized transformation $\frac{X-\mu}{\sigma} \sim N(0, 1)$.

Table 1: Some characteristic values corresponding to FAV of normal distribution

Sample sizes (n)	Subsample sizes (k)								
	2	3	4	5	6	7	8	9	10
Mean[$\widehat{FAV}$]									
10	-1.880	-1.870	-1.867	-1.856	-1.857	-1.867	-1.880	-1.946	-2.061
20	-1.828	-1.847	-1.847	-1.831	-1.820	-1.813	-1.806	-1.795	-1.794
30	-1.790	-1.805	-1.826	-1.818	-1.810	-1.816	-1.795	-1.795	-1.786
40	-1.743	-1.767	-1.780	-1.790	-1.801	-1.798	-1.799	-1.793	-1.793
50	-1.726	-1.728	-1.742	-1.765	-1.779	-1.783	-1.788	-1.787	-1.791
Variance[$\widehat{FAV}$]									
10	0.64	0.54	0.47	0.42	0.40	0.38	0.35	0.34	0.34
20	0.26	0.26	0.25	0.24	0.22	0.21	0.21	0.20	0.20
30	0.15	0.16	0.15	0.15	0.15	0.15	0.14	0.14	0.14
40	0.12	0.10	0.11	0.11	0.11	0.11	0.11	0.11	0.10
50	0.09	0.09	0.08	0.08	0.08	0.08	0.08	0.08	0.08
Cdf[Mean[$\widehat{FAV}$]]									
10	0.0301	0.0307	0.0310	0.0317	0.0316	0.0310	0.0301	0.0258	0.0197
20	0.0338	0.0324	0.0323	0.0336	0.0344	0.0349	0.0354	0.0363	0.0364
30	0.0368	0.0356	0.0339	0.0345	0.0352	0.0347	0.0363	0.0364	0.0370
40	0.0406	0.0386	0.0375	0.0367	0.0359	0.0361	0.0360	0.0365	0.0364
50	0.0422	0.0420	0.0408	0.0388	0.0376	0.0373	0.0369	0.0370	0.0367
Mean[Cdf[$\widehat{FAV}$]]									
10	0.0651	0.0625	0.0598	0.0581	0.0566	0.0551	0.0523	0.0457	0.0368
20	0.0504	0.0491	0.0484	0.0490	0.0494	0.0495	0.0497	0.0503	0.0501
30	0.0469	0.0462	0.0439	0.0449	0.0455	0.0447	0.0463	0.0461	0.0466
40	0.0497	0.0460	0.0451	0.0442	0.0435	0.0435	0.0435	0.0439	0.0436
50	0.0488	0.0486	0.0468	0.0447	0.0434	0.0432	0.0427	0.0427	0.0424

4. Selection of the Subsample Size

As mentioned in the section 2, the procedure for FAV calculation in EPA guideline uses only the four points with P_r closest to 0.05. In this report, n and k denote the sample size and the number of a subset of the data near the fifth percentile, respectively.

In combination of the sample size of n and the subsample size of k, the averages of estimated FAV over the 10,000 simulations are shown in Tables 1, 2, 3, 4, and 5 for the standardized normal, logistic, uniform, Laplace, and triangular distributions, which have mean zero and variance one. The subsample size of $k = 2(1)10$ closest to the fifth percentile from the sample size of $n = 10(10)50$ are considered. Other characteristics (Variance, cumulative probability of average of the FAV, and its mean of cumulative probability of the FAV) of the estimate of the FAV are also presented in these Tables.

A small subsample size has the advantage of reducing the effects of derivation from the assumed distribution and a large subsample size has the advantage of reducing the variance of estimates when the distribution assumptions are correct. In these five tables the means of cumulative probabilities of estimates for the FAV in the cases of $k = 2$ and $n = 20(10)50$ are appropriate for the FAV. EPA guidelines proposed that $k = 4$. From these simulation results it is proposed that k is equal to 2 for a practical point of view.

Table 2: Some characteristic values corresponding to FAV of logistic distribution

Sample sizes (n)	Subsample sizes (k)								
	2	3	4	5	6	7	8	9	10
	Mean[$\widehat{FAV}$]								
10	-1.951	-1.929	-1.870	-1.865	-1.830	-1.835	-1.837	-1.898	-2.043
20	-1.933	-1.940	-1.924	-1.894	-1.881	-1.854	-1.838	-1.823	-1.805
30	-1.847	-1.879	-1.895	-1.893	-1.885	-1.873	-1.865	-1.854	-1.850
40	-1.766	-1.816	-1.843	-1.862	-1.868	-1.872	-1.863	-1.869	-1.854
50	-1.737	-1.758	-1.785	-1.819	-1.838	-1.853	-1.863	-1.845	-1.852
	Variance[$\widehat{FAV}$]								
10	1.09	0.85	0.71	0.64	0.57	0.51	0.47	0.45	0.44
20	0.50	0.47	0.45	0.40	0.38	0.35	0.34	0.31	0.29
30	0.24	0.27	0.28	0.28	0.27	0.26	0.25	0.24	0.23
40	0.19	0.17	0.18	0.19	0.19	0.19	0.19	0.19	0.18
50	0.14	0.14	0.13	0.14	0.14	0.15	0.15	0.15	0.15
	Cdf[Mean[$\widehat{FAV}$]]								
10	0.0282	0.0294	0.0326	0.0328	0.0349	0.0346	0.0345	0.0310	0.0240
20	0.0291	0.0288	0.0296	0.0312	0.0319	0.0335	0.0344	0.0354	0.0365
30	0.0339	0.0321	0.0312	0.0313	0.0317	0.0324	0.0328	0.0335	0.0337
40	0.0391	0.0358	0.0341	0.0330	0.0327	0.0324	0.0330	0.0326	0.0335
50	0.0411	0.0396	0.0378	0.0356	0.0344	0.0335	0.0330	0.0340	0.0336
	Mean[Cdf[$\widehat{FAV}$]]								
10	0.0674	0.0640	0.0642	0.0621	0.0627	0.0594	0.0580	0.0518	0.0406
20	0.0487	0.0476	0.0480	0.0487	0.0489	0.0501	0.0505	0.0508	0.0514
30	0.0455	0.0441	0.0434	0.0434	0.0437	0.0443	0.0443	0.0450	0.0446
40	0.0494	0.0445	0.0430	0.0417	0.0413	0.0413	0.0418	0.0414	0.0422
50	0.0489	0.0474	0.0447	0.0426	0.0415	0.0407	0.0400	0.0413	0.0406

5. Exact Evaluation for Estimator of the FAV

Another advantage of using $k = 2$ is that we can evaluate the estimator of the FAV analytically.

5.1 Standard Normal Distribution

Let $X_1, X_2, \ldots, X_n$ be a random sample from a population with pdf $f(x)$. Further, let $X_{1:n} \leq X_{2:n} \leq \cdots \leq X_{n:n}$ be the order statistics obtained from this sample.

Let us consider the standard normal population with pdf

$$f(x) = \frac{1}{\sqrt{2\pi}} e^{-\frac{x^2}{2}}, \qquad -\infty < x < \infty,$$

and cdf

$$F[x] = \Phi(x).$$

From the definition we have the moments of $X_{i:n}$ as

$$\mu_{i:n} = E[X_{i:n}] = \frac{n!}{(i-1)!(n-i)!} \int_{-\infty}^{\infty} x[\Phi(x)]^{i-1}[1 - \Phi(x)]^{n-i} f(x) dx,$$

where $(1 \leq i \leq n)$.

Bose and Gupya(1959) derived exact explicit expressions for the mean of order statistics for sample sizes up to 5. In the case of large sample sizes, it

Table 3: Some characteristic values corresponding to FAV of uniform distribution

Sample sizes (n)	Subsample sizes (k)								
	2	3	4	5	6	7	8	9	10
					Mean[$\widehat{FAV}$]				
10	-1.615	-1.682	-1.746	-1.800	-1.855	-1.914	-1.971	-2.034	-2.093
20	-1.557	-1.572	-1.589	-1.607	-1.628	-1.642	-1.669	-1.693	-1.716
30	-1.556	-1.553	-1.556	-1.561	-1.572	-1.580	-1.589	-1.600	-1.614
40	-1.559	-1.551	-1.550	-1.548	-1.552	-1.555	-1.562	-1.566	-1.573
50	-1.558	-1.557	-1.550	-1.547	-1.545	-1.547	-1.548	-1.550	-1.555
					Variance[$\widehat{FAV}$]				
10	0.12	0.13	0.14	0.15	0.16	0.18	0.19	0.21	0.22
20	0.02	0.03	0.03	0.03	0.03	0.04	0.04	0.05	0.05
30	0.01	0.01	0.01	0.01	0.01	0.02	0.02	0.02	0.02
40	0.01	0.01	0.01	0.01	0.01	0.01	0.01	0.01	0.01
50	0.01	0.01	0.01	0.01	0.01	0.01	0.01	0.01	0.01
					Cdf[Mean[$\widehat{FAV}$]]				
10	0.0338	0.0145	0.0000	0.0000	0.0000	0.0000	0.0000	0.0000	0.0000
20	0.0506	0.0463	0.0413	0.0361	0.0300	0.0261	0.0183	0.0112	0.0046
30	0.0509	0.0515	0.0507	0.0493	0.0461	0.0439	0.0412	0.0381	0.0342
40	0.0499	0.0524	0.0527	0.0530	0.0521	0.0510	0.0491	0.0480	0.0460
50	0.0501	0.0505	0.0526	0.0534	0.0540	0.0535	0.0530	0.0525	0.0512
					Mean[Cdf[$\widehat{FAV}$]]				
10	0.0557	0.0456	0.0386	0.0340	0.0303	0.0271	0.0239	0.0223	0.0193
20	0.0506	0.0471	0.0438	0.0405	0.0372	0.0364	0.0325	0.0299	0.0276
30	0.0509	0.0515	0.0507	0.0494	0.0464	0.0447	0.0426	0.0404	0.0376
40	0.0499	0.0524	0.0527	0.0530	0.0521	0.0510	0.0491	0.0481	0.0463
50	0.0501	0.0505	0.0526	0.0534	0.0540	0.0535	0.0530	0.0525	0.0512

is difficult to derive exact explicit ones of the mean of $X_{i:n}$. The values of $\mu_{1:n}$ for all i and for $n = 2(1)100(25)250(50)400$ have been tabulated to five decimals by Harter(1961).

We evaluate $\mu_{i:n}$ by using *Mathematica*, numerically.

5.2 Standardized Logistic Distribution

Let X be a random variable with the standardized logistic pdf

$$f(x) = \frac{\pi e^{-\frac{\pi}{\sqrt{3}}x}}{\sqrt{3}(1 + e^{-\frac{\pi}{\sqrt{3}}x})^2}, \qquad -\infty < x < \infty.$$

Birnbaum and Dudman (1963) derived the explicit formula of the mean of $X_{i:n}$ as follows:

$$\mu_{i:n} = \frac{\sqrt{3}}{\pi}(\psi(i) - \psi(n - i + 1)),$$

where $\psi(x)$ is the digamma function.

5.3 Standardized Uniform Distribution

Let X be a random variable with the standardized uniform pdf

$$f(x) = \frac{1}{\sqrt{12}}, \qquad -\sqrt{3} < x < \sqrt{3}.$$

Table 4: Some characteristic values corresponding to FAV of Laplace distribution

Sample	Subsample sizes (k)								
sizes (n)	2	3	4	5	6	7	8	9	10
					Mean[$\widehat{FAV}$]				
10	-2.019	-1.982	-1.900	-1.845	-1.807	-1.775	-1.751	-1.818	-2.029
20	-2.006	-2.005	-1.988	-1.972	-1.920	-1.906	-1.877	-1.851	-1.829
30	-1.919	-1.961	-1.975	-1.973	-1.969	-1.954	-1.940	-1.916	-1.895
40	-1.805	-1.867	-1.911	-1.941	-1.942	-1.942	-1.943	-1.937	-1.917
50	-1.776	-1.797	-1.843	-1.896	-1.908	-1.918	-1.932	-1.928	-1.942
					Variance[$\widehat{FAV}$]				
10	1.61	1.29	1.06	0.93	0.79	0.74	0.62	0.58	0.60
20	0.75	0.72	0.66	0.63	0.58	0.54	0.52	0.48	0.46
30	0.39	0.41	0.43	0.42	0.41	0.41	0.38	0.37	0.36
40	0.29	0.26	0.28	0.30	0.30	0.30	0.30	0.30	0.27
50	0.21	0.21	0.20	0.21	0.22	0.23	0.24	0.23	0.23
					Cdf[Mean[$\widehat{FAV}$]]				
10	0.0288	0.0303	0.0340	0.0368	0.0388	0.0406	0.0420	0.0383	0.0284
20	0.0293	0.0294	0.0301	0.0307	0.0331	0.0337	0.0351	0.0365	0.0376
30	0.0331	0.0312	0.0306	0.0307	0.0309	0.0315	0.0322	0.0333	0.0343
40	0.0390	0.0356	0.0335	0.0321	0.0321	0.0321	0.0320	0.0323	0.0332
50	0.0405	0.0394	0.0369	0.0342	0.0337	0.0332	0.0325	0.0327	0.0321
					Mean[Cdf[$\widehat{FAV}$]]				
10	0.0698	0.0667	0.0688	0.0688	0.0684	0.0688	0.0668	0.0594	0.0458
20	0.0498	0.0495	0.0492	0.0495	0.0515	0.0515	0.0530	0.0535	0.0543
30	0.0455	0.0437	0.0434	0.0432	0.0430	0.0441	0.0439	0.0452	0.0463
40	0.0496	0.0446	0.0425	0.0415	0.0414	0.0415	0.0414	0.0416	0.0420
50	0.0487	0.0474	0.0440	0.0414	0.0409	0.0405	0.0400	0.0402	0.0394

The mean of $X_{i:n}$ is, due to Arnold *et al.* (1992), p.128,

$$\mu_{i:n} = \sqrt{12}\left(\frac{i}{n+1} - \frac{1}{2}\right).$$

5.4 Standardized Laplace Distribution

Let X be a random variable with the standardized Laplace pdf

$$f(x) = \frac{1}{\sqrt{2}}e^{-\sqrt{2}|x|}, \qquad -\infty < x < \infty.$$

Govindarajula (1966) derived the explicit formula of the mean of $X_{i:n}$ as follows:

$$\mu_{i:n} = 2^{n-\frac{1}{2}}\left\{\sum_{j=0}^{i-1}\binom{n}{j}\sum_{m=n-i+1}^{n-j}\frac{1}{m} - \sum_{j=i}^{n}\binom{n}{j}\sum_{m=i}^{j}\frac{1}{m}\right\}.$$

5.5 Standardized Triangular Distribution

Let X be a random variable with the standardized triangular pdf

$$f(x) = \frac{\sqrt{6} - |x|}{6}, \qquad -\sqrt{6} < x < \sqrt{6}.$$

Table 5: Some characteristic values corresponding to FAV of triangular distribution

Sample sizes (n)	Subsample sizes (k)								
	2	3	4	5	6	7	8	9	10
				Mean$[\widehat{FAV}]$					
10	-1.803	-1.842	-1.846	-1.855	-1.860	-1.879	-1.912	-1.971	-2.084
20	-1.745	-1.758	-1.761	-1.768	-1.766	-1.773	-1.775	-1.779	-1.770
30	-1.731	-1.740	-1.750	-1.747	-1.748	-1.759	-1.753	-1.750	-1.755
40	-1.718	-1.719	-1.720	-1.730	-1.734	-1.734	-1.733	-1.734	-1.737
50	-1.708	-1.712	-1.708	-1.717	-1.718	-1.723	-1.723	-1.722	-1.724
				Variance$[\widehat{FAV}]$					
10	0.37	0.34	0.31	0.30	0.29	0.29	0.29	0.29	0.29
20	0.12	0.12	0.12	0.12	0.13	0.13	0.13	0.13	0.13
30	0.07	0.06	0.06	0.06	0.07	0.07	0.07	0.07	0.07
40	0.06	0.05	0.05	0.05	0.05	0.05	0.05	0.05	0.05
50	0.05	0.05	0.04	0.04	0.04	0.04	0.04	0.04	0.04
				Cdf$[$Mean$[\widehat{FAV}]]$					
10	0.0348	0.0307	0.0304	0.0295	0.0290	0.0271	0.0240	0.0190	0.0111
20	0.0414	0.0399	0.0395	0.0387	0.0389	0.0381	0.0379	0.0374	0.0384
30	0.0430	0.0420	0.0407	0.0411	0.0410	0.0397	0.0405	0.0408	0.0402
40	0.0446	0.0445	0.0444	0.0432	0.0427	0.0427	0.0428	0.0427	0.0422
50	0.0458	0.0453	0.0458	0.0448	0.0446	0.0440	0.0439	0.0441	0.0438
				Mean$[$Cdf$[\widehat{FAV}]]$					
10	0.0657	0.0588	0.0566	0.0547	0.0533	0.0514	0.0478	0.0431	0.0351
20	0.0512	0.0497	0.0498	0.0490	0.0496	0.0488	0.0488	0.0481	0.0493
30	0.0487	0.0474	0.0461	0.0465	0.0466	0.0453	0.0462	0.0467	0.0460
40	0.0499	0.0486	0.0483	0.0470	0.0465	0.0465	0.0467	0.0466	0.0463
50	0.0497	0.0491	0.0491	0.0478	0.0477	0.0470	0.0469	0.0472	0.0469

We derive the explicit expression of the mean of $X_{i:n}$ as follows:

$$\mu_{i:n} = \frac{\sqrt{6}n!}{(i-1)!(n-i)!}\left\{\left(\frac{1}{2}\right)^{i-1}(A1 - A2) + \left(\frac{1}{2}\right)^{n-i}(B1 - B2)\right\},$$

where $A1$, $A2$, $B1$, and $B2$ are as follows:

$$A1 = \frac{2^{i+\frac{1}{2}}}{1+2i}{}_2F_1(i+\frac{1}{2}, n+\frac{3}{2}; i+\frac{3}{2}; -1),$$

$$A2 = \frac{2^{i-1}}{i}{}_2F_1(i, n+1; i+1; -1),$$

$$B1 = \frac{2^{-i}}{1-i+n}{}_2F_1(1, 1-i; 2-i+n; -1), and$$

$$B2 = \frac{2^{1-i}}{3-2i+2n}{}_2F_1(1, 1-i; \frac{5}{2}-i+n; -1),$$

where ${}_2F_1(a, b; c; z)$ is the Gauss hypergeometric function.

Proof: By definition of the first moment of order statistics $Y_{i:n}$ with pdf $f(y) = 1 - |y|$, where $-1 < y < 1$,

$$\mu_{i:n} = \frac{n!}{(i-1)!(n-i)!}\left\{\int_{-1}^{0} y\left[\frac{1}{2}(1+y)^2\right]^{i-1}\left[1-\frac{1}{2}(1+y)^2\right]^{n-i}(1+y)dy\right.$$

$$\left.+\int_{0}^{1} y\left[1-\frac{1}{2}(1-y)^2\right]^{i-1}\left[\frac{1}{2}(1-y)^2\right]^{n-i}(1-y)dy\right\},$$

$$= \frac{n!}{(i-1)!(n-i)!}\left\{\left(\frac{1}{2}\right)^{i-1}\left[\int_{0}^{1} y^{2i}\left(1-\frac{1}{2}y^2\right)^{n-i}dy\right.\right.$$

$$\left.-\int_{0}^{1} y^{2i-1}\left(1-\frac{1}{2}y^2\right)^{n-i}dy\right]$$

$$+\left(\frac{1}{2}\right)^{n-i}\left[\int_{0}^{1}\left(1-\frac{1}{2}y^2\right)^{i-1}y^{2n-2i+1}dy\right.$$

$$\left.\left.-\int_{0}^{1}\left(1-\frac{1}{2}y^2\right)^{i-1}y^{2n-2i+2}dy\right]\right\}.$$

From the formula 3.197.3, due to Gradshteyn and Ryzhik (2000), p.314, we get the following equation:

$$\int_{0}^{1} y^{\lambda-1}(1-\beta y)^b dy = B(\lambda,1)_2F_1(\lambda,-b;\lambda+1;\beta),$$

where $\lambda > 0$, $|\beta| < 1$, $B(\lambda,1) = \frac{\Gamma(\lambda)\Gamma(1)}{\Gamma(\lambda+1)}$, and $\Gamma(y)$ is the gamma function. Therefore, we get the following formula:

$$\int_{0}^{1} y^a\left(1-\frac{y^2}{2}\right)^b dy = \frac{{}_2F_1(\frac{a+1}{2},-b;\frac{a+3}{2};\frac{1}{2})}{a+1},$$

$$= \frac{2^{\frac{a+1}{2}}{}_2F_1(\frac{a+1}{2},\frac{a+3+2b}{2};\frac{a+3}{2};-1)}{a+1},$$

where $a > -1$. By using this one, we can calculate the mean of $Y_{i:n}$. This completes the proof.

5.6 FAV

Let $X_1, X_2, \ldots, X_n$ be a random sample from a symmetric population. Further, let $X_{1:n} \le X_{2:n} \le \cdots \le X_{n:n}$ be the order statistics obtained from these samples. We can rewrite the estimator of the FAV based on the subsample size of 2 as

$$\widehat{FAV} = \overline{X_{m=2}} + \frac{\sqrt{0.05}-\overline{\sqrt{p}}}{\sqrt{S_{\sqrt{p}}^2}}\sqrt{S_{m=2}^2},$$

where

$$\overline{X_{m=2}} = \frac{1}{2}(X_{r:n} + X_{r+1:n}),$$

$$\overline{\sqrt{p}} = \frac{1}{2}(\sqrt{P_r} + \sqrt{P_{r+1}}),$$

$$P_r = \frac{r}{n+1},$$

$$\sqrt{S^2_{\sqrt{p}}} = \sqrt{\frac{1}{2}\sum_{i=r}^{r+1} P_r - \overline{\sqrt{p}}^2}, and$$

$$\sqrt{S^2_{m=2}} = \frac{1}{2}(X_{r+1:n} - X_{r:n}).$$

For $n < 39$, r will be 1; for $40 \leq n < 59$, $r = 2$; for $60 \leq n < 79$, $r = 3$; for $80 \leq n < 99$, $r = 4$; for $100 \leq n < 119$, $r = 5$ in the case of subsample size of 2.

It should be mentioned that $(\sqrt{0.05} - \overline{\sqrt{p}})/\sqrt{S^2_{\sqrt{p}}}$ is a constant value and $\sqrt{S^2_{m=2}}$ is easy for evaluation in the case of $k=2$. In other words, $\widehat{FAV}$ is expressed as a linear combination of $X_{i:n}$ and $X_{i+1:n}$. Therefore, we can evaluate the estimator of the FAV for five distributions except for the standard normal distribution, analytically. If the subsample size of k is greater than 2, it is difficult to evaluate $\sqrt{S^2_{m=k}} = \sqrt{\sum_{i=1}^{k}(X_{r+i-1:n} - \overline{X_{m=k}})^2/k}$, analytically, where $\overline{X_{m=k}} = \sum_{i=1}^{k} X_{r+i-1:n}/k$.

In Table 6, we present the exact values of $E[\widehat{FAV}]$ and its cumulative of $E[\widehat{FAV}]$ for five distributions. The values of $E[\widehat{FAV}]$ for each distributions correspond to ones of the average of estimates for the simulated FAV in Tables 1, 2, 3, 4, and 5. We rewrite them with parentheses in Table 6.

6. Conclusions

We would like to make two conclusions.

Firstly, we strongly recommend the two points with the cumulative probabilities closest to 0.05 to estimate the FAV. One of the most important properties of using the subsample size of $k = 2$ is its statistical performance under the logistic, Laplace, uniform, and triangular distributions. We can evaluate the estimator of the FAV only by using the subsample size of 2, analytically. There are many cases that the number of measurements is small in the field of environment chemistry. From the simulation results, the one where the subsample size is small has the advantage that a calculation is easy.

Finally, following EPA guidelines in combination of the sample size of n and the subsample size of k, we can not determine the unique rank of r, for example, $n = 39$ and $k = 2$, and so on. Therefore, we should modify the

Table 6: $E[\widehat{FAV}]$ and its cumulative probability for five distributions in the case of subsample size of 2

Sample sizes (n)	normal (numerical)	logistic	uniform	Laplace	triangular
			$E[\widehat{FAV}]$		
10	-1.874(-1.880)[*]	-1.942(-1.951)	-1.614(-1.615)	-2.022(-2.019)	-1.811(-1.803)
20	-1.840(-1.828)	-1.921(-1.933)	-1.557(-1.557)	-2.012(-2.006)	-1.755(-1.745)
30	-1.790(-1.790)	-1.847(-1.847)	-1.554(-1.556)	-1.917(-1.919)	-1.732(-1.731)
40	-1.740(-1.743)	-1.764(-1.766)	-1.558(-1.559)	-1.809(-1.805)	-1.718(-1.718)
50	-1.725(-1.726)	-1.742(-1.737)	-1.557(-1.558)	-1.781(-1.776)	-1.710(-1.708)
			Cumulative of $E[\widehat{FAV}]$		
10	0.0305	0.0287	0.0342	0.0286	0.0339
20	0.0329	0.0297	0.0505	0.0291	0.0401
30	0.0367	0.0339	0.0513	0.0333	0.0429
40	0.0409	0.0392	0.0501	0.0387	0.0446
50	0.0423	0.0407	0.0505	0.0403	0.0455

[*]The values in parentheses present the means of the estimated FAVs shown in the Tables 1, 2, 3, 4, and 5.

procedure for calculating the FAV or newly propose the method for estimating the FAV. One of the solutions on the above mentioned issue is using the following rank determining rule:

Based on the data set size (n), determine ranks (r) with the logit of cumulative probabilities $(log\frac{P_r}{1-P_r})$ closest to $log\frac{0.05}{1-0.05}$.

References:

Arnold, B. C., Balakrishnan, N., and Nagaraja, H. N. (1992). *A first course in order statistics*, John Wiley & Sons, Inc. New York.

Bose, R. C. and Gupta, S. S. (1959). Moments of order statistics from a normal population, *Biometrika*, **46**, 433-440.

Erickson, R. J. and Stephan, C. E. (1988). *Calculation of the final acute value for water quality criteria for aquatic organisms*, U.S. Environmental Protection Agency, EPA/600/3-88/018.

Govindarajula, Z. (1966). Best linear estimates under symmetric censoring of the parameters of a double exponential population, *J. Amer. Statist. Assoc.*, **61**, 248-258.

Gradshteyn, I. S. and Ryzhik, I. M. (2000). *Table of Integrals, Series, and Products*, Academic Press San Diego.

Harter, H. L. (1961). Expected values of normal order statistics, *Bimometrika*, **48**, 151-165.

On Usefulness of Maximum Likelihood Estimator Using Incomplete Data

Hironori Fujisawa

The Institute of Statistical Mathematics
4-6-7, Minami-Azabu, Minato-ku
Tokyo 106-8569, Japan

Summary: We often encounter with missing data which consist of a complete part and an incomplete part. In this case, there are two types of maximum likelihood estimator. One is the conventional maximum likelihood estimator based on all of the data and another is based on the complete part only, which neglects the incomplete part. Let n and n_* be the sample sizes of the complete part and the incomplete part, respectively. It is well-known that the former is asymptotically better than the latter when n_*/n is constant. However, other cases have not been well-known. This paper shows that the former is better than the latter in other cases in view of higher-order. In addition, this paper illustrates some exact comparison.

1. Introduction

We often encounter with missing data which consist of a complete part and an incomplete part. For example, we observe the complete part $(x_1, y_1), \ldots,$ (x_n, y_n) and the incomplete part $x_1^*, \ldots, x_{n_*}^*$ for the bivariate random variate (x, y). Such a situation has been studied by numerous researchers (see e.g. Little and Rubin (1987)).

Focus on the maximum likelihood estimator for an objective parameter u. Two types of estimators are naturally suggested. One is the conventional maximum likelihood estimator based on all of the data, say $\hat{u}$, and the other is the simple maximum likelihood estimator based only on the complete part, say $\check{u}$. This paper compares $\hat{u}$ with $\check{u}$.

Let n and n_* be the sample sizes of the complete part $z = (x, y)$ and the incomplete part x^*, respectively, It is well-known that $\hat{u}$ is asymptotically better than $\check{u}$ when we assume that $w = n_*/n$ is constant, more generally, that $w = n_*/n$ converges to a positive constant (or infinity) as n and n_* go to infinity (see e.g. Kano *et al.* (1993)). Let $\mathrm{MSE}(\hat{u}, u) = (\hat{u} - u)(\hat{u} - u)'$ and $\mathrm{MSE}(\check{u}, u) = (\check{u} - u)(\check{u} - u)'$. It holds that $\mathrm{MSE}(\hat{u}, u) \leq \mathrm{MSE}(\check{u}, u)$ up to the order $o(n^{-1})$. However, the case that w converges to zero as n goes to infinity or as n and n_* go to infinity has not been well-known. This paper focuses on such a contiguous case. This case means that the sample size of the complete part is much larger than the sample size of the incomplete part. In the following, the terms 'as n goes to infinity' and 'as n and n_* go to infinity' will often be omitted.

The contiguous case is different from the case that w is constant because $\mathrm{MSE}(\hat{u}, u)$ is asymptotically the same as $\mathrm{MSE}(\breve{u}, u)$ up to the order $o(n^{-1})$. The difference appears in higher-order terms. This paper shows that $\mathrm{MSE}(\hat{u}, u) \leq \mathrm{MSE}(\breve{u}, u)$ up to the order $o(n^{-2})$. Similar results can be extended to more general missing situations and to other risks, e.g., Kullback-Leibler risk. In a special situation, it is also possible to compare $\hat{u}$ with $\breve{u}$ exactly in view of the Kullback-Leibler risk.

2. Mean Square Error

Higher-order asymptotic properties of $\breve{u}$ are well-known (see e.g. Amari (1985)). The mean square error can be approximated by

$$\mathrm{MSE}(\breve{u}, u) = \mathrm{E}(\breve{u} - u)(\breve{u} - u)' = \frac{1}{n}G^{-1} + \frac{1}{n^2}H + o\left(\frac{1}{n^2}\right), \qquad (1)$$

where G is the Fisher information matrix given by $G = \mathrm{E}[-\partial^2 \log f(z; u)/\partial u \partial u']$, $f(z; u)$ is the probability density function of the complete part $z = (x, y)$, and H is the appropriate matrix. The exact form of H is omitted in this paper because the form is not simple and not essential.

Consider the case where $w = n_*/n$ is constant. Higher-order asymptotic properties of $\hat{u}$ is similar to that of $\breve{u}$, which is demonstrated by replacing $\log f(z; u)$ with $\log f(z; u) + w \log f_*(x^*; u)$, where $f_*(x^*; u)$ is the probability density function of the incomplete part x_*. This can be explained as follows. Let $n = rd$ and $n_* = r_* d$, where d is the greatest common measure of n and n_*. Then, it holds that $w = r_*/r$ and d goes to infinity as n goes to infinity. The $\breve{u}$ is the maximizer of

$$\sum \log f(z_i; u) = \sum_{k=1}^{d} \left(\sum_{j=1}^{r} \log f(z_{r(k-1)+j}; u) \right),$$

and the $\hat{u}$ is the maximizer of

$$\sum \log f(z_i; u) + \sum \log f_*(x_i^*; u)$$
$$= \sum_{k=1}^{d} \left(\sum_{j=1}^{r} \log f(z_{r(k-1)+j}; u) + \sum_{j=1}^{r_*} \log f(x_{r_*(k-1)+j}^*; u) \right).$$

Consequently, asymptotic properties of $\breve{u}$ and $\hat{u}$ are based on $r \log f(z; u)$ and $r \log f(z; u) + r_* \log f_*(x^*; u) = r(\log f(z; u) + w \log f_*(x^*; u))$, respectively.

The mean square error is given by

$$\text{MSE}(\hat{u}, u) = \text{E}(\hat{u} - u)(\hat{u} - u)' = \frac{1}{n}G_w^{-1} + \frac{1}{n^2}H_w + o\left(\frac{1}{n^2}\right), \qquad (2)$$

where $G_w = \text{E}[-\partial^2\{\log f(z; u) + w \log f_*(x^*; u)\}/\partial u \partial u']$ and H_w is the appropriate matrix which is continuous on w. It may be noted that the above asymptotic expansion of $\text{MSE}(\hat{u}, u)$ can be used even if w is not constant.

3. Comparison of Two Types

It is clear that $G_w \geq G$, which implies that $\hat{u}$ is asymptotically better than $\breve{u}$ if w is constant. It may be noted that G_w converges to G if w goes to zero, which implies that $\hat{u}$ is asymptotically equivalent to $\breve{u}$. Consider this case in view of higher-order. Remember the form (2) and note that H_w converges to H. It holds that

$$\text{MSE}(\hat{u}, u) = \text{E}(\hat{u} - u)(\hat{u} - u)' = \frac{1}{n}G_w^{-1} + \frac{1}{n^2}H + o\left(\frac{1}{n^2}\right). \qquad (3)$$

The first-order of the above is G^{-1}/n, which is the same as the first-order of (1). However, we can see that $\text{MSE}(\hat{u}, u) \leq \text{MSE}(\breve{u}, u)$ up to the order $o(n^{-2})$ since $G_w \geq G$.

It may be noted that if $w = o(n^{-1})$, then $G_w = G + o(n^{-1})$ so that $\text{MSE}(\hat{u}, u)$ is the same as $\text{MSE}(\breve{u}, u)$ up to the order $o(n^{-2})$. However, this case does not happen because the form w is n_*/n.

Consider another risk. Let the Kullback-Leibler divergence be denoted by

$$D(u, v) = \text{E}\left[\log \frac{f(z; u)}{f(z; v)}\middle| f(z; u)\right].$$

The asymptotic expansion of $D(u, \breve{u})$ was derived from the proof of Theorem 1 of Komaki (1996). The formula obtained implies that the risk $\text{E}[D(u, \breve{u})]$ can be expressed by the same form as (1). Consequently, it is seen by the same way as the above that $\text{E}[D(u, \hat{u})] \leq \text{E}[D(u, \breve{u})]$ up to the order $o(n^{-2})$.

4. Exact Comparison

Let $p(x; \theta)$ be one of the exponential families given by

$$p(x; \theta) = \exp\{\theta' t(x) - \psi(\theta) + h(x)\}.$$

Assume that x and y are mutually independent and that the probability density functions of x and y are $p(x;\theta_1)$ and $p(y;\theta_2)$, where $\theta_j = \theta_j(u)$ for $j = 1,2$ and u is the objective parameter. Assume that the dimension of $\theta = (\theta_1, \theta_2)$ is the same as that of u, which enables us to regard θ as a free parameter. It then holds that $\hat{\theta} = \theta(\hat{u})$ is also the maximum likelihood estimator and that $\hat{\theta}_1$ and $\hat{\theta}_2$ are the functions of $x_1, \ldots, x_n, x_1^*, \ldots, x_{n_*}^*$ and $y_1, \ldots, y_n$, respectively. Let $R(\hat{u}, u) = \mathrm{E}[D(\hat{u}, u)]$. We can easily see in virtue of independency that

$$R(\hat{u}, u) = R(\hat{\theta}_1, \theta_1) + R(\hat{\theta}_2, \theta_2).$$

The following hold:

$$R(\hat{\theta}_1, \theta_1) \leq R(\check{\theta}_1, \theta_1), \qquad R(\hat{\theta}_2, \theta_2) = R(\check{\theta}_2, \theta_2), \tag{4}$$

which implies that $R(\hat{u}, u) \leq R(\check{u}, u)$. The proof is given in the next section. This property means that $\hat{u}$ is exactly better than $\check{u}$ in view of the risk R under the above assumptions.

An example satisfying the assumptions is the following. Let X and Y be the random variates exponentially distributed with parameters λ and μ, and assume that X and Y are mutually independent. We can observe $Z = \min\{X, Y\}$ and δ where $\delta = 1$ if $Z = X$ and $\delta = 0$ if $Z = Y$, where δ is sometimes missing. The objective parameter is $u = (\lambda, \mu)$ or $u = (\mathrm{E}(Z), \mathrm{E}(\delta)) = (1/(\lambda + \mu), \lambda/(\lambda + \mu))$. We see that $\dim u = \dim \theta = 2$. The former case was analyzed by Miyakawa (1984), who showed that the incomplete part is useful when the objective parameter is $u = (\lambda, \mu)$ in view of mean square error.

Weakening the assumption might be difficult. For such the a discussion, see Fujisawa (2000). He has focused on the case where λ or μ is known in the above example and the bivariate normal distribution with an intraclass correlation (Kariya *et al.* (1983); Konishi and Shimizu (1994)). In addition, it will be difficult that a similar exact result generally holds in view of the mean square error because the exponential family is compatible with the Kullback-Leibler risk, not with the mean square error.

5. Proof

This section proves (4). The latter equality is clear. The following is the proof of the former inequality.

It follows from p.95 of Kullback (1959) that

$$R(\tilde{\theta}_1, \theta_1) = E\left[\frac{1}{n}\sum_{i=1}^{n}\log\frac{p(x_i;\hat{\theta}_1)}{p(x_i;\theta_1)}\right]. \tag{5}$$

If we show that

$$R(\hat{\theta}_1, \theta_1) = E\left[\frac{1}{n}\sum_{i=1}^{n}\log\frac{p(x_i;\hat{\theta}_1)}{p(x_i;\theta_1)}\right],$$

then the proof is complete because

$$\sum_{i=1}^{n}\log p(x_i;\hat{\theta}_1) \le \sum_{i=1}^{n}\log p(x_i;\tilde{\theta}_1).$$

This idea is derived from Yanagimoto (1991). It holds by the same way as (5) that

$$R(\hat{\theta}_1, \theta_1) = E\left[\frac{1}{n+n_*}\left\{\sum_{i=1}^{n}\log\frac{p(x_i;\hat{\theta}_1)}{p(x_i;\theta_1)} + \sum_{i=1}^{n_*}\log\frac{p(x_i^*;\hat{\theta}_1)}{p(x_i^*;\theta_1)}\right\}\right].$$

Let $C = E[\log\{p(x_1;\hat{\theta}_1)/p(x_1;\theta_1)\}]$. Since the maximum likelihood estimator $\hat{\theta}_1$ is invariant under permutation of variates, we see that

$$R(\hat{\theta}_1, \theta_1) = C = E\left[\frac{1}{n}\sum_{i=1}^{n}\log\frac{p(x_i;\hat{\theta}_1)}{p(x_i;\theta_1)}\right].$$

References:

Amari, S. (1985). *Differential-Geometrical Methods in Statistics*. Springer-Verlag, New York.

Fujisawa, H. (2000). On usefulness of incomplete data for maximum likelihood estimator. *Hiroshima University Technical Report*, 99-19.

Kano, Y., Bentler, P.M., and Mooijaart, A. (1993). Additional information and precision of estimators in multivariate structural models. *In Stat.Sci.& Data Anal.*, Matsusita, K. et al. (eds.), 187–196, VSP, Utrecht.

Kariya, T., Krishnaiah, P.R., and Rao, C.R. (1983). Inference on parameters of multivariate normal populations when some data is missing. *In Developments in Statistics, Vol.4"*, Krishnaiah, P.R. (eds.), 137–184, Academic Press, New York.

Komaki, F. (1996). On asymptotic properties of predictive distributions. *Biometrika*, **83**, 299–313.

Konishi, S. and Shimizu, K. (1994). Maximum likelihood estimation of an intraclass correlation in a bivariate normal distribution with missing observations. *Commun. Statist. Theory Meth.*, **23**, 1593–1604.

Kullback, S. (1959). *Information Theory and Statistics*. Wiley, New York

Little, R.J.A. and Rubin, D.B. (1987). *Statistical Analysis with Missing Data*. Wiley, New York

Miyakawa, M. (1984). Analysis of incomplete data in competing risks model. *IEEE Trans. Reliab.*, **33**, 293–296.

Yanagimoto, T. (1991). Estimating a model through the conditional MLE. *Ann. Inst. Statist. Math.*, **43**, 735–746.

Sensitivity Analysis in Semiparametric Regression Models[*]

Wing-Kam Fung[1], Zhong-Yi Zhu[2], Bo-Cheng Wei[3]

[1]Department of Statistics and Actuarial Science, University of Hong Kong,
Pokfulam Road, Hong Kong, China
[2]Department of Statistics, East China Normal University, China
[3]Department of Mathematics, Southeast University, China

Summary: Research in semiparametric regression models has received attention in recent years. However, there is little work in the sensitivity analysis for such models. In this paper, we investigate the statistical diagnostics in semiparametric regression models. The case deletion influence diagnostics are constructed. An outlier diagnostic of the case deletion model is studied and it is shown to be equivalent to that of the mean-shift outlier model. The popular Cook's distance is constructed which can be expressed in terms of the leverage measure and the residual. The proposed diagnostics are illustrated using a real data set.

1. Introduction

Diagnostic methods for nonparametric regression models, in particular, for smoothing splines have received attention; see Eubank (1984, 1985), Silverman (1985), Thomas (1991) and Kim (1996), who studied the basic diagnostic building blocks such as the residual and the leverage, and the local influence for the choice of the smoothing parameter. However, there does not seem to have any work on diagnostics for semiparametric regression models. In this paper, we present the case deletion diagnostics for semiparametric regression models.

Suppose that the scalar response y_i follows the model

$$y_i = x_i^T \beta + g(t_i) + \epsilon_i, \quad 1 \le i \le n, \tag{1.1}$$

where β is a p-vector of regression coefficients, x_i is a p-vector of explantory variables, t_i is a scalar ($a \le t_1, \cdots, t_n \le b$), and $t_i's$ are not all identical, g is a smooth curve and the errors ϵ_i are uncorrelated with zero mean and constant variance σ^2. The nonparametric part is to be estimated assuming only that g is an element of $W_2^2[a, b]$ of functions f that have first order continuous derivatives and square-integratable second order derivatives f'' in $[a, b]$.

There are many possible estimators for β and g in (1.1); see Green & Silverman (1994). The popular estimators, based on the assumptions stated

[*]The project is supported by NSFC (19631040), NSFJ of China and the Hong Kong RGC CERG Grant HKU7134/98H.

above, are $\hat{g}(t)$ and $\hat{\beta}$ which minimize the penalized sum of squares

$$\sum_{i=1}^{n} \{y_i - x_i^T\beta - g(t_i)\}^2 + \lambda \int g''(t)^2 dt, \quad \lambda > 0. \tag{1.2}$$

Discussions of partial spline models and their statistical applications may be found in Green & Silverman (1994). In this paper, the choice of the smoothing parameter λ is accomplished by minimizing the generalized cross-validation citerion $GCV(\lambda)$.

The paper is organized as follows: Section 2 introduces the penalized least square estimates and related notations for model (1.1). Section 3 presents several diagnostic measures based on the case deletion method, and we show that the case deletion model and the mean-shift outlier model are equivalent in the estimation of regression coefficients of the partial spline models (1.1). In Section 4, a numerical example is given to illustrate our results.

2. Partial Spline Models and Estimation

We first introduce the penalized least square estimates (PLSE) for β and g for model (1.1). Let the ordered, distinct values among $t_1, \cdots, t_n$ be denoted by $s_1, \cdots, s_q$. The connection between $t_1, \cdots, t_n$ and $s_1, \cdots, s_q$ is captured by means of the $n \times q$ incidence matrix N, with entries $N_{ij} = 1$ if $t_i = s_j$ and 0 otherwise. It follows that $2 \leq q$ from the assumption that the $t_i's$ are not all identical.

Let g be the vector of value $a_i = g(s_i)$. Model (1.2) can be written as

$$L(\theta) = \|Y - X\beta - Ng\|^2 + \lambda \int g''(t)^2 dt, \quad \lambda > 0,$$

where $\theta = (\beta^T, g^T)^T$, and Y, X, g are the corresponding matrices or vectors. Conceptually, the minimization of $L(\theta)$ can be considered in two steps, first minimizing $L(\theta)$ subject to $g(s_j) = a_j$, $j = 1, \cdots, q$, and then minimizing the result over the choice of g and β (Green and Silverman, 1994).

The problem of minimizing $\int g''(t)^2 dt$ subject to g interpolating given points $g(s_j) = a_j$ is given by Green and Siliverman (1994), and the minimizing curve g is a natural cubic spline with knots $\{s_j\}$. There exists a matrix K only depending on the knots $\{s_j\}$, such that the minimized value of $\int g''(t)^2 dt$ is $g^T Kg$ (Green & Silverman, 1994, p.66). For this g, $L(\theta)$ takes the value

$$L(\theta) = \|Y - X\beta - Ng\|^2 + \lambda g^T Kg, \quad \lambda > 0. \tag{2.1}$$

By simple calculus it follows that (2.1) is minimized when β and g satisfy the equation:

$$\begin{pmatrix} X^T X & X^T N \\ N^T X & N^T N + \lambda K \end{pmatrix} \begin{pmatrix} \beta \\ g \end{pmatrix} = \begin{pmatrix} X^T \\ N^T \end{pmatrix} Y. \tag{2.2}$$

This equation results in

$$\hat{\beta} = \{X^T(I - S)X\}^{-1} X^T(I - S)Y, \tag{2.3}$$

$$\hat{g} = (N^T N + \lambda K)^{-1} N^T (Y - X\hat{\beta}), \tag{2.4}$$

where $S = N(N^T N + \lambda K)^{-1} N^T$. After some calculations from knots $\{s_j\}$ and $\hat{g}$, we can get a natural cubic spline estimate $\hat{g}(t)$ of curve $g(t)$. Under the regularity assumptions stated by Heckman (1986), $\hat{\beta}$ is consistent and asymptotically normal. We have the generalized likelihood ratio statistic

$$PLR(\beta) := \{L(\beta, \hat{g}(\beta)) - L(\hat{\beta}, \hat{g})\}/\sigma^2$$

$$= (\beta - \hat{\beta})^T (X^T (I - S)X)(\beta - \hat{\beta})/\sigma^2 \xrightarrow{\mathcal{L}} \chi^2(p), \tag{2.5}$$

where " $\xrightarrow{\mathcal{L}}$ " denotes convergence in distribution, and $\hat{g}(\beta) = (N^T N + \lambda K)^{-1} N^T (Y - X\beta)$.

¿From (2.3) and (2.4), we can get the fitted value $\hat{Y} = X\hat{\beta} + N\hat{g} = HY$, where $H = (h_{ij})$ is the hat matrix which is given as

$$H = S + (I - S)X\{X^T (I - S)X\}^{-1} X^T (I - S).$$

The residual vector is evaluated as

$$\hat{e} = Y - \hat{Y} = (I - H)Y,$$

and the estimate of σ^2 is $\hat{\sigma}^2 = \|\hat{e}\|^2/(n - \text{tr}(H))$.

3. Case Deletion Diagnostics
3.1 Estimates under Case Deletion

To study the influence of observation (y_i, x_i^T, t_i) on estimates $\hat{\beta}$ and $\hat{g}$, it is common to compare the estimates $\hat{\theta} = (\hat{\beta}^T, \hat{g}^T)^T$ with the estimates $\hat{\theta}_{(i)} = (\hat{\beta}_{(i)}^T, \hat{g}_{(i)}^T)^T$ which correspond to the case deletion model with the $i - th$ case deleted:

$$y_j = x_j^T \beta + g(t_j) + \epsilon_j, \quad j \neq i, \quad j = 1, \cdots, n. \tag{3.1}$$

To compute $\hat{\theta}_{(i)}$ for all i and compare them with $\hat{\theta}$ would be consuming when the sample size n is large. Fortunately, the following theorem gives an updating formula under case deletion to avoid direct model estimation for each of the n cases. This result is essential for our case deletion diagnostics.

Theorem 1. Using the notation stated above, we have

$$\hat{\beta}_{(i)} = \hat{\beta} - \frac{\{X^T (I - S)X\}^{-1} X^T (I - S)d_i \hat{e}_i}{1 - h_{ii}}, \tag{3.2}$$

$$\hat{g}_{(i)} = \hat{g} - \frac{(N^T N + \lambda K)^{-1} N^T [I - X\{X^T (I - S)X\}^{-1} X^T (I - S)]d_i \hat{e}_i}{1 - h_{ii}}, \tag{3.3}$$

where d_i is an $n \times 1$ vector with 1 at $i - th$ position and zeros elsewhere, $\hat{e}_i$ is the ith element of the residual vector $\hat{e}$, and h_{ii} is the ith diagonal element of the hat matrix.

Proof. Let $Y^* = (y_1^*, y_2^* \cdots y_n^*)^T$ and

$$y_j^* = \begin{cases} y_i^* & j = i \\ y_j & j \neq i \end{cases},$$

where $y_i^* = x_i^T \hat{\beta}_{(i)} + \hat{g}_{(i)}(t_i)$. For any parameter β and smooth curve g, we have by the definition of $\hat{\beta}_{(i)}$ and $\hat{g}_{(i)}$,

$$\|Y^* - X\beta - Ng\|^2 + \lambda \int g''(t)^2 dt$$

$$= (y_i^* - x_i^T \beta - g(t_i))^2 + \sum_{j \neq i}(y_j - x_j^T \beta - g(t_j))^2 + \lambda \int g''(t)^2 dt$$

$$\geq \sum_{j \neq i}(y_j - x_j^T \beta - g(t_j))^2 + \lambda \int g''(t)^2 dt$$

$$\geq \|Y^* - X\hat{\beta}_{(i)} - N\hat{g}_{(i)}\|^2 + \lambda \int \hat{g}''(t)^2 dt.$$

It follows that $(\hat{\beta}_{(i)}, \hat{g}_{(i)}^T)^T$ minimizes

$$\|Y^* - X\beta - Ng\|^2 + \lambda \int g''(t)^2 dt.$$

So that

$$\begin{aligned} \hat{\beta}_{(i)} &= \{X^T(I-S)X\}^{-1}X^T(I-S)Y^* \\ &= \hat{\beta} - \{X^T(I-S)X\}^{-1}X^T(I-S)d_i(y_i - y_i^*), \\ \hat{g}_{(i)} &= (N^T N + \lambda K)^{-1}N^T(Y^* - X\hat{\beta}_{(i)}) \\ &= \hat{g} - (N^T N + \lambda K)^{-1}N^T[I - X\{X^T(I-S)X\}^{-1} \\ &\quad X^T(I-S)]d_i(y_i - y_i^*). \end{aligned}$$

On the other hand, we can verify

$$\begin{aligned} y_i - y_i^* &= d_i^T(Y - X\hat{\beta}_{(i)} - N\hat{g}_{(i)}) \\ &= d_i^T(Y - HY^*) = d_i^T[Y - HY + H(Y - Y^*)] \\ &= d_i^T(Y - HY) + d_i^T H d_i(y_i - y_i^*). \end{aligned}$$

Solving for $y_i - y_i^*$ we get

$$y_i - y_i^* = \hat{e}_i/(1 - h_{ii}),$$

from which Theorem 1 follows. □

If the nonparametric component g is not present in the model, (3.2) reduces to the well-known updating formula of Cook (1977) for linear regression. Moreover, the updating formula of Eubank (1985) is also a special case of Theorem 1 here.

3.2 Cook's Distance

Cook (1977) introduced what is now known as the Cook's distance for the measurement of influential observations in linear regression models. Kim (1996) showed how the Cook's distance can be defined for smoothing spline models. An analogous definition of it can be given for the partial spline models (1.1) as follows.

The (genearlized) Cook's distance is defined as the standardized norm of $\hat{\theta} - \hat{\theta}_{(i)}$ with respect to a certain weight matrix $M > 0$, i.e.

$$D_i(\theta) = \|\hat{\theta} - \hat{\theta}_{(i)}\|_M^2 = (\hat{\theta} - \hat{\theta}_{(i)})^T M (\hat{\theta} - \hat{\theta}_{(i)}).$$

Let $G = \partial^2 L(\theta)/\partial\theta\partial\theta^T$. It is natural to choose $M = G/(2\hat{\sigma}^2)$. From Theorem 1 and after some calculations, we can get

$$D_i(\theta) = \frac{h_{ii}r_i^2}{1 - h_{ii}}, \tag{3.4}$$

where $r_i = \hat{e}_i/[\hat{\sigma}(1 - h_{ii})]$ is the studentized residual. (3.4) is in the same form as the Cook's distance in regression (Cook, 1977), apart from a constant proportion.

The Cook's distances for β and g can be respectively defined as

$$D_i(\beta) = (\hat{\beta} - \hat{\beta}_{(i)})^T [(I_p, 0)G^{-1}(I_p, 0)^T]^{-1}(\hat{\beta} - \hat{\beta}_{(i)})/\hat{\sigma}^2,$$

and

$$D_i(g) = (\hat{g} - \hat{g}_{(i)})^T [(0, I_q)G^{-1}(0, I_q)^T]^{-1}(\hat{g} - \hat{g}_{(i)})/\hat{\sigma}^2.$$

By Theorem 1, we can get

$$D_i(\beta) = \frac{d_i^T(H - S)d_i r_i^2}{1 - h_{ii}}; \quad D_i(g) = \frac{d_i^T(H - X(X^T X)^{-1}X^T)d_i r_i^2}{1 - h_{ii}}.$$

3.3 Likelihood Distance

Cook and Weisberg (1982, p.183) suggested using the (generalized) likelihood distance as a general approach to influence. The distance for θ is defined as $LD_i(\theta) = [L(\hat{\theta}_{(i)}) - L(\hat{\theta})]/\hat{\sigma}^2$. Using Theorem 1 and after some calculation, we obtain that

$$LD_i(\theta) = D_i(\theta).$$

Likelihood distances, $LD_i(\beta|g)$ and $LD_i(g|\beta)$, for the parametric part β and the nonparametric part g respectively can be defined in similar ways; see Cook and Weisberg (1982, p184). The following relationship can be derived,

$$LD_i(\beta|g) = D_i(\beta) \quad \text{and} \quad LD_i(g|\beta) = D_i(g).$$

Thus, the likelihood distance is essentially equivalent to the Cook's distance.

3.4 Mean-Shift Outlier Model and Test for Outlier

The above case deletion model has been commonly used for constructing effective diagnostic statistics because of its simplicity. Another popoular diagnostic model is the mean-shift outlier model (Cook & Weisberg, 1982, p.20) which can be represented as

$$
\begin{cases}
y_j = x_j^T \beta + g(t_j) + \epsilon_j, & j \neq i, \; j = 1, \cdots, n; \\
y_i = x_i^T \beta + g(t_i) + \gamma + \epsilon_i,
\end{cases}
\tag{3.5}
$$

where γ is an extra parameter to indicate the presence of outlier. This model is usually easier to formulate than the case deletion model. Let $\hat{\beta}_{mi}$, $\hat{g}_{mi}$ and $\hat{\gamma}_{mi}$ be the PLSE for model (3.5). We have the following relationship.

Theorem 2. The PLSE of (β, g) under (3.1) without the ith case is the same as the estimate under (3.5) with the full data set, that is

$$
\hat{\beta}_{(i)} = \hat{\beta}_{mi}, \qquad \text{and} \qquad \hat{g}_{(i)} = \hat{g}_{mi}.
$$

Proof. By the definition of $\hat{\beta}_{(i)}$, $\hat{g}_{(i)}$, it follows that $\hat{\beta}_{(i)}$, $\hat{g}_{(i)}$ minimize

$$
L_{(i)}(\theta) = \sum_{j \neq i}(y_j - x_j^T \beta - g(t_j))^2 + \lambda \int g''(t)^2 dt.
$$

For model (3.5), $\hat{\beta}_{mi}$, $\hat{g}_{mi}$, and $\hat{\gamma}_{mi}$ minimize

$$
L_{mi}(\theta, \gamma) = \sum_{j \neq i}(y_j - x_j^T \beta - g(t_j))^2 + (y_i - x_i^T \beta - g(t_i) - \gamma)^2 + \lambda \int g''(t)^2 dt.
\tag{3.6}
$$

So $\hat{\gamma}_{mi}$ satisfies

$$
\frac{\partial L_{mi}(\theta, \gamma)}{\partial \gamma} = -2(y_i - x_i^T \beta - g(t_i) - \gamma) = 0,
$$

and we get $\hat{\gamma}_{mi} = y_i - x_i^T \beta - g(t_i)$. Substituting this into (3.6), we can easily obtain $L_{mi}(\theta, \hat{\gamma}_{mi}) = L_{(i)}(\theta)$. Thus, $\hat{\beta}_{(i)} = \hat{\beta}_{mi}$ and $\hat{g}_{(i)} = \hat{g}_{mi}$. $\square$

The equivalence between the case deletion model and the mean-shift outlier model in regression was observed in Cook and Weisberg (1982). Wei (1998) has extended the result to a broader class of parametric models. Theorem 2 shows that the phenomenon is also true for semiparametric models.

A likelihood ratio test for outliers based on the mean-shift outlier model can be constructed for the hypotheses $H_0 : \gamma = 0$, and $H_1 : \gamma \neq 0$. The test statistic is taken as

$$
GLR_i(0) = \{L_{mi}(\hat{\theta}, 0) - L_{mi}(\hat{\theta}_{mi}, \hat{\gamma}_{mi})\}/\hat{\sigma}^2,
$$

which as in (2.5), is asymptotically distributed as χ_1^2.

Theorem 3. The likelihood ratio test for H_0: $\gamma = 0$ versus H_1 : $\gamma \neq 0$ for model (3.5) is

$$GLR_i = \frac{\hat{e}_i^2}{\hat{\sigma}^2(1 - h_{ii})} = r_i^2. \tag{3.7}$$

Proof. To prove the theorem, we evaluate $L_{mi}(\hat{\theta}_{mi}, \hat{\gamma}_{mi})$ at $(\hat{\theta}, 0)$ based on the Taylor expansion. It follows from (3.6) that

$$GLR_i = -\{\dot{L}_{mi}^T \delta + \delta^T J \delta / 2\}/\hat{\sigma}^2, \tag{3.8}$$

where $\dot{L}_{mi} = (\partial L_{mi}/\partial \theta^T, \partial L_{mi}/\partial \gamma)^T$, $\delta = ((\hat{\beta}_{mi} - \hat{\beta})^T, \hat{\gamma}_{mi})^T$,

$$J = \begin{pmatrix} \dfrac{\partial^2 L_{mi}(\theta, \gamma)}{\partial\theta\partial\theta^T} & \dfrac{\partial^2 L_{mi}(\theta, \gamma)}{\partial\theta\partial\gamma} \\[2mm] \dfrac{\partial^2 L_{mi}(\theta, \gamma)}{\partial\gamma\partial\theta^T} & \dfrac{\partial^2 L_{mi}(\theta, \gamma)}{\partial\gamma^2} \end{pmatrix} = 2\begin{pmatrix} X^T X & X^T N & X^T d_i \\ N^T X & N^T N + \lambda K & N^T d_i \\ d_i^T X & d_i^T N & 1 \end{pmatrix}.$$

It follows from Theorems 1 and 2 that

$$\hat{\theta}_{mi} - \hat{\theta} = -\begin{pmatrix} X^X & X^T N \\ N^T X & N^T N + \lambda K \end{pmatrix}^{-1} \begin{pmatrix} X^T \\ N^T \end{pmatrix} d_i \hat{e}_i / (1 - h_{ii}),$$

$$\dot{L}_{mi}^T \hat{\gamma}_{mi} = -2\hat{e}_i^2 / (1 - h_{ii}).$$

Substituting these into (3.8) yields (3.7). □

4. Mouthwash Data

The data are taken from an experiment to determine whether or not a common brand of analgesic is effective in treating a type of gum disease (Speckman, 1988). The data set consists of measurements from 30 subjects. The response variable y is the week-three SBI, a measurement indicating gum shrinkage for which lower numbers indicate better health. Two explanatory variables namely, $t =$ base-line SBI and x ($x = 0$ or 1 for treatment or control groups), are employed. These data have been studied in the context of kernel smoothing by Speckman (1988). Here, the data are analysed with the partial spline model (2.1) with $\hat{\lambda} = 0.0041$ obtained from GCV. The estimated treatment effect is $\hat{\beta} = 0.046$.

Figures 1 and 2 give the index plots for the leverage values and the studentized residuals. It is clear that observation 30 is a high leverage point, but there does not seem to have any outliers in the model. Figures 3 and 4 plot the Cook's distances $D_i(\beta, g)$ for the full parameters and $D_i(g)$ for g. Case 30 exerts the largest influence on (β, g) and g. The two figures, however, essentially give the same information, which suggests that the observations may not be influential in the estimation of the regression parameter β. This can be observed from the plot of $D_i(\beta)$. For brevity, the plot is omitted.

In conclusion, the proposed diagnostics are able to identify the influential observations in the data set.

240

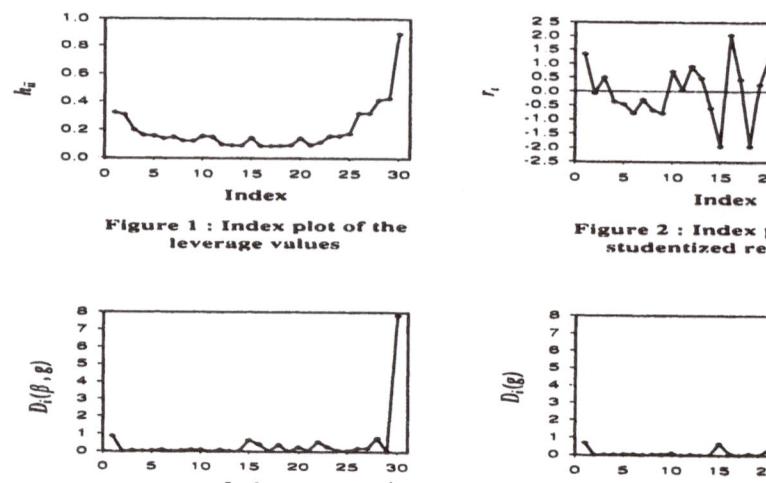

Figure 1 : Index plot of the
leverage values

Figure 2 : Index plot of the
studentized residuals

Figure 3 : Index plot of the
Cook's distance for (β, g)

Figure 4 : Index plot of the
Cook's distance for g

References:

Cook, R.D. (1977). Detection of influential observations in linear regression. *Technometrics*, **19**, 15-18.

Cook, R.D. and Weisberg, S. (1982). *Residual and Influence in Regression* New York: Chapman and Hall.

Eubank, R.L. (1984). The hat matrix for smoothing splines. *Statistics and Probability Letters*, **2**, 9-14.

Eubank, R.L. (1985). Diagnostics for smoothing splines. *Journal of Royal Statistical Society, B*, **47**, 332-341.

Green, P.J. and Silverman, B.W. (1994). *Nonparametric Regression and Generalized Linear Models*. London: Chapman and Hall.

Heckman, N. E. (1986). Spline smoothing in a partly linear model. *Journal of Royal Statistical Society, B*, **48**, 244-248.

Kim, C. (1996). Cook's distance in spline smoothing. *Statistics and Probability Letters*, **31**, 139-144.

Silverman, B.W. (1985). Some aspects of the spline smoothing approach to nonparametric regression curve fitting (with discussion). *Journal of Royal Statistical Society*, **47**, 1-52.

Speckman, P. (1988). Kernel smoothing in partial linear models. *Journal of Royal Statistical Society, B*, **50**, 413-436.

Thomas, W. (1991). Influence diagnostics for the cross-validated smoothing parameter in spline smoothing. *Journal of American Statistical Association*, **86**, 693-698.

Wei, B.C. (1998). *Exponential Family Nonlinear Models*. Singapore: Springer-Verlag.

Cohort Analysis of Data Obtained Using a Multiple Choice Question

Takashi Nakamura

The Institute of Statistical Mathematics
4-6-7, Minami-Azabu, Minato-ku
Tokyo 106-8569, Japan

Summary: Cohort analysis is a method of estimating the age, period, and cohort effects from time-series social survey data. A Bayesian multinomial logit cohort (BMLC) model is presented to analyze categorical data obtained using a multiple choice question. The model is identical to the existing Bayesian logit cohort (BLC) model when the number of response categories equals two. Data on religious attitudes from a study of the Japanese national character is analyzed both by using the BLC model with each category separated, and by using the BMLC model on all categories simultaneously, to show the utility of the latter model.

1. Introduction

Cohort analysis is a method of estimating the age, period (calendar year), and cohort (birth time) effects from a set of social survey data classified by age groups and survey periods and is widely employed in various fields (Mason and Fienberg, 1985). It has confronted, however, an identification problem in that the age, period, and cohort effects cannot be decomposed uniquely without some prior information. In order to overcome the problem, Nakamura (1982, 1986, 1995) introduced a Bayesian cohort model with a gradually-changing-parameter assumption and a procedure of selecting the optimal model based on the Akaike Bayesian information criterion, ABIC (Akaike, 1980).

A Bayesian multinomial logit cohort model is presented here to analyze data obtained using a multiple choice question which asks a respondent to choose only one from the given response categories. The model treats simultaneously all categories of a multiple choice question to avoid inconsistent results when we use the logit model with each category separated.

2. Bayesian Logit Cohort (BLC) Model

Let π_{ijr} denote the response probability of the rth category of a multiple choice question for the ith age group in the jth survey period. For simplicity, we drop the subscript r for the time being since the logit model focuses only on a certain category.

Let π_{ij} denote the response probability of some category of a question for the ith age group in the jth survey period, then the logit cohort model

242

decomposes the logit, η_{ij}, of π_{ij} into the form

$$\eta_{ij} \equiv \log[\pi_{ij}/(1-\pi_{ij})] = \beta_0 + \beta_i^A + \beta_j^P + \beta_k^C, \tag{1}$$
$$i = 1,\ldots,I; \ j = 1,\ldots,J; \ k = 1,\ldots,K$$

where β_0, β_i^A, β_j^P and β_k^C are the effects of the grand mean, the ith age, the jth period, and the kth cohort, respectively. Further, I, J, and K are the numbers of age groups, survey periods, and cohorts, respectively. We call the above the age-period-cohort model, or the APC model for short. Note that the relations $k = j-i+I$ and $K = I+J-1$ hold for a standard cohort table, in which the range in years covered by each age group equals the interval in years between successive survey periods. The β's in (1) are subject to the zero-sum constraints:

$$\sum_{i=1}^{I} \beta_i^A = \sum_{j=1}^{J} \beta_j^P = \sum_{k=1}^{K} \beta_k^C = 0. \tag{2}$$

For convenience, we represent the model (1) with the constraints (2) in terms of vectors and matrices as follows:

$$\eta = X\beta, \tag{3}$$

where
$$\eta = (\eta_{11}, \eta_{21}, \ldots, \eta_{IJ})',$$
$$\beta = (\beta_0, \beta_*')',$$
$$\beta_* = (\beta_1^A, \ldots, \beta_{I-1}^A, \beta_1^P, \ldots, \beta_{J-1}^P, \beta_1^C, \ldots, \beta_{K-1}^C)',$$
$$X = \left(1, X_*\right),$$
$$1 = (1, \ldots, 1)',$$

and X_* is an appropriate design matrix expressing the relationship between the cells in a cohort table and the three factors.

Let m_{ij} denote the sample size and y_{ij} denote the observed positive response count of the (i,j) cell. Assume the product binomial distribution for $y = (y_{11}, \ldots, y_{IJ})'$, then the likelihood of the model (1) is given by

$$f(\beta | y) = \prod_{i=1}^{I} \prod_{j=1}^{J} \binom{m_{ij}}{y_{ij}} \pi_{ij}^{y_{ij}} (1-\pi_{ij})^{m_{ij}-y_{ij}}. \tag{4}$$

In order to overcome the identification problem in the decomposition (1), we make the assumption that successive parameters change gradually. This can be realized by reducing the sum of squares of the first-order differences of the parameters,

$$\frac{1}{\sigma_A^2}\sum_{i=1}^{I-1}(\beta_i^A - \beta_{i+1}^A)^2 + \frac{1}{\sigma_P^2}\sum_{j=1}^{J-1}(\beta_j^P - \beta_{j+1}^P)^2 + \frac{1}{\sigma_C^2}\sum_{k=1}^{K-1}(\beta_k^C - \beta_{k+1}^C)^2, \tag{5}$$

or expressed statistically by a prior distribution such that

$$\pi(\boldsymbol{\beta}_* \mid \sigma_A^2, \sigma_P^2, \sigma_C^2) = (2\pi)^{-\frac{M}{2}} |\boldsymbol{D}_*' \boldsymbol{\Sigma}^{-1} \boldsymbol{D}_*|^{\frac{1}{2}} \exp\left\{ -\frac{1}{2} \boldsymbol{\beta}_*' \boldsymbol{D}_*' \boldsymbol{\Sigma}^{-1} \boldsymbol{D}_* \boldsymbol{\beta}_* \right\} \quad (6)$$

where σ_A^2, σ_P^2, and σ_C^2 are the unknown hyperparameters,

$$\boldsymbol{\Sigma} = \operatorname{diag}(\sigma_A^2, \ldots, \sigma_A^2, \sigma_P^2, \ldots \sigma_P^2, \sigma_C^2, \ldots, \sigma_C^2),$$

$\boldsymbol{D}_*$ is an appropriate design matrix expressing the first-order differences of the parameters, and $M = I + J + K - 3$.

To estimate the parameter $\boldsymbol{\beta}$, we first determine the hyperparameters by taking the minimum of

$$\text{ABIC} = -2 \log \int f(\boldsymbol{\beta} \mid \boldsymbol{y}) \cdot \pi(\boldsymbol{\beta}_* \mid \sigma_A^2, \sigma_P^2, \sigma_C^2) \, d\boldsymbol{\beta}_* + 2(h+1) \quad (7)$$

where h is the number of the hyperparameters. ABIC is a kind of the Akaike Information Criterion, AIC, and is based on the marginal log likelihood of hyperparameters of a Bayesian model.

Once the hyperparameters $\hat{\sigma}_A^2$, $\hat{\sigma}_P^2$, and $\hat{\sigma}_C^2$, are obtained, solve the maximization problem

$$\log f(\boldsymbol{\beta} \mid \boldsymbol{y}) \cdot \pi(\boldsymbol{\beta}_* \mid \hat{\sigma}_A^2, \hat{\sigma}_P^2, \hat{\sigma}_C^2) \longrightarrow \max \quad (8)$$

which provides the maximum posterior estimate $\hat{\boldsymbol{\beta}}$.

We can consider and compare the variant models such as the model with only the grand mean effect (the β_0 model), the A, the P, the C, the AP, the AC, the PC, and the APC models in (1). For instance, the AC model is expressed by $\eta_{ij} = \beta_0 + \beta_i^A + \beta_k^C$.

3. Application of the BLC Model

The BLC models are applied to data from the nationwide survey on the Japanese national character conducted every five years since 1953 by the Institute of Statistical Mathematics (Sakamoto et al., 2000). The data analyzed are concerned with the religious attitudes asking "Without reference to any of established religions, do you think a religious attitude is important, or not important?" We call this question item "Spirituality" hereafter. The response categories are "Important," "Not important," and "Other+D.K." Note that the entire sample was asked this question only four times, that is, in 1983, 1988, 1993, and 1998.

Table 1 compares the eight BLC models. As an example, for males who choose the category "Important," the AP model is selected as the best one since it gives the minimum ABIC value. Similarly, the C model is selected as the best one for both the category "Not important" and "Other+D.K." The results for Japanese females are omitted in this paper.

Table 1: Comparison of the eight variant BLC models for the category "Important" concerning "Spirituality" for Japanese males.

Model	ABIC	ΔABIC	h	σ_A^2	σ_P^2	σ_C^2
AP	67.4085	--	2	1	1	--
APC	69.4480	2.0395	3	1	1	1/128
AC	71.4947	4.0862	2	1/4	--	1
C	72.8247	5.4162	1	--	--	1
PC	73.2962	6.8877	2	--	1/16	1
A	96.6756	29.2671	1	1	--	--
P	207.7126	140.3042	1	--	1/2	--
β_0	224.1461	156.7376	0	--	--	--

Figure 1 shows the parameter estimates of the best models for the three categories for males. In this case, the inconsistent results among the categories can be seen in that the period and the age effects exist for the category "Important" while no period effects exist for the other two categories, and that no cohort or age effects exist for the category "Important" to compensate for the cohort effects for the other two categories. Of course, the sum of the estiamtes $\hat{\pi}_{ijr}$ of the response probabilities over the categories is not guaranteed to equal unity ($\sum_r \hat{\pi}_{ijr} \neq 1$).

These phenomena happen to occur because the BLC model tries to find the most parsimonious one for each category regardless of the other categories of a question.

4. Bayesian Multinomial Logit Cohort (BMLC) Model

Again let π_{ijr} denote the response probability of the rth category of a multiple choice question for the ith age group in the jth survey period and y_{ijr} the corresponding observed response count ($r = 1, \ldots, R$ where R is the number of categories). Note that $\sum_r \pi_{ijr} = 1$ and $\sum_r y_{ijr} = m_{ij}$ for all i and j.

The multinomial logit cohort model decomposes the multinomial logit, η_{ijr}, of π_{ijr} into the form

$$\eta_{ijr} \equiv \log\left[\pi_{ijr}^R \Big/ \prod_{r=1}^R \pi_{ijr}\right] = \beta_{0r} + \beta_{ir}^A + \beta_{jr}^P + \beta_{kr}^C, \tag{9}$$

$$i = 1, \ldots, I; \; j = 1, \ldots, J; \; k = 1, \ldots, K; \; r = 1, \ldots, R.$$

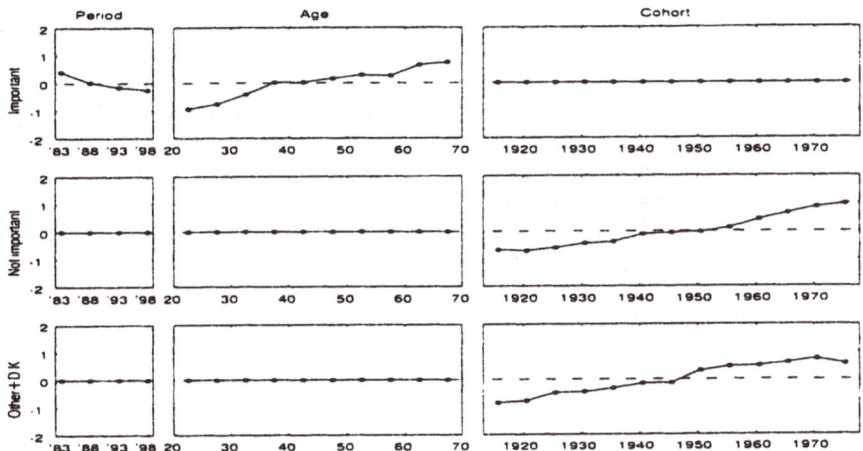

Figure 1: Plot of the parameter estimates of the best models for the three categories of "Spirituality" for Japanese males, based on the BLC model applied to each category separately.

Note that π_{ijr} can be expressed inversely by

$$\pi_{ijr} = \exp(\eta_{ijr}/R)\Big/\sum_{r=1}^{R}\exp(\eta_{ijr}/R) .$$

The zero-sum constraints are

$$\sum_{i=1}^{I}\beta_{ir}^{A} = \sum_{j=1}^{J}\beta_{jr}^{P} = \sum_{k=1}^{K}\beta_{kr}^{C} = 0, \qquad r = 1,\ldots,R; \qquad (10)$$

$$\sum_{r=1}^{R}\beta_{0r} = \sum_{r=1}^{R}\beta_{ir}^{A} = \sum_{r=1}^{R}\beta_{jr}^{P} = \sum_{r=1}^{R}\beta_{kr}^{C} = 0,$$

$$i = 1,\ldots.I; \ j = 1,\ldots,J; \ k = 1,\ldots,K. \qquad (11)$$

The gradually-changing-parameter assumption for the model (9) is realized by reducing the sum

$$\sum_{r\in R.}\left\{\frac{1}{\sigma_{A,r}^{2}}\sum_{i=1}^{I-1}(\beta_{ir}^{A}-\beta_{i+1,r}^{A})^{2}\right.$$

$$\left. +\frac{1}{\sigma_{P,r}^{2}}\sum_{j=1}^{J-1}(\beta_{jr}^{P}-\beta_{j+1,r}^{P})^{2}+\frac{1}{\sigma_{C,r}^{2}}\sum_{k=1}^{K-1}(\beta_{kr}^{C}-\beta_{k+1,r}^{C})^{2}\right\} \qquad (12)$$

where $R_{\bullet}$ is any subset composed of $(R-1)$ elements chosen from the index set $\{1,2,\ldots,R\}$.

Table 2: Top 20 BMLC models for the question "Spirituality" for Japanese males.

Model	ABIC	ΔABIC	h	Important $(\sigma_A^2/\sigma_P^2/\sigma_C^2)$	Not important $(\sigma_A^2/\sigma_P^2/\sigma_C^2)$	Other+D.K. $(\sigma_A^2/\sigma_P^2/\sigma_C^2)$
AP/C/APC	111.9808	–	5	(4 / 2 / –)	(– / – / 1)	(4 / 2 / 1)
AC/C/AC	113.3897	1.4089	5	(1 / – / *)	(– / – / 1)	(1 / – / 2)
APC/C/APC	113.5980	1.6172	6	(1 / 1 / *)	(– / – / 1)	(1 / 1 / 1)
APC/C/AP	113.7537	1.7729	5	(2 / 1 / 1)	(– / – / 1)	(2 / 1 / –)
AP/AC/APC	113.9870	2.0062	6	(* / 2 / –)	($\frac{1}{128}$/ – / 1)	(4 / 2 / 1)
AP/PC/APC	113.9949	2.0141	6	(4 / * / –)	(– / $\frac{1}{128}$/ 1)	(4 / 2 / 1)
APC/AP/C	114.7540	2.7731	5	(1 / 1 / 1)	(1 / 1 / –)	(– / – / 1)
APC/PC/AC	115.3020	3.3211	6	(1 / $\frac{1}{16}$/ *)	(– / $\frac{1}{16}$ / 1)	(1 / – / 2)
AP/AP/AP	115.3402	3.3594	6	(* / * / –)	(1 / 1 / –)	(2 / 1 / –)
AC/AC/AC	115.4111	3.4303	6	(* / – / *)	($\frac{1}{128}$/ – / 1)	(1 / – / 2)
AC/PC/APC	115.4147	3.4339	6	(1 / ·· / *)	(– / $\frac{1}{128}$/ 1)	(1 / $\frac{1}{128}$/ 2)
C/C/C	115.4197	3.4388	4	(– / – / *)	(– / – / 1)	(– / – / 1)
APC/AC/APC	115.6078	3.6270	7	(* / 1 / *)	($\frac{1}{128}$/ – / 1)	(1 / 1 / 1)
APC/PC/APC	115.6230	3.6422	7	(1 / * / *)	(– / $\frac{1}{128}$/ 1)	(1 / 1 / 1)
APC/PC/AP	115.7605	3.7796	6	(2 / * / 1)	(– / $\frac{1}{128}$/ 1)	(2 / 1 / –)
APC/AC/AP	115.7642	3.7833	6	(* / 1 / 1)	($\frac{1}{128}$/ – / 1)	(2 / 1 / ··)
APC/AP/AC	115.8790	3.8982	6	(* / 1 / 1)	(1 / 1 / –)	($\frac{1}{2}$ / ·· / 1)
AP/APC/APC	116.0150	4.0342	7	(* / 2 / –)	($\frac{1}{128}$/$\frac{1}{128}$/ 1)	(4 / * / 1)
AC/AC/C	116.3551	4.3743	5	($\frac{1}{4}$ / – / *)	($\frac{1}{4}$ / – / 1)	(– / – / 1)
PC/C/PC	116.4644	4.4836	5	(– / $\frac{1}{2}$ / *)	(– / – / 1)	(– / $\frac{1}{2}$ / 1)

Letting $\eta = (\eta_{111}, \ldots, \eta_{11R}, \ldots, \eta_{IJ1}, \ldots, \eta_{IJR})'$ and defining y, X, β, D_*, and Σ appropriately, we can represent the model (9) and the prior density $\pi(\beta_* \mid \Sigma)$ as the same forms (3) and (6), respectively, where $M = (R-1)(I+J+K-3)$. The likelihood of the model is given by

$$f(\beta \mid y) = \prod_{i=1}^{I} \prod_{j=1}^{J} \frac{m_{ij}!}{y_{ij1}! \cdots y_{ijR}!} \pi_{ij1}^{y_{ij1}} \cdots \pi_{ijR}^{y_{ijR}} \tag{13}$$

based on the product multinomial distribution. The estimation of β can be conducted in a similar manner as explained in Section 2. Actually, ABIC is evaluated approximately as a function of the hyperparameters:

$$\begin{aligned} \text{ABIC}(\Sigma) \simeq{} & 2y'(\log y - \log \hat{\mu}) + \hat{\beta}_*' D_*' \Sigma^{-1} D_* \hat{\beta}_* \\ & - \log |D_*' \Sigma^{-1} D_*| + \log |R^{-2} X_*' \hat{V}^{-1} X_* + D_*' \Sigma^{-1} D_*| \\ & + 2(h + R - 1) \end{aligned} \tag{14}$$

where

$$\hat{\mu} = (m_{11}\hat{\pi}_{11}', \ldots, m_{IJ}\hat{\pi}_{IJ}')',$$

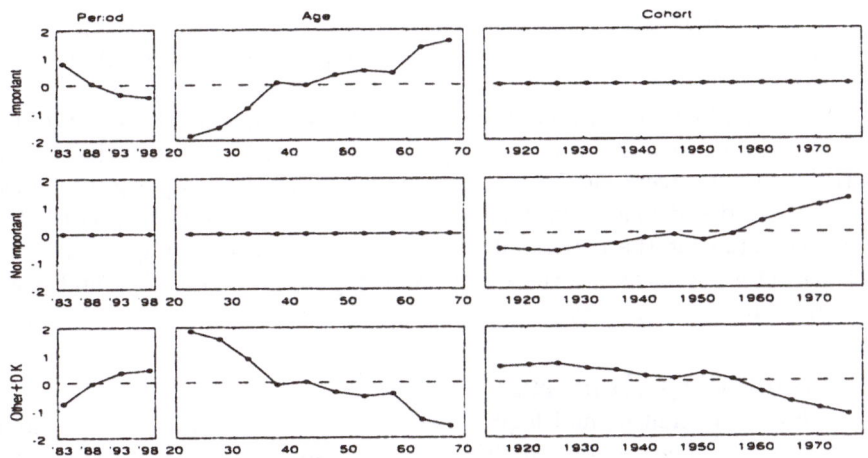

Figure 2: Plot of the parameter estimates of the best model for "Spirituality" for Japanese males, based on the BMLC model applied to all the categories simultaneously.

$$\widehat{\pi}_{ij} = (\widehat{\pi}_{ij1}, \ldots, \widehat{\pi}_{ijR})',$$

$$\widehat{V} = \bigoplus_{j,i} m_{ij}[(\mathrm{diag}\widehat{\pi}_{ij}) - \widehat{\pi}_{ij}\widehat{\pi}'_{ij}].$$

The estimate $\widehat{\beta}$ is obtained by maximizing $\log f(\beta \mid y) \cdot \pi(\beta_* \mid \Sigma)$ for fixed hyperparameters in Σ. When the number of categories $R = 2$, the Bayesian multinomial logit cohort (BMLC) model in this section coincides with the Bayesian logit cohort (BLC) model in Section 2.

We can also consider and compare many variant models. The numbers of the BMLC models are 8 for $R = 2$, 125 for $R = 3$, 1728 for $R = 4$, 19683 for $R = 5$, and $(2^R - R)^3$ in general regardless of the way to choose the subset R_*. For examples, the APC/APC/APC model is the full cohort model for $R = 3$; the AP/C/APC model indicates that the age and period effects exist for both the first and the last categories and that the cohort effects exist for both the second and the last categories. Note that, for example, age effects cannot appear in just one category.

5. An Application of the BMLC Model

The BMLC models are applied to the same data in Section 3 on all three categories of "Spirituality" simultaneously.

Table 2 lists the top 20 BMLC models having smaller ABIC values than others. The symbol '*' in parenthesis in the table indicates that the model does not include the gradually-changing-parameter assumption to the effects for the corresponding category and the symbol '–' indicates that the model does not have the effects for the corresponding category. The AP/C/APC model is considered to be the best one for "Spirituality" for males.

Figure 2 plots the parameter estimates of the best one, the AP/C/APC model. The consistent results among the categories can be seen here. The period and age effects for the category "Important" are compensated by the reverse period and cohort effects, respectively, for the category "Other+D.K." It is the same for the cohort effects except between "Not important" and "Other+D.K." Of course, the constranits $\sum_r \hat{\pi}_{ijr} = 1$ for all i and j are guaranteed.

6. Concluding Remarks

A Bayesian multinomial logit cohort (BMLC) model that treats simultaneously all categories of a multiple choice question is developed to avoid inconsistent results when we use the logit model with each category separated. If the results by the BLC model are inconsistent among categories, we need the BMLC model although the number of models to be compared is large and the amount of computation is huge when the number of categories $R > 4$. A more efficient computational algorithm is needed.

References:

Akaike, H.1980 Likelihood and the Bayes Procedure Bayesian Statistics Bernardo, J. M., et al. (eds.) 143–166 University Press Valencia

Mason, M. M. and Fienberg, S. E. (Eds.)1985Cohort Analysis in Social Research, Beyond the Identification ProblemSpringer-Verlag, New York

Nakamura, T.1982 A Bayesian cohort model for standard cohort table analysis Proc. Inst. Statist. Math.29277–97 (in Japanese)

Nakamura, T.1986 Bayesian cohort models for general cohort table analyses Ann. Inst. Statist. Math.32353–370 Nakamura, T.1995 Bayesian logit cohort models with age-by-period interaction effects and an over-dispersion parameter and their application to the data from the study of the Japanese national character Proc. Inst. Statist. Math.43199–119 (in Japanese)

Sakamoto, Y., Tsuchiya, T., Nakamura, T., Maeda, T., and Fouse, D. B.2000 A Study of the Japanese National Character: the Tenth Nationwide Survey (1998) Research Report (Inst. Statist. Math.)85

Mining Web Navigation Path Fragments

Wolfgang Gaul and Lars Schmidt-Thieme

Institut für Entscheidungstheorie und Unternehmensforschung,
University of Karlsruhe, D-76128 Karlsruhe, Germany
{ Wolfgang.Gaul, Lars.Schmidt-Thieme } @wiwi.uni-karlsruhe.de

Summary: For many web usage mining applications like, e.g., user segmentation, it is crucial to compare navigation paths of different users. We model user navigation path fragments by generalized subsequences that take into consideration local deviations but still sketch the global user navigational behavior. This paper presents a new algorithm of apriori type for mining all generalized subsequences of user navigation paths with prescribed minimal occurrence from a given database.

1. Introduction

E-commerce needs web usage mining that aims at considering different phases of consumer behavior, extending the focus from classical buying behavior analysis to data mining of different kinds of contacts with (potential) customers. User navigation paths in the web or even fragments of visits of websites establish an important source of information. For most higher level analytical tasks and applications like user segmentation, recommender systems etc., paths of different users have to be compared. Most path distances can be viewed as ordinary distance measures on a feature space of path fragments. As this space turns out to be high-dimensional and sparsely populated, dimension reduction schemes are needed. One such scheme consists in selecting the subspace spanned by frequent path fragments.

There are different kinds of fragments: The simplest kinds of fragments are occurrences of single pages or sets of pages in a user path. Frequent page sets can be mined by the standard apriori algorithm (see Agrawal and Srikant (1994)). As subsets neglect the sequential structure of user paths, better choices for path fragments are subsequences. Frequent *contiguous subsequences* can be mined by a well known variant of the apriori algorithm (see Agrawal and Srikant (1995) with modifications by Srikant and Agrawal (1996)). Borges and Levene (1998 and 1999) have developed algorithms for sequence mining on aggregated data. As a third kind of path fragments *generalized subsequences* containing wildcards have been proposed in the web mining literature (see Spiliopoulou (1999)). Generalized subsequences sketch the global navigational behavior of users. Several algorithms exist to mine frequent generalized subsequences of a specified type (called templates, i.e., subsequences with prescribed positions of wildcards, see Spiliopoulou (1999) and Buechner et al. (1999)). Other authors following a broader approach have constructed algorithms to find frequent subsequences of pages with attached attributes and relations (called generalized episodes, see Mannila and Toivonen (1996)). While those algorithms are perfectly suited for use in interactive analysis, a general algorithm mining *all* frequent generalized subsequences (of a given minimal support) still is missing. In this paper we describe a new algorithm that fills this gap.

2. Formal Background

Let R be an arbitrary set of resources extracted from a webserver's logfile, where the navigational behavior of anonymous visitors has been recorded, and $R^* := \bigcup_{i \in \mathbb{N}} R^i \cup \{\emptyset\}$ the set of finite sequences of elements of R (with $\emptyset$ as the empty sequence), here used to model user paths. For a sequence $x \in R^*$ the *length* $|x|$ is the number of symbols in the sequence ($|x| := n$ for $x \in R^n$, $|\emptyset| := 0$). Let $x, y \in R^*$ be two such sequences. We say that x *is a subsequence of* y ($x \leq y$), if there is an index $i \in \{0, \ldots, |y| - |x|\}$ with $x_j = y_{i+j}$ $\forall j = 1, \ldots, |x|$. x is a *strict subsequence of* y ($x < y$), if it is a subsequence of y but not equal to y ($x \leq y \wedge x \neq y$).

A pair of sequences $x, y \in R^*$ *is overlapping on* $k \in \mathbb{N}_0$ *elements*, if the last k elements of x are equal to the first k elements of y ($x_{last-k+i} = y_i$ $\forall i = 1, \ldots k$). For such a pair of sequences $x, y \in R^*$ overlapping on k elements we define the *k-telescoped concatenation of* x *and* y to be

$$
\begin{aligned}
x +_k y &:= (x_1, \ldots, x_{last-k}, y_1, \ldots, y_{last}) \\
&= (x_1, \ldots, x_{last}, y_{k+1}, \ldots, y_{last}).
\end{aligned}
$$

Note that any two sequences are 0-overlapping and the 0-telescoped concatenation of two sequences is just their arrangement one behind the other. For a pair of sets of sequences $X, Y \subseteq R^*$ we denominate the set of k-overlapping pairs $x \in X, y \in Y$ by $X \ominus_k Y$ and the set of k-telescoped sequences of all k-overlapping pairs shortly as the *set of k-telescoped sequences of X and Y*:

$$
\begin{aligned}
X +_k Y &:= +_k(X \ominus_k Y) \\
&= \{x +_k y \mid x \in X, y \in Y \text{ are over-} \\
&\qquad \text{lapping on } k \text{ elements}\}.
\end{aligned}
$$

Now let S be a finite list of such sequences $x \in R^*$ (allowing multiplicities if different users take the same path). For an arbitrary sequence $x \in R^*$ we denominate the relative frequency of sequences of S containing x as subsequence as *support of x with respect to S*:

$$
\sup_S(x) := \frac{|\{s \in S \mid x \leq s\}|}{|S|}
$$

The task of *searching all frequent subsequences* in the given list of sequences S means to find all sequences $x \in R^*$ with at least a given minimal support, i.e. with $\sup_S(x) \geq$ minsup and minsup $\in \mathbb{R}^+$ a given constant. As the support of subsequences of a sequence is greater than or equal to the support of the sequence itself, one can build frequent subsequences recursively starting from the sequences of length $n = 1$. With all sequences of length 1 as initial *set of candidates* the algorithm performs two steps: first, it computes the support values of all candidates and selects those candidates as frequent subsequences that satisfy the minimal support constraint; second, it builds a new set of candidates of length $n + 1$ for the next step by trying to join frequent subsequences of length n in the following manner: two sequences c and d of length n are joined to a sequence of length $n + 1$ if they overlap on $n - 1$ elements, i.e. $(c_2, \ldots, c_n) = (d_1, \ldots, d_{n-1})$; the joined sequence is $c +_{n-1} d$. Algorithm 1 gives the formal description of this procedure.

Algorithm 1 Apriori algorithm adapted for sequences

Require: set of items R (resources), list S of (finite) sequences (user paths) of elements of R, minimal support value minsup $\in \mathbb{R}^+$.

Ensure: set of frequent subsequences $F := \bigcup_{n \in \mathbb{N}} F_n$ of the sequences of S with support of at least minsup.

$C := \{\{r\} \mid r \in R\}$ set of initial candidates,

$n := 1$.

while $C \neq \emptyset$ **do**

    compute $\sup_S(c) \quad \forall c \in C$ by counting the number of occurrences of each c in S (one loop through S).

    $F_n := \{c \in C \mid \sup_S(c) \geq \text{minsup}\}$

    $C := F_n +_{n-1} F_n$ {compute new candidate sequences with length n+1}

    $n := n + 1$

end while

Please note, that for the special case of sequences describing paths on a graph, in the first join step only 0-overlapping pairs of sequences of length 1, i.e., pairs of nodes of the graph, have to be considered that are linked by an edge. — This adaption of the classical apriori algorithm for sets (see Agrawal and Srikant (1994)) to sequences has first been published by Agrawal and Srikant (1995) (with modifications by Srikant and Agrawal (1996)). It has been used for finding subsequences in web mining paths by Chen et al. (1996) and other authors afterwards (Viveros et al. (1997), Chen et al. (1998), Cooley et al. (1999)).

3. Mining Frequent Generalized Subsequences

By a *generalized sequence in R* we mean a (finite ordinary) sequence in the symbols $R \cup \{*\}$ with an additional symbol $* \notin R$ called *wildcard*, such that no two wildcards are adjacent:

$$R^{\text{gen}} := \{x \in (R \cup \{\star\})^* \mid \not\exists i \in \mathbb{N} : x_i = x_{i+1} = \star\}$$

The wildcard symbol $\star$ is used to model partially indeterminate sequences, matching arbitrary subsequences. This notion of generalized sequence first has been introduced in web mining literature by Spiliopoulou and Faulstich (1998). For a generalized sequence $x \in R^{\text{gen}}$ we define its *length* $|x|$ as the length of the sequence in the symbols $R \cup \{\star\}$, i.e., $|x| := n$, if $x \in (R \cup \{\star\})^n$. Now let $x, y \in R^{\text{gen}}$ be two generalized sequences. We say that x *matches* y or y *generalizes* x ($y \vdash x$), if there exists a mapping

$$m : \{1, \ldots, |x|\} \to \{1, \ldots, |y|\}$$

(called *matching*) with the following properties:

1. m maps indices of elements of x to indices of elements of y that coincide or to a wildcard ($y_{m(i)} = x_i$ or $y_{m(i)} = \star$).

2. m covers all indices of y of non-wildcard elements ($y_i \in R \Rightarrow m^{-1}(i) \neq \emptyset$).

3. m is weakly monotonic increasing.

4. m is even strictly monotonic at places where its image does not belong to a wildcard
   $(m(i) = m(i + 1) \Rightarrow y_{m(i)} = \star)$.

Please note that as the set of ordinary sequences R^* is a subset of the set of generalized sequences R^{gen}, this also defines the notion of an ordinary sequence matching a generalized sequence. Obviously matchings are not uniquely determined by two generalized sequences x and y. A trivial example is $\star A \star \vdash AA$ with the two matchings $m_1 = \{(1,1), (2,2)\}$ and $m_2 = \{(1,2), (2,3)\}$. Finally we carry over the notions of subsequence and of k-telescoped concatenation from ordinary sequences to generalized sequences without any change. Note the difference between $A \star C$ not being a subsequence of $ABCD$ but matching a subsequence of it (i.e. $A \star C \vdash ABC$ and $ABC \leq ABCD$).

Again, let S be a finite list of ordinary sequences (user paths) $x \in R^*$. For an arbitrary generalized sequence $x \in R^{\text{gen}}$ we denominate the relative frequency of sequences containing a subsequence which matches x as *support of x with respect to S*:

$$\text{sup}_S(x) := \frac{|\{s \in S \mid \exists y \leq s : x \vdash y\}|}{|S|}$$

Now, *mining frequent generalized subsequences* is the label for the task to find all generalized sequences with at least a given minimal support. As we

Table 1: Construction of closed generalized subsequences of length ≥ 4.

	sequence	length		sequence	length
	ab...cd	n+1		a*b...cd	n+1
=	ab...c	n	=	a*b...c	n
$+_{n-1}$	b...cd	n	$+_{n-2}$	b...cd	n-1
	ab...c*d	n+1		a*b...c*d	n+1
=	ab...c	n-1	=	a*b...c	n-1
$+_{n-2}$	b...c*d	n	$+_{n-3}$	b...c*d	n-1

are looking for subsequences anyway, we can narrow our view to *closed generalized subsequences*, i.e. generalized subsequences without leading or trailing wildcard ($x \in R^{\mathrm{gen}}$ with $x_1, x_{\mathrm{last}} \in R$), that we call *path fragments*.

Up to now no general algorithm for finding all frequent generalized subsequences in a list of sequences is known. Spiliopoulou (1999) has invented an algorithm for finding frequent generalized subsequences in a limited subspace of the search space: her generalized sequence miner (GSM) looks for generalized sequences of a given length and wildcards at given positions (such subspaces are described by so called *templates*: see Buechner et al. (1999) for another approach using templates to limit the search space; templates are useful in the framework of interactive tools like WUM, see Spiliopoulou and Faulstich (1998) and Spiliopoulou et al. (1999)).

We present a modification of the apriori algorithm for sequences to path fragments, resulting in a general algorithm for finding frequent generalized subsequences. The idea is pretty simple. As we are looking only at closed generalized sequences, the support of any subsequence of such a closed generalized sequence again is greater than or equal to the support of the sequence itself. Now, as adjacent wildcards are not allowed, we can get every path fragment of length $n + 1$ (for $n \geq 3$) as junction of two overlapping path fragments of the kind described in table 1.

Thus, we only have to modify the join step of the apriori algorithm for building new candidates of length $n + 1$ in such a way that we not only use the frequent (closed generalized) subsequences of length n but also those of length $n - 1$ from the step before, and try all possible combinations. Closed generalized subsequences of length 3 containing a wildcard have the form $(x, *, y)$ with $x, y \in R$, shorter closed generalized subsequences cannot contain wildcards.

Algorithm 2 gives the exact formulation of the necessary comparisons. Of course, the computation of the support values of the candidate generalized sequences also has to be modified. The performance characteristics of the algorithm is the same as for the apriori algorithm for ordinary sequences: to find sequences of length n, n loops through the database have to be accomplished.

Algorithm 2 Apriori algorithm adapted for generalized sequences

---

Require: set of items R (resources), list S of (finite) sequences (user paths) of elements of R, minimal support value minsup $\in \mathbb{R}^+$.

Ensure: set of frequent (closed) generalized subsequences $F := \bigcup_{n \in \mathbb{N}} F_n$ of the sequences of S with support of at least minsup.

$C := \{\{r\} \mid r \in R\}$ set of initial candidates,

$n := 1$, $F_0 := \emptyset$.

while $C \neq \emptyset$ or $F_{n-1} \neq \emptyset$ **do**

    compute $\sup_S(c)$ $\forall c \in C$ by counting the number of occurrences of each c in S (one loop through S).

    $F_n := \{c \in C \mid \sup_S(c) \geq \text{minsup}\}$

    $C := F_n +_{n-1} F_n$ {compute new candidate sequences with length n+1}

    if $n = 2$ **then** {introduce wildcards}

        $C := C \cup \{(x, \star, y) \mid x, y \in F_{n-1}\}$

    else if $n > 2$ **then** {additional joins considering wildcards}

        $C \quad := C$

            $\cup \{x +_{n-2} y \mid (x, y) \in F_n \ominus_{n-2} F_{n-1},$

                  $x_2 = \star\}$

            $\cup \{x +_{n-2} y \mid (x, y) \in F_{n-1} \ominus_{n-2} F_n,$

                  $y_{\text{last} - 1} = \star\}$

            $\cup \{x +_{n-3} y \mid (x, y) \in F_{n-1} \oplus_{n-3} F_{n-1},$

                  $x_2 = y_{\text{last} - 1} = \star\}$

    end if

    $n := n + 1$

end while

As algorithms of the apriori type return all subsequences of the frequent sequences found, one often prunes the result set by removing all subsequences of a frequent sequence contained in the result set, thus retaining only the "maximal" subsequences:

$$F' := \{c \in F \mid \not\exists d \in F : c < d\}$$

For generalized subsequences the algorithm also returns all generalizations of all subsequences found. Reasonably one prunes the result set further, by removing all generalizations of a sequence contained in the result set, thus retaining only the "most concrete" subsequences:

$$F'' := \{c \in F \mid \not\exists d \in F : c \vdash d\}$$

We call these two pruning steps *subsequence pruning* and *generalization pruning*, respectively.

4. Example and Experiments

Figure 1 shows a simple example web site and some paths traveled on

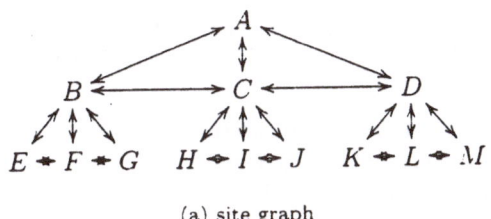

(a) site graph

nr	path
1	ABEF(EB)CHIJ
2	ACBE(BC)HI(HC)D
3	BCJ(C)HI
4	ABG(B)E(B)CH(C)I(C)D
5	ABEFG(FEB)CH(C)JI
6	ACJ(C)D(C)B(C)HI
7	BEFG(FEB)CHIJ(IHC)DKLM
8	ABF(B)CIH(I)J
9	ADK(D)L(D)AB(A)CHI
10	ABEFG(FEBA)CJ(C)HI(HC)D
11	ABCD(C)HIJ(IHCD)M
12	CBF(BC)H(C)DK(DC)I(CDKDCHCB)E

(b) analyzed paths

Figure 1: Example web site and example set of paths.

the site. Looking for ordinary frequent subsequences by applying the apriori algorithm for sequences (algorithm 1) does not give very useful results here: one finds the sequences CHI with a support of 8/12 and BCH with a support of 7/12. The first sequence containing more than three resources appears at support 5/12: EBCH.

Searching for frequent generalized sequences with algorithm 2 results in the set of three sequences with high support: B*C*H*I with support 12/12 and two lightly more specialized sequences B*CH*I and BC*H*I with support 11/12 and 10/12 respectively. Of course, the algorithm finds all literal subsequences of these sequences as well as all more general sequences (like B*H*I etc.), but these less useful subsequences are pruned by the two pruning steps (subsequence pruning and generalization pruning) presented at the end of section . — Already this simple example gives some insights into the good properties of generalized sequences: first, they are more robust than ordinary sequences against artifacts coming from navigation path construction steps; second, they can cope with local deviations of the navigation paths,

256

thus resulting in longer paths with higher support values, i.e. they better sketch user navigational behavior in the large, contrary to local descriptions by ordinary subsequences.

We tested our algorithm for path fragments systematically with synthetic data. The data has been created by randomly instantiating a set of navigation templates. Each template describes the navigational behavior of a user segment by a generalized sequence (that may be open) and a distribution of the lengths of the replacements of the wildcards as well as its relative size by a weight. A navigation template is instantiated by randomly filling in the wildcards with concrete resource sequences.

Figure 2 shows the experimental results. We created datasets of different sizes from a set of 5 templates with relative sizes 0.3, 0.3, 0.2, 0.1, and 0.1, and $N(4,2)$-distributed replacement lengths on a site of 100 resources. — We implemented the Apriori algorithm for sets and our adaptation for path fragments using prefix trees (see, e.g., Mueller (1995)) in Java. All experiments have been run on the IBM JVM 1.3 on an Athlon-600 Linux-PC with 256 MB RAM.

Figure 2a shows the execution time for datasets of different sizes. As expected the execution time increases linear in the size of the dataset. The apriori algorithm for sets is only between 1 and 5% faster on this dataset. In figure 2b the dependency of the execution time on the minimum support is depicted in comparison to the execution times of the Apriori algorithm for sets. In both cases one can clearly see the steps in the performance curve at the support values that correspond to the weights of the user segments (0.3, 0.2, and 0.1). Figure 2c shows the execution times that the algorithms spends on the individual passes (i.e., for sequences of degree 1, 2, etc.). As support values for single items are computed in parallel with reading the data, the execution time for pass 1 includes the time for file I/O. For minimum support 0.2 most items are not frequent, so the algorithm has not much to check. For minimum support 0.1 all items are frequent and the main part of the computation time is spend on sequences of length 2 (20000 candidates have to be checked). For an even lower minimum support of 0.05 the frequent sequences of length 2 are common enough to yield a large pool of candidates of length 3 (ca. 15000 candidates).

5. Applications to Association Rules and Recommender Systems

As the retrieval of frequent (generalized) subsequences is the hard part of the generation of association rules, we can easily apply our algorithm to find association rules with prescribed minimal support and confidence. An *association rule* is (described by) a pair of (generalized) sequences $x, y \in R^{gen}$ with the meaning that if x (the body of the rule) has occurred then — under conditions to be explained in the following — y (the head of the rule) will occur,

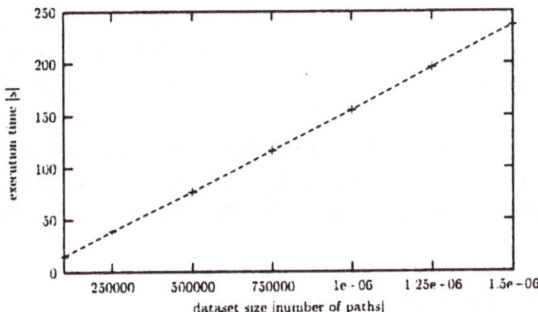

(a) Execution time depending on dataset size.

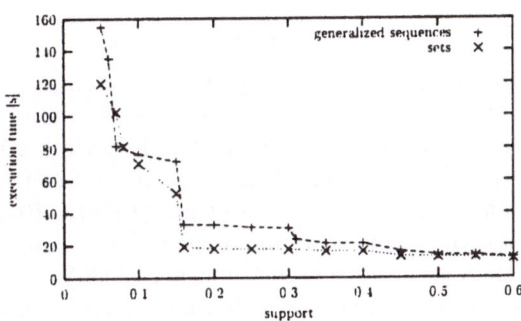

(b) Execution time depending on minimum support
(size of dataset: 500000).

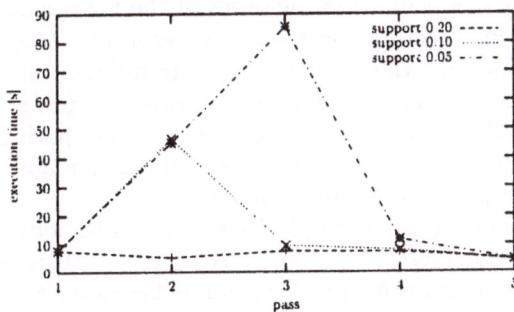

(c) Execution time depending on pass (size of dataset:
500000).

Figure 2: Experimental results.

too, where occurrence is related to the underlying list S of sequences. We suggest different interpretations of the rule notation that all have their origin in the web site traversal behavior of users as reconstructed via path completion and depicted in S. First, $x \to y \cong x +_1 y = (x_1, \ldots, x_{\text{last}} = y_1, \ldots, y_{\text{last}})$ which best corresponds to the usage of ordinary navigation paths. Second, $x \rightsquigarrow y \cong x \star y = (x_1, \ldots, x_{\text{last}}, \star, y_1, \ldots, y_{\text{last}})$, i.e., a wildcard is used to combine x and y. Both cases can be handled with the tools described so far. In addition to

$$\text{sup}_S(x \to y) := \text{sup}_S(x +_1 y) \quad \text{or}$$
$$\text{sup}_S(x \rightsquigarrow y) := \text{sup}_S(x \star y)$$

we need the confidence

$$\text{conf}_S(x \to y) := \frac{\text{sup}_S(x +_1 y)}{\text{sup}_S(x)} \quad \text{or}$$
$$\text{conf}_S(x \rightsquigarrow y) := \frac{\text{sup}_S(x \star y)}{\text{sup}_S(x)}$$

a number that counts the occurrence of $x +_1 y$ or $x \star y$ given x.

From early papers on web usage mining, the idea of feeding back the usage information extracted from the logfiles to the hyperlink structure of the underlying website has been suggested as an application of the results found by various data analysis tasks (see Yan et al. (1996)). Recently this idea has been revived by the name of *recommender system* making use of frequent item sets and association rules (see Mobasher (2000)). The paths of active users are compared to the left sides (the bodies) of a rule set previously extracted from the logfiles and (parts of) the right sides (the heads) of the matching rules with highest confidence are recommended via dynamically included direct hyperlinks.

Using only sets of resources (and not sequences) as the base for recommendations has the drawback of neglecting the order of the navigation patterns, and, thus, may result in directing users back to resources, they might no longer have an interest in. On the other hand, using ordinary subsequences as base for recommendations retains the order information, but only catches local navigational behavior. Generalized subsequences, i.e. path fragments, combine the strengths of the two methods, retaining order and not being bound to local behavior (by allowing deviations).

Let us go back to our simple example from above and look at user 9. Imagine he has already done ADK(D)L(D)AB(A)C. Using frequent ordinary subsequences we cannot recommend a next resource, because no subsequence of the frequent literal subsequences (CHI, BCH, BCHI and EBCH) can be found in his partial path. But the subsequence B⋆C of the frequent generalized subsequence B⋆C⋆H⋆I matches a subsequence of the tail B(A)C of his partial path. Thus, using the association rules B⋆C⤳H and B⋆C⤳I (both with support and confidence 1) we can recommend H and I for subsequent browsing, exactly the resources, he visits afterwards.

6. Outlook

Path fragments described by frequent closed generalized subsequences are ideal candidates to describe the user navigation path space, and thus the basis of distance computations of user paths, which in turn are necessary for clustering user paths. We will report about user path clustering resting upon path fragments in an upcoming paper.

References:

Agrawal, R. and Srikant, R. (1994): Fast Algorithms for Mining Association Rules. In: Bocca, J.B., Jarke, M., and Zaniolo, C. (eds.): *Proceedings of the 20th International Conference on Very Large Data Bases (VLDB'94)*, September 12-15, 1994, Santiago de Chile, Morgan Kaufmann, Chile. 487–499.

Agrawal, R. and Srikant, R. (1995): Mining Sequential Patterns. In: Yu, P.S., and Chen, A.L.P. (eds.): *Proceedings of the Eleventh International Conference on Data Engineering*, March 6-10, 1995, Taipei, Taiwan, IEEE Computer Society. 3–14.

Borges, J. and Levene, M. (1998): Mining Association Rules in Hypertext Databases. In: Agrawal, R. (ed.): *Proceedings / The Fourth International Conference on Knowledge Discovery and Data Mining*, August 27 - 31, 1998, New York, New York, Menlo Park, Calif., 149–153.

Borges, J. and Levene, M. (1999): Data Mining of User Navigation Patterns. In: *Proceedings of the Workshop on Web Usage Analysis and User Profiling (WEBKDD'99)*, August 15, 1999, San Diego, CA, Springer, 31–36.

Buechner, A.G., Baumgarten, M., Anand, S.S., Mulvenna, M.D., and Hughes, J.G. (1999): Navigation Pattern Discovery from Internet Data. In: *Proceedings of the Workshop on Web Usage Analysis and User Profiling (WEBKDD'99)*, August 15, 1999, San Diego, CA. Springer. 25-30.

Chen, M.-S., Park, J.S., and Yu, P.S. (1996): Data Mining for Path Traversal Patterns in a Web Environment. In: *Proceedings of the 16th International Conference on Distributed Computing Systems (ICDCS)*, May 27-30, 1996, Hong Kong, IEEE Computer Society, 385–392.

Chen, M.-S., Park, J.S., and Yu, P.S. (1998): Efficient Data Mining for Path Traversal Patterns. *IEEE Transactions on Knowledge & Data Engineering 10/2 (1998)*, 209–221.

Cooley, R., Mobasher, B., and Srivastava, J. (1999): Data Preparation for Mining World Wide Web Browsing Patterns. In *9th International Conference on Tools with Artificial Intelligence (ICTAI '97)*, November 3-8, 1997, Newport Beach, CA.

Mannila, H., and Toivonen, H. (1996): Discovering generalized episodes using minimal occurrences. In: *The Second International Conference on Knowledge Discovery and Data Mining (KDD '96)*, Portland, Oregon, August 2-4 1996, 146–151.

Mobasher, B. (2000): Mining Web Usage Data for Automatic Site Personalization. To appear in *Studies in Classification, Data Analysis, and Knowledge Organization 2000*.

Mueller, A. (1995): Fast Sequential and Parallel Algorithms for Association Rule Mining: A Comparison. Department of Computer Science, University of Maryland-College Park. CS-TR-3515.

Spiliopoulou, M. and Faulstich, L.C. (1998): WUM: A Tool for Web Utilization Analysis. In: Atzeni, P., Mendelzon, A., and Mecca, G. (eds.): *The World Wide Web and Databases, International Workshop WebDB'98*, Valencia, Spain, March 27-28, 1998, LNCS 1590, Springer, 184–203.

Spiliopoulou, M., Faulstich, L.C., and Winkler, K. (1999): A Data Miner Analyzing the Navigational Behavior of Web Users. In: *Proc. of the Workshop on Machine Learning in User Modeling of the ACAI'99 Int. Conf.*, Creta, Greece, July 1999.

Spiliopoulou, M. (1999): The Laborious Way from Data Mining to Web Mining. *Int. Journal of Comp. Sys., Sci. & Eng. 14 (1999)*, Special Issue on "Semantics of the Web", 113-126.

Srikant, R. and Agrawal, R. (1996): Mining Sequential Patterns: Generalizations and Performance Improvements. In: Apers, P.M.G., Bouzeghoub, M., and Gardarin, G. (eds.): *Advances in Database Technology - EDBT'96, 5th International Conference on Extending Database Technology*, Avignon, France, March 25-29, 1996, Proceedings. LNCS 1057, Springer.

Viveros, M.S., Elo-Dean, S., Wright, M.A., and Duri, S.S. (1997): Visitor's Behavior: Mining Web Servers. In: *Proceedings of the 1st International Conference on the Practical Application of Knowledge Discovery and Data Mining*, Blackpool 1997, 257–269.

Yan, T.W., Jacobsen, M., Garcia-Molina, H., and Dayal, U. (1996): From User Access Patterns to Dynamic Hypertext Linking. In: *Fifth International World Wide Web Conference* May 6-10, 1996, Paris, France.

Measures and Admissibilities for the Structure of Clustering

Akinobu Takeuchi[1], Hiroshi Yadohisa[2], and Koichi Inada[2]

[1] College of Social Relations, Rikkyo (St. Paul's) University,
3-34-1, Nishi-Ikebukuro, Tokyo 351-0034, Japan
[2] Department of Mathematics and Computer Science, Kagoshima
University, 1-21-35, Korimoto, Kagoshima 890-0065, Japan

Summary: The problem of selecting a clustering algorithm from the myriad of algorithms has been discussed in recent years. Many researchers have attacked this problem by using the concept of admissibility (e.g. Fisher and Van Ness, 1971, Yadohisa, et al., 1999). We propose a new criterion called the "structured ratio" for measuring the clustering results. It includes the concept of the well-structured admissibility as a special case, and represents some kind of "goodness-of-fit" of the clustering result. New admissibilities of the clustering algorithm and a new agglomerative hierarchical clustering algorithm are also provided by using the structured ratio. Details of the admissibilities of the eight popular algorithms are discussed.

1. Introduction

Several criteria for measuring the results of clustering algorithms have been proposed. Examples are the cophenetic correlation coefficient (Sokal and Rohlf, 1962), sum of squares (Hartigan, 1967), and Minkowski metrics (Jardine and Sibson, 1971). Takeuchi, et al. (1999) proposed the distortion ratio based on the concept of space distortion introduced by Lance and Williams (1967).

The well-structured criterion proposed by Rubin (1967) is another measure and is based on the dispersion of clusters. He defined data as well-structured (l-group) if there exist clusters $C_1, C_2, \ldots, C_l$ such that all within-cluster distances are smaller than the smallest between-cluster distance. Using this concept, Fisher and Van Ness (1971) proposed a new clustering algorithm admissibility called the well-structured admissible.

In this paper, we propose a new criterion for measuring clustering results called the "structured ratio". It includes the well-structured concept as a special case, and represents some kind of goodness-of-fit of a clustering result. New admissibilities and a new agglomerative hierarchical clustering algorithm (AHCA) are also provided by using the structured ratio, and details of the admissibilities of the eight popular algorithms are discussed.

Cluster I at stage m ($1 \leq m < N$) is denoted as $C_I(m)$. We denote the dissimilarity between objects p and q by d_{pq}, the dissimilarity between $C_I(m)$ and $C_J(m)$ by d_{IJ}, and the number of objects to be clustered by N.

We use the standard set theoretic notation $p \in C_I(m)$ to indicate that object p belongs to $C_I(m)$; the number of objects belonging to $C_I(m)$ is denoted by n_I. To simplify notation, we define $_{n_I}C_2 = 0$ if $n_I = 1$. Additionally, we suppose that $d_{pq}, d_{IJ} > 0$ and "tie" is broken though the analysis in some way.

We assume that clusters $C_I(m)$ are obtained using some AHCAs. From this assumption, the number of the clusters at stage m is $N - m$. When $C_T(m)$ and $C_K(m)$ are combined at stage m and $C_T(m)$ is not a singleton, it is assumed that $C_T(m)$ was formed from $C_I(t)$ and $C_J(t)$, which were combined at stage t $(1 < t < m)$, and that $C_K(m)$ is a singleton or was formed from $C_{I'}(t')$ and $C_{J'}(t')$, which were combined at stage t' $(1 \le t' < t)$. Hereafter, we assume this relationship between the two combined clusters, without loss of generality, and we assume $d_{IJ} < d_{IK} \le d_{JK}$.

We abbreviate the single linkage algorithm as SL, the complete linkage algorithm as CL, the weighted average algorithm (WPGMA) as WA, the median algorithm (WPGMC) as MD, the group average algorithm (UPGMA) as GA, the centroid algorithm (UPGMC) as CE, the minimum variance algorithm (Ward's method) as WD, and the flexible algorithm with $\beta = -0.25$ (see Gordon, 1996) as FX.

2. Structured Measures

Here we define the "structured ratio" as an extension of the well-structured concept that was first proposed by Rubin (1967). We define W_h as the dispersion within a cluster and B_h as the dispersion between clusters.

Definition 1: The structured ratio at stage m $(< N - 1)$ is defined as:

$$SR_h(m) = W_h(m)/B_h(m), \tag{1}$$

where $W_h(m)$ and $B_h(m)$ are within cluster and between cluster dispersions at stage m, respectively. We define several measures of within cluster and between cluster dispersion. For example, for $I \ne J$, which we assume hereafter,

$$W_1(m) = \max_I \max_{p,q \in C_I(m)} d_{pq}, \quad B_1(m) = \min_{I,J} \min_{p \in C_I(m), q \in C_J(m)} d_{pq},$$

$$W_2(m) = \sum_I \left(\max_{p,q \in C_I(m)} d_{pq} \right) \Big/ (N - m),$$

$$B_2(m) = \sum_{I,J} \left(\min_{p \in C_I(m), q \in C_J(m)} d_{pq} \right) \Big/ {_{N-m}C_2},$$

$$W_3(m) = \max_I \left(\sum_{p,q \in C_I(m)} d_{pq}/{_{n_I}C_2} \right), \quad B_3(m) = \min_{I,J} \left(\sum_{p \in C_I(m), q \in C_J(m)} d_{pq}/n_I n_J \right),$$

$$W_4(m) = \sum_I \left(\sum_{p,q \in C_I(m)} d_{pq}/n_I C_2 \right) \Big/ (N - m),$$

$$B_4(m) = \sum_{I,J} \left(\sum_{p \in C_I(m), q \in C_J(m)} d_{pq}/n_I n_J \right) \Big/ {}_{N-m}C_2,$$

$$W_5(m) = \sum_I \left(\sum_{p,q \in C_I(m)} d_{pq} \right) \Big/ \sum_I n_I C_2,$$

$$B_5(m) = \sum_{I,J} \left(\sum_{p \in C_I(m), q \in C_J(m)} d_{pq} \right) \Big/ \sum_{I,J} n_I n_J.$$

Using the same dispersion measures, we define another ratio for representing the structure of clustering results, while the structured ratio is defined for each combination.

Definition 2: The total structured ratio is defined as:

$$TSR_h(N - L) = \sum_{m=1}^{N-L} SR_h(m)/(N - L), \tag{2}$$

where L $(1 < L < N)$ is the number of clusters selected.

The total structured ratio can be used to measure the structure of clustering algorithms. If we would like to measure the final results of a clustering algorithm, $SR_h(N - L)$ may be more appropriate than the total structured ratio.

Since the structured ratio is the ratio of dispersion within a cluster to dispersion between clusters, a smaller value is preferable in terms of the concept of structure. However, the value of the structured ratio depends heavily on the dispersion measure. The characterization of dispersion measures still remains to be completed. However, we can obtain some useful information from the structured ratio using the following properties.

Property 1: $W_1(m)$ and $B_1(m)$ are monotone increasing functions.

Property 2: If $B_1(m+1)/B_1(m) \le W_1(m+1)/W_1(m)$ for all m, then SR_1 is a monotone increasing function.

Property 3: For any m $(< N - 1)$, the following inequalities hold;

$$SR_4(m) \le SR_2(m) \le SR_1(m), \quad SR_4(m) \le SR_3(m) \le SR_1(m).$$

Property 4: If

$$W_3 \le \min_I \max_{p,q \in C_I(m)} d_{pq} \quad \text{and} \quad \max_{I,J} \left(\min_{\substack{p \in C_I(m) \\ q \in C_J(m)}} d_{pq} \right) \le B_3$$

hold, then the following inequalities hold.

$$SR_4(m) \leq SR_3(m) \leq SR_2(m) \leq SR_1(m).$$

Property 5: The following equation hold for all m $(< N - 1)$;

$$\sum_I n_I C_2 W_5(m) + \sum_{I,J} n_I n_J B_5(m) = \sum_{p,q} d_{pq}.$$

3. ζ-structured Admissibility

In this section, we propose some admissibilities of the clustering by using the structured ratio and the total structured ratio defined in the previous section. Using the structured ratio, we redefine the condition of the well-structured (L-group) admissible first proposed by Fisher and Van Ness (1971), as follows.

An algorithm is well-structured (L-group) admissible if and only if the following equation is satisfied for any well-structured (L-group) data;

$$SR_1(N - L) < 1.$$

We defined an admissibility including this as a special case.

Definition 3: Suppose an algorithm classifies objects to L $(1 < L < N)$ clusters at stage m $(= N - L)$. If the following inequality is satisfied, the algorithm is ζ-structured (L-group) admissible;

$$SR_h(m) < \zeta. \tag{3}$$

The ζ-structured (L-group) admissible is defined at one combined stage. Next we define admissibilities for the entire set of combined stages.

Definition 4: Suppose an algorithm classifies objects to L $(1 < L < N)$ clusters at stage m $(= N - L)$. If the following inequality is satisfied for all n $(\leq m)$, the algorithm is ζ-structured (perfect) admissible;

$$SR_h(n) < \zeta. \tag{4}$$

Similarly, we can define admissibility by using the total structured ratio.

Definition 5: Suppose an algorithm classifies objects to L $(1 < L < N)$ clusters. If the following inequality is satisfied, the algorithm is ζ-total structured admissible;

$$TSR_h(N - L) < \zeta. \tag{5}$$

As is obvious from these definitions, the concept of ζ-structured admissible is determined only at the stage when the data is separated into L clusters.

This admissibility is a looser condition than the ζ-structured (perfect) admissible. In fact, for a small value of L, it is necessary to select quite a large value of ζ when the algorithm is ζ-structured (perfect) admissible.

These admissibilities satisfy the following properties.

Property 6: If an AHCA is ζ-structured (perfect) admissible, then the algorithm is ζ-structured (L-group) admissible and ζ-total structured admissible.

Property 7: If an AHCA is 1-structured (L-group) admissible, then the algorithm is well-structured (L-group) admissible, as proposed by Fisher and Van Ness (1971).

4. New Algorithm

In this section, we defined a new AHCA which has the minimum $SR_h(m)$ at the stage $m(\geq 2)$.

Definition 6: An AHCA that combines $C_I(m-1)$ and $C_J(m-1)$ to make a new cluster at stage $m(\geq 2)$ by minimizing $SR_h(m)$ is called MSR_h algorithm. When $m = 1$, it is supposed that MSR_h algorithm combines clusters which attains the between cluster (objects) distance minimum.

We note that the structured ratio is not a monotone (increasing) function with respect to m. Therefore dendrograms of this algorithm cannot be drawn. Additionally, the algorithm is not invariant on monotone transformations of distances used.

5. A Numerical Example

Here, we analyze an artificial dataset in two-dimensional space (see, Figure 1 and Table 1). These distances between objects are calculated by squared Euclidean distance. We anticipate that this data can be separated into three clusters.

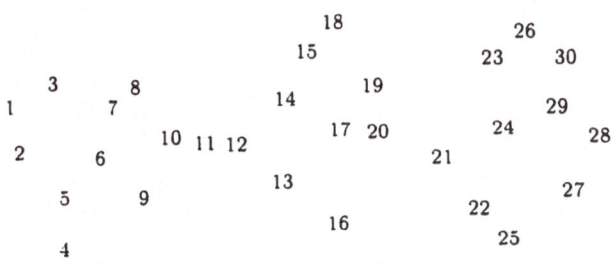

Figure 1: Scatter plots of 30 objects

Table 1: Coordinates of 30 objects in two-dimensional space

No	Coordinates	No	Coordinates	No	Coordinates
1	0.43, 3.22	11	0.55, 2.62	21	0.97, 3.51
2	1.12, 1.39	12	1.12, 2.05	22	1.57, 2.55
3	1.72, 3.21	13	2.01, 3.46	23	2.12, 2.05
4	2.52, 2.81	14	2.95, 2.74	24	3.34, 2.72
5	3.92, 2.25	15	3.95, 3.31	25	4.22, 3.91
6	4.62, 1.72	16	4.64, 2.91	26	4.55, 4.28
7	5.05, 3.48	17	5.12, 2.88	27	5.92, 2.55
8	6.38, 1.90	18	6.55, 3.82	28	6.68, 2.92
9	6.75, 1.51	19	6.95, 4.15	29	7.55, 2.12
10	7.88, 2.81	20	7.35, 3.20	30	7.46, 3.82

Here, we analyze the data using eight popular AHCAs and the MSR_1 algorithm.

The SR_1 of the results of these algorithms are represented on the ordinate in Figures 2 and 3. The values of $SR_1(27)$ and $TSR_1(27)$ are shown in these figures and the abscissa shows the combined stage. We select the 27th stage because there are three clusters that combine at this stage.

The structured admissibilities for $\zeta = 10, 15, 20$ or $\zeta = 4.5, 5, 10$ are indicated in Table 2. Generally, from the definitions, the structured admissibilities are sensitive concepts in contrast to the total structured admissibilities. For example, all algorithms are not 10-structured (3-group or perfect) admissible, but they are 10-total structured admissible. By changing the value of ζ, we can control the condition of the structured admissibilities. For example, the structured (perfect) admissibilities are changed from 'No' to 'Yes' by decreasing the value of ζ at the assessment of CL, GA, CE, WD, FX, and MSR_1 algorithms, respectively.

Table 2: ζ-structured admissibilities of the AHCAs

Admissible	SL	CL	WA	MD	GA	CE	WD	FX	MSR_1
10-structured (3-group)	No	No	No	No	No	No	No	No	No
15-structured (3-group)	No	Yes	No	No	No	No	No	No	No
20-structured (3-group)	No	Yes	Yes	Yes	Yes	Yes	Yes	Yes	Yes
10-structured (perfect)	No	No	No	No	No	No	No	No	No
15-structured (perfect)	No	No	No	No	No	No	No	No	No
20-structured (perfect)	No	Yes	No	No	Yes	Yes	Yes	Yes	Yes
4.5-total structured	No	Yes	No	No	Yes	No	Yes	Yes	No
5-total structured	No	Yes	Yes	Yes	Yes	Yes	Yes	Yes	Yes
10-total structured	Yes	Yes	Yes	Yes	Yes	Yes	Yes	Yes	Yes

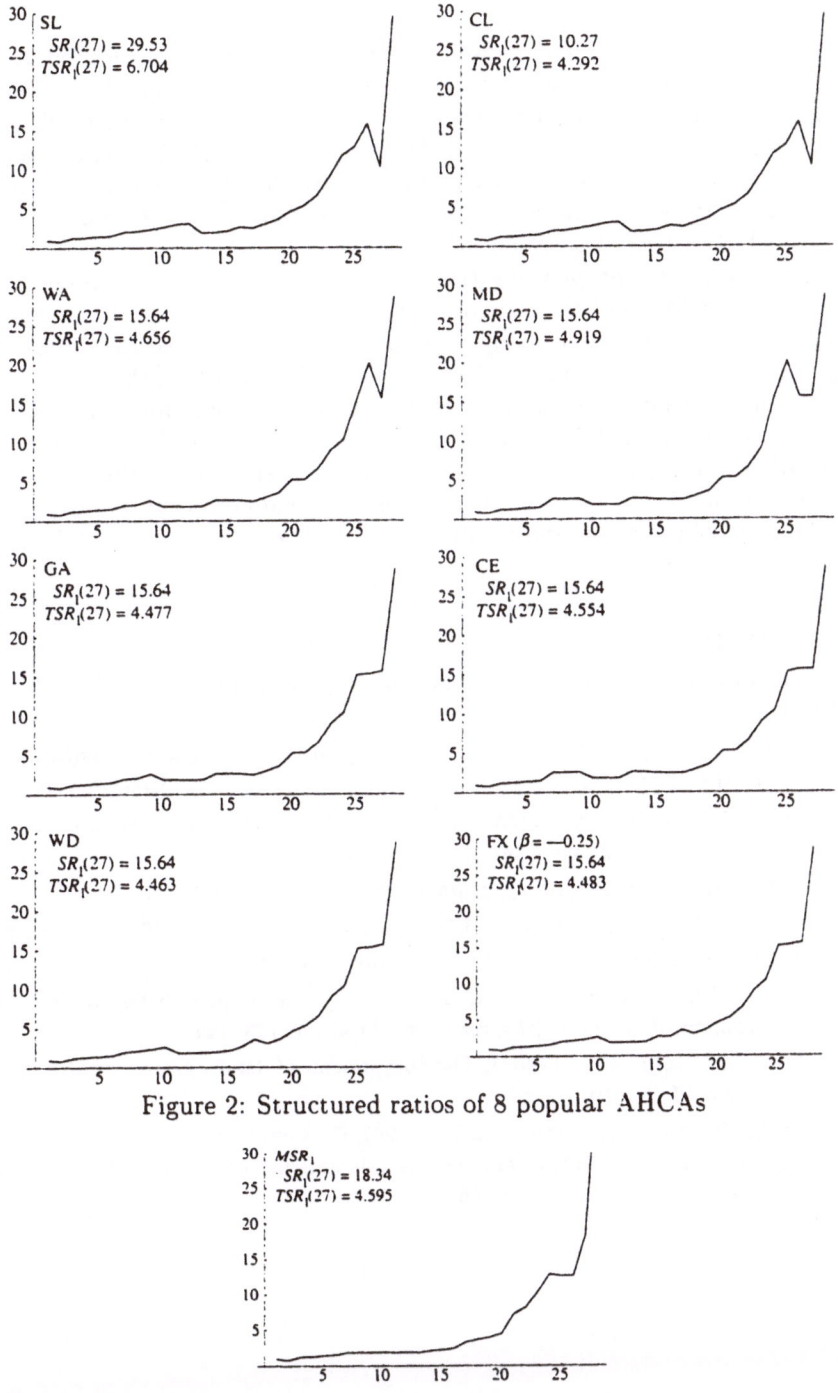

Figure 2: Structured ratios of 8 popular AHCAs

Figure 3: The structured ratio of MSR_1

6. Discussion

We consider the concept of well-structured, in which the desired classification condition has similar objects classified to the same cluster with small within-cluster dispersion, and dissimilar objects classified to different clusters with large between-cluster dispersion. However, the concept of well-structured is very strict, and is determined for specific data. The equation of the condition for a well-structured is only satisfied by a large L in data, and popular AHCAs are not satisfied with small values of L.

The structured concept using the structured ratio that we proposed can be used for any data and can control the condition of judgement. In addition, our concept can select from many criteria what is most suited for the user's purpose, not only the particular criterion. Thus, the concept of the structured includes the existing well-structured concept and can be used for more general and extensive cases. Additionally, this concept can numerically measure the degree of structure, so it can be used in a manner similar to admissibilities of an algorithm. By using this concept, we believe analysts can better select algorithms to obtain a desired result.

References:

Fisher, L. and Van Ness, J. (1971), Admissible clustering procedures, *Biometrika*, **58**, 91–104.

Gordon, A. D. (1996), Hierarchical classification, In: *clustering and classification*, P. Arabie, L. Hubert and G. Soete (Eds.), World Scientific, New Jersey, 65–121.

Hartigan, J. A. (1967), Representation of similarity matrices by trees, *Journal of the American Statistical Association*, **62**, 1140–1158.

Jardine, N. and Sibson, R. (1971), *Mathematical taxonomy*, London, Wiley.

Lance, G. N. and Williams, W. T. (1967), A general theory of classificatory sorting strategies: 1. hierarchical systems, *The Computer Journal*, **9**, 373–380.

Rubin, J. (1967), Optimal classification into groups: an approach for solving the taxonomy problem, *Journal of Theoretical Biology*, **15**, 103–144.

Sokal, R. R. and Rohlf, F. J. (1962), The comparison of dendrograms by objective methods, *Taxon*, **11**, 33–40.

Yadohisa, H., Takeuchi, A. and Inada, K. (1999), Developing criteria for measuring space distortion in combinatorial cluster analysis and methods for controlling the distortion, *Journal of Classification*, **16**, 45–62.

Multivariate Regressions, Genetic Algorithms, and Information Complexity: A Three Way Hybrid

Peter Bearse[1] and Hamparsum Bozdogan[2]

[1]Department of Economics, University of North Carolina at Greensboro,
Greensboro, NC 27402-6165, USA
E-Mail: bearse@uncg.edu
[2]Department of Statistics,336 SMC, The University of Tennessee,
Knoxville, TN 37996-0532, USA
E-Mail: bozdogan@utk.edu

Summary: We develop a computationally feasible intelligent data mining and knowledge discovery technique to select the best subset of predictors in multivariate regression (MR) models. Our approach integrates novel statistical modeling procedures based on the information-theoretic measure of complexity (ICOMP) criterion with the genetic algorithm (GA). When ICOMP is used as the fitness function, the GA, which by itself is an extremely clever non-local optimization algorithm, becomes an intelligent statistical model selection device capable of pruning combinatorially large numbers of sub-models to obtain an optimal or near-optimal subset MR model of multivariate data. We demonstrate our approach by determining the best predictors of taste and odor in a Japanese rice wine (i.e., sake) data set.

1. Introduction

Over the past thirty years, the statistical literature has placed more and more emphasis on model selection. Following the seminal work of Akaike (1973), the question is posed as one of choosing the best approximating model from a portfolio of competing statistical models by a suitable model evaluation criterion given a data set. The model that optimizes the criterion is deemed best.

A difficulty encountered with this approach is that the space of models one would like to consider is often quite large. For instance, in a regression setting with p responses and q predictors, exhaustive approaches to selecting the subset model with the best predictors would require the evaluation of 2^{pq} models. With even moderate values for p and q, this can be a daunting combinatorial task, and, with the advent of the high dimensional data sets now readily available in both business and science, it is an increasingly relevant one. Indeed, what is often required is the ability to identify the best-fitting model among a portfolio of competing models for a given complex data structure in real time.

In this paper, we propose an approach capable of doing just this. While our general approach is applicable to a variety of linear and nonlinear statistical modeling settings, we focus in this paper on Gaussian multivariate regression (MR) models. In section 2, we set-up the MR model using a vectorized notation that is convenient for the purposes of our genetic algorithm (GA)-based approach to subset MR model selection. In section 3, we briefly discuss Bozdogan's (2000) information complexity (ICOMP) criterion. In section 4, we describe our approach to selecting MR models using the GA with ICOMP as the fitness function. In section 5, we demonstrate our approach by using it to choose the best predictors of the perceived quality of Japanese rice wine (i.e., sake).

2. Multivariate Regression (MR)

We observe n observations of a p-dimensional response vector

$$\mathbf{y}_t = (\; y_{t1}, \quad y_{t2}, \quad \cdots, \quad y_{tp} \;)', \quad t = 1, 2, \ldots, n, \tag{1}$$

and a corresponding q-dimensional vector of predictors

$$\mathbf{x}_t = (\; x_{t1}, \quad x_{t2}, \quad \cdots, \quad x_{tq} \;)', \quad t = 1, 2, \ldots, n. \tag{2}$$

The responses are related to the predictors by the regression equation

$$\mathbf{y}_t = \mathbf{B}'\mathbf{x}_t + \varepsilon_t, \quad \varepsilon_t \sim i.i.d. \; N_p(0, \Sigma), \tag{3}$$

$$\varepsilon_t = (\; \varepsilon_{t1}, \quad \varepsilon_{t2}, \quad \cdots, \quad \varepsilon_{tp} \;)', \tag{4}$$

$$E(\varepsilon_{ti}| x_{\tau j}) = 0 \quad for \; all \; t, \; i, \; \tau, \; and \; j, \tag{5}$$

and

$$\mathbf{B} = [b_{ij}], \quad i = 1, 2, \ldots, q; \; j = 1, 2, \ldots, p \tag{6}$$

so that b_{ij} is the coefficient relating the ith predictor to the jth response. In compact matrix form, this Gaussian MR model is written as

$$\mathbf{Y} = \mathbf{XB} + \mathbf{E} \tag{7}$$

where

$$\underset{(n \times p)}{\mathbf{Y}} = (\; \mathbf{y}_1, \quad \mathbf{y}_2, \quad \cdots, \quad \mathbf{y}_n \;)', \tag{8}$$

$$\underset{(n \times q)}{\mathbf{X}} = (\; \mathbf{x}_1, \quad \mathbf{x}_2, \quad \cdots, \quad \mathbf{x}_n \;)', \tag{9}$$

and

$$\underset{(n \times p)}{\mathbf{E}} = (\; \varepsilon_1, \quad \varepsilon_2, \quad \cdots, \quad \varepsilon_n \;)'. \tag{10}$$

When no restrictions are imposed on the elements of $\mathbf{B}$, we call (7) a *saturated* MR. A *subset* MR model is obtained when we impose that some of the elements of $\mathbf{B}$ are equal to zero. For instance, imposing $b_{ij} = 0$ is equivalent to excluding the ith predictor (i.e., x_i) from the jth equation (i.e., the equation determining y_j).

It is convenient for our purposes to write the MR model using the vectorized notation

$$vec(\mathbf{Y}) = vec(\mathbf{XB}) + vec(\mathbf{E}) \tag{11}$$

$$= (\mathbf{I}_p \otimes \mathbf{X}) \, vec(\mathbf{B}) + vec(\mathbf{E}) \tag{12}$$

or, equivalently,

$$\underset{(np \times 1)}{\mathbf{y}} = \underset{(np \times pq)(pq \times 1)}{\mathbf{X}_{sup}} \underset{(np \times 1)}{\beta} + \underset{(np \times 1)}{\mathbf{e}}, \quad \mathbf{e} \sim N_{np}(\mathbf{0}, \Omega) \tag{13}$$

where

$$\mathbf{y} = vec(\mathbf{Y}), \quad \mathbf{X}_{sup} = (\mathbf{I}_p \otimes \mathbf{X}) \tag{14}$$

$$\beta = vec(\mathbf{B}), \quad \mathbf{e} = vec(\mathbf{E}), \tag{15}$$

and

$$\Omega = \Sigma \otimes \mathbf{I}_n. \tag{16}$$

Let

$$\beta = \underset{(pq \times m)(m \times 1)}{\mathbf{R}} \gamma, \quad m \le pq. \tag{17}$$

and impose zero restrictions on the elements of β by appropriate choice of $\mathbf{R}$. The column dimension of $\mathbf{R}$, denoted m, is equal to the number of unrestricted elements in β. That is,

$$m = pq - (\text{\# of zero restrictions on } \beta). \tag{18}$$

When the ith element of β is restricted to zero, the ith row of $\mathbf{R}$ contains all zeros. Otherwise, when the ith element of β is unrestricted, the ith row of $\mathbf{R}$ has a one in its jth column and zeros elsewhere where j equals the number of nonzero (i.e., unrestricted) elements of β in the rows prior to i. That is, if we let

$$\beta = (\beta_1, \beta_2, \cdots, \beta_{pq})' \tag{19}$$

and define the indicator

$$\chi_i = \begin{cases} 1 & if \quad \beta_i \text{ is not restricted to zero} \\ 0 & \text{otherwise} \end{cases}, \tag{20}$$

then the ith row of $\mathbf{R}$ contains a one in its jth column, where

$$j = \sum_{k \le i} \chi_i, \tag{21}$$

and zeros in its remaining $(m-1)$ columns. The m-vector γ contains the unrestricted coefficients in β.

Every subset MR model can be represented through an appropriate choice of $\mathbf{R}$. To see this, substitute (17) into (13) to obtain

$$\begin{aligned} \mathbf{y} &= \mathbf{X}_{sup}\beta + \mathbf{e} \tag{22} \\ &= \mathbf{X}_{sup}\mathbf{R}\gamma + \mathbf{e} \tag{23} \end{aligned}$$

or, equivalently,

$$y = \mathbf{X}^{\bullet}_{sup}\gamma + \mathbf{e} \tag{24}$$

where the columns of

$$\mathbf{X}^{\bullet}_{sup} = \mathbf{X}_{sup}\mathbf{R} \tag{25}$$

contain the predictors for the subset model.

When written as in (24), the subset MR model can be viewed as a multiple regression model subject to both heteroskedasticity and autocorrelation since, in general, at least some diagonal elements of

$$\Omega \equiv \mathbf{cov}\,(\mathbf{e}) \tag{26}$$

will be unequal and at least some off-diagonal elements of Ω will be nonzero. However, our formulation is equivalent to Zellner's (1962) seemingly unrelated regressions (SUR), and Ω is completely determined by the observational error covariance matrix

$$\Sigma \equiv \mathbf{cov}\,(\varepsilon_t)\,. \tag{27}$$

Since Σ contains at most $p\,(p + 1)\,/2$ distinct elements, consistent estimation of Ω is possible without further restrictions.

We use a two-step method to estimate the unrestricted coefficients γ in the subset MR model. In the first step, we construct the least squares (LS) estimator

$$\widehat{\gamma} = \left(\mathbf{X}^{\bullet\prime}_{sup}\mathbf{X}^{\bullet}_{sup}\right)^{-1}\mathbf{X}^{\bullet\prime}_{sup}\mathbf{y} \tag{28}$$

and the LS residuals

$$\widehat{\mathbf{e}} = \mathbf{y} - \mathbf{X}^{\bullet}_{sup}\widehat{\gamma}. \tag{29}$$

Let $\widehat{e}_i$ denote the ith element of $\widehat{\mathbf{e}}$ and define

$$\widehat{\mathbf{E}} = \begin{bmatrix} \widehat{e}_1 & \widehat{e}_{n+1} & \cdots & \widehat{e}_{n(p-1)+1} \\ \widehat{e}_2 & \widehat{e}_{n+2} & \cdots & \widehat{e}_{n(p-1)+2} \\ \vdots & \vdots & \ddots & \vdots \\ \widehat{e}_n & \widehat{e}_{2n} & \cdots & \widehat{e}_{np} \end{bmatrix} = \begin{bmatrix} \widehat{\varepsilon}_{1,1} & \widehat{\varepsilon}_{1,2} & \cdots & \widehat{\varepsilon}_{1,p} \\ \widehat{\varepsilon}_{2,1} & \widehat{\varepsilon}_{2,2} & \cdots & \widehat{\varepsilon}_{2,p} \\ \vdots & \vdots & \ddots & \vdots \\ \widehat{\varepsilon}_{n,1} & \widehat{\varepsilon}_{n,2} & \cdots & \widehat{\varepsilon}_{n,p} \end{bmatrix}. \tag{30}$$

A consistent estimator of Ω is given by

$$\widehat{\Omega} = \widehat{\Sigma} \otimes \mathbf{I}_n \tag{31}$$

where

$$\widehat{\Sigma} = \frac{1}{n}\widehat{\mathbf{E}}'\widehat{\mathbf{E}}. \tag{32}$$

In the second step, we construct the feasible generalized least squares (FGLS) estimator

$$\tilde{\gamma} = \left(\mathbf{X}_{sup}^{*\prime} \widehat{\Omega}^{-1} \mathbf{X}_{sup}^{*} \right)^{-1} \mathbf{X}_{sup}^{*\prime} \widehat{\Omega}^{-1} \mathbf{y}. \tag{33}$$

The FGLS residuals are given by

$$\tilde{\mathbf{e}} = \mathbf{y} - \mathbf{X}_{sup}^{*} \tilde{\gamma} \tag{34}$$

and a consistent estimator of the covariance matrix of $\tilde{\mathbf{e}}$ is given by

$$\tilde{\Sigma} = \frac{1}{n} \tilde{\mathbf{E}}' \tilde{\mathbf{E}} \tag{35}$$

where, analogous to above, $\tilde{\mathbf{E}}$ is the reshaped version of $\tilde{\mathbf{e}}$. A consistent estimator of the covariance matrix of $\tilde{\gamma}$ is given by

$$\widehat{\mathrm{Cov}}\,(\tilde{\gamma}) = \left(\mathbf{X}_{sup}^{*\prime} \tilde{\Omega}^{-1} \mathbf{X}_{sup}^{*} \right)^{-1} \tag{36}$$

where $\tilde{\Omega} = \tilde{\Sigma} \otimes \mathbf{I}_n$.

3. Information Complexity (ICOMP)

A general form of Bozdogan's (2000) ICOMP criterion is given by

$$ICOMP = -2 \log L + 2C_1 \left(\widehat{F}^{-1} \right) \tag{37}$$

where $\log L$ is the fitted log likelihood, $\widehat{F}^{-1}$ is the estimated inverse Fisher information matrix, and

$$C_1 \left(\widehat{F}^{-1} \right) = \frac{\dim \left(\widehat{F}^{-1} \right)}{2} \log \left(\frac{tr \left(\widehat{F}^{-1} \right)}{\dim \left(\widehat{F}^{-1} \right)} \right) - \frac{1}{2} \log \left| \widehat{F}^{-1} \right| \tag{38}$$

is Van Emden's (1971) maximal covariance complexity index. In the spirit of Akaike's (1973) information criterion (AIC), ICOMP prefers models that, other things equal, minimize Kullback and Leibler's (1951) KL discrepancy between the true (but unknown) data generating process and the fitted process. Unlike AIC, ICOMP also considers the KL information divergence against independence (see Kullback (1968), Harris (1978), and Theil and Fiebig (1984)) of the parameter estimates. Consequently, when evaluating the parameterization of models, ICOMP penalizes not merely number of parameters but also their interactions/redundancy. Models with smaller ICOMP scores are deemed better.

For the case of a Gaussian subset MR model, the log likelihood evaluated at the FGLS estimates is

$$\log L = -\frac{np}{2} \log (2\pi) - \frac{n}{2} \log \left| \tilde{\Sigma} \right| - \frac{np}{2},$$

and the estimated inverse Fisher information matrix is

$$
\widehat{F}^{-1} \equiv
\begin{bmatrix}
\underset{(m \times m)}{\widehat{\mathrm{Cov}}\,(\widetilde{\gamma})} & \mathbf{0} \\
\mathbf{0} & \underset{\left(\frac{p(p+1)}{2} \times \frac{p(p+1)}{2}\right)}{\frac{2}{n}\mathbf{D}_p^+ \left(\widetilde{\Sigma} \otimes \widetilde{\Sigma}\right) \mathbf{D}_p^{+\prime}}
\end{bmatrix},
\tag{39}
$$

where $\mathbf{D}_p^+$ is the Moore-Penrose inverse of the duplication matrix $\mathbf{D}_p$ (i.e., the unique $p^2 \times p\,(p+1)\,/2$ matrix satisfying $\mathbf{D}_p vech\,(\Sigma) = vec\,(\Sigma)$). An equivalent computationally convenient expression for ICOMP(IFIM) — in the sense that all the required inputs are available as a part of the standard output of most regression packages — is

$$
ICOMP \;=\; np\left[1 + \log\,(2\pi)\right] + n\log\left|\widetilde{\Sigma}\right|
\tag{40}
$$

$$
+ s\log\left(\frac{tr\left(\widehat{\mathrm{Cov}}\,(\widetilde{\gamma})\right) + \frac{1}{2n}G}{s}\right)
\tag{41}
$$

$$
- \log\left|\widehat{\mathrm{Cov}}\,(\widetilde{\gamma})\right| - p\log\,(2) + \frac{p\,(p+1)}{2}\log\,(n)
\tag{42}
$$

$$
- (p+1)\log\left|\widetilde{\Sigma}\right|
\tag{43}
$$

where

$$
G \equiv tr\left(\widetilde{\Sigma}^2\right) + \left(tr\widetilde{\Sigma}\right)^2 + 2\sum_{j=1}^{p} \left(\widetilde{\sigma}_j^2\right)^2,
\tag{44}
$$

and $\widetilde{\sigma}_j^2$ is the jth diagonal element of $\widetilde{\Sigma}$.

4. Genetic Algorithm (GA) for Statistical Model Selection

While information criteria, such as ICOMP, provide a means of discriminating between good and bad models of data, we often face the computational problem of a combinatorially large number of subset MR in many applied settings. To make this problem tractable while still employing non-local search techniques and system-wide estimation methods, we develop a genetic algorithm (GA) for subset MR modeling. In this approach, a model selection criterion, such as ICOMP, determines the *fitness function* for the GA. In so doing, the GA, which by itself is an extremely clever non-local optimization algorithm, becomes an intelligent *statistical* model selection device capable of pruning combinatorially large numbers of sub-models to obtain an optimal or near-optimal subset MR representation of multivariate data.

The GA, which is attributed to Holland (1975), mimics aspects of biological evolution to determine optimal or near-optimal solutions to difficult optimization problems with rugged fitness landscapes. Following Bearse and Bozdogan (1998), to implement the GA, we first need a coding scheme for each subset MR model. Each subset MR model is coded as a string, where each locus in the string is a binary variable indicating the presence (1) or absence (0) of a given predictor variable. Every string has the same length, but each can contain a different binary coding representing a different combination of predictor variables. For example, in the case where $p = 2$ and $q = 3$, the unrestricted MR model is

$$
\begin{aligned}
y_{t1} &= b_{11}x_{t1} + b_{21}x_{t2} + b_{31}x_{t3} + \varepsilon_{t1} \\
y_{t2} &= b_{12}x_{t1} + b_{22}x_{t2} + b_{32}x_{t3} + \varepsilon_{t2}
\end{aligned}
\tag{45}
$$

and the string 100011 represents a model where the response variable y_1 includes only the predictor variable x_1 and the response variable y_2 includes only the predictor variables x_2 and x_3. We draw an initial population of N subset MR models by randomly initializing N binary strings, each of length pq. The choice of N is typically determined experimentally.

The GA also requires a fitness function. The fitness function provides the criteria to map the members of a candidate population into the domain of performance. It can be the hardest part of a GA since it is often difficult to induce the desired behavior on the basis of fitness, and its appropriate choice is problem dependent. In our case, we base the fitness function on ICOMP. Since smaller ICOMP values indicate better models, we define fitness as the negative of the relevant criterion value. Thus, models with lower criterion scores have higher fitness values. This marriage of the GA with an information theoretic model selection criterion expands the role of the GA to *statistical* modeling.

Given the initial population of models, the GA explores the model space through an evolutionary process. This process continues for a number of generations (i.e., iterations) which can be pre-determined or subject to a termination criterion. Moving from generation to generation, new models are created through the operations of natural selection, crossover, and mutation.

The natural selection role of the GA is a mechanism by which we can expect members of the current generation with better fitness values to be better-represented in the mating pool used to determine the next generation's models. We use a form of ranking selection. We compute the ICOMP score of each subset MR model in a population of size N. We then sort these models in ascending order by the rank of the negative of their ICOMP scores. The model with the worst (i.e., largest) criterion value is ranked 1 while that with the best is ranked N. We then constructed a "weighted roulette wheel" with N bins, one for each subset MR model, where

$$
Bin\ Width\ for\ Model\ with\ Rank\ i = \frac{i}{N(N+1)/2}
\tag{46}
$$

We then draw N uniform (0,1) random numbers and include a model in the mating pool each time one of the random numbers falls in its bin.

This mating pool is subjected to a crossover operation that recombines subset MR models of the current generation (i.e., the parents) into new subset MR models for the next generation (i.e., the offspring). Crossover provides a means of combining aspects of existing models in the current population. To

Table 1: Subset MR Model Chosen by GA with ICOMP as the Fitness Function

Predictor	Response	
	sake taste (y_1)	sake odor (y_2)
pH (x_1)	0	0
acidity 1 (x_2)	1	0
acidity 2 (x_3)	0	0
sake-meter (x_4)	0	1
direct reducing sugar (x_5)	0	0
total sugar (x_6)	0	0
alcohol content (x_7)	0	0
formol-nitrogen (x_8)	0	0
constant (x_9)	0	0

perform crossover, the subset MR models included in the mating pool are randomly paired. Each locus in the binary coding scheme is swapped with the corresponding locus of its mate with crossover rate (probability) p_c. The overall probability of at least one locus crossing over in a given mating pair is given by

$$p_c^* = 1 - (1 - p_c)^{pq}. \tag{47}$$

Whenever $p_c \in (0, 1)$, offspring are expected to differ from the mating pool.

The resulting N offspring models are then subjected to mutation. Mutation is a means of creating new combinations of variables not available in the current population, thereby allowing the GA to explore models not attainable through crossover alone. We permit mutation by specifying a mutation rate (probability) p_m at which a locus changes from 0 to 1 or 1 to 0. That is, for each offspring model, mutation allows predictor variables to be added or removed randomly.

5. Example

Here we consider a data set on Japanese rice wine (i.e., sake) containing 30 observations (i.e., $n = 30$) on two responses (i.e., $p = 2$) and nine predictors (i.e., $q = 9$). The variable definitions are provided in Table 1. Since the saturated (i.e., unrestricted) MR model contains 9 predictors for each of its 2 responses, there are

$$2^{18} = 262,144 \tag{48}$$

subset MR models (including the empty and the saturated models). The ICOMP score for the saturated MR model is 144.4794 with a C_1 complexity of 58.4120.

We applied our GA-based procedure to choose a subset MR model for this data set. We set the GA to run for 50 generations with population size 76. The initial population size was also equal to 76. Consequently, a run of the GA entails evaluating

$$50 \cdot 76 + 76 = 3,876 \tag{49}$$

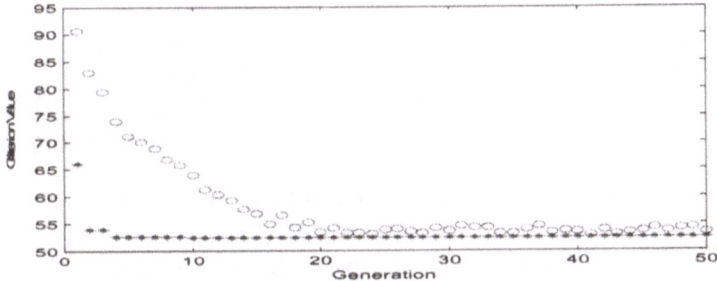

Figure 1: Summary of a GA Run for the Rice Data ('*' = Minimum ICOMP(IFIM), '0' = Average ICOMP(IFIM))

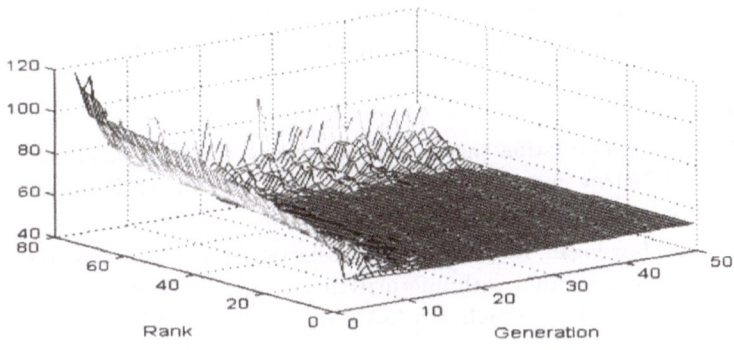

Figure 2: All Models Evaluated by GA

subset models which is 1.48% of the model space. We imposed an elitist rule to ensure that the GA never discards the best model found. The crossover rate p_c was set to 0.5 and the mutation rate p_m to 0.01. Figure 1 plots a summary of the ICOMP scores found over the evolution of our GA-based procedure. Figure 2 plots the ICOMP score for each of the 3,876 models evaluated by the GA.

Again referring to Table 1, we see that our GA-based approach chooses a subset MR model including only the predictor x_2 (acidity 1) in the equation for y_1 (sake taste), and only the predictor x_4 (sake-meter) in the equation for y_2 (sake odor). The ICOMP(IFIM) score for this subset MR model is 52.1637 with a C_1 complexity of 0.9311, which is a drastic improvement, according to ICOMP, over the saturated model. In particular, the saturated model is much more complex, according to C_1, than the best subset model.

In the typical applied setting, it is not feasible to evaluate each subset MR model in a reasonable amount of time. However, in the context of this small example, we can. We find that the model chosen by this run of the GA is indeed the one that minimizes ICOMP.

6. Conclusion

In this paper, we have outlined a new and novel approach to identifying the optimal predictors in multivariate regression. Our approach uses a genetic algorithm (GA) with Bozdogan's (2000) information complexity (ICOMP) as the fitness function. In the context of a small example, we demonstrated the feasibility of our approach.

Since our GA approach is stochastic, it will be interesting to see how it performs in Monte Carlo environments. In particular, in setting where it is feasible to combinatorially determine the best subset MR model, how often does our GA-based approach find it? In settings where this is not possible, how does our GA-based approach fare relative to other combinatorial optimization methods? To what extent does its performance depend on the parameters chosen (e.g., number of generations, population size, crossover and mutation rates)? Additionally, while the framework described here applies directly to subset MR models, it is easily adapted to many other important modeling settings.

We plan to explore these interesting questions in future work.

References:

Akaike, H. (1973), Information theory and an extension of the maximum likelihood principle, In B.N. Petrov and F. Csaki (eds), Second International Symposium on Information Theory, Academiai Kiado, Budapest, 267-81.

Bearse, P.M. and Bozdogan, H. (1998), Subset selection in vector autoregressive models using the genetic algorithm with informational complexity as the fitness function, Systems Analysis, Modeling, Simulation (SAMS), Vol. 31, pp. 61-91.

Bozdogan, H. (2000), Akaike's Information Criterion and Recent Developments in Information Complexity, Journal of Mathematical Psychology, 44, 62-91.

Harris, C.J. (1978), An information theoretic approach to estimation, In M.J. Gregson (ed.), Recent Theoretical Developments in Control, London: Academic Press, pp. 563-590.

Holland, J.H. (1975), Adaptation in natural and artificial systems (Ann Arbor, MI USA: University of Michigan Press). Second Edition, 1992, MIT Press.

Kullback, S. (1968), Information Theory and Statistics, New York: Dover Publications.

Kullback, S. and Leibler, R. (1951), On information and sufficiency, Annals of Mathematical Statistics, Vol. 22, pp. 79-86.

Theil, H. and Feibig, D.G. (1984), Exploiting continuity: maximum entropy estimation of continuous distributions, Cambridge, MA: Ballinger Publishing Company.

Van Emden, M. H. (1971), An Analysis of Complexity, Mathematical Centre Tracts, 35, Amsterdam.

Zellner, A. (1962), An efficient method of estimating seemingly unrelated regression and tests for aggregation bias, Journal of the American Statistical Association, Vol. 57, pp. 348-368.

Statistical Software SAMMIF for Sensitivity Analysis in Multivariate Methods

Yuichi Mori[1], Shingo Watadani[2], Yoshiro Yamamoto[3], Tomoyuki Tarumi[4] and Yutaka Tanaka[4]

[1] Dept. of Socio-Information, Okayama University of Science
1-1 Ridai-cho, Okayama 700-0005, Japan
[2] Dept. of Liberal Arts and Science, Kurashiki University of Science and the Arts
2640 Nishinoura Tsurajima-cho, Kurashiki, 712-8505, Japan
[3] Dept. of Management and Information sciences, Tama University,
4-1-1 Hijirigaoka, Tama-shi, Tokyo 206-0022, Japan
[4] Dept. of Environmental and Mathematical Science, Okayama University
3-1-1 Tsushima-Naka, Okayama 700-8530, Japan

Summary: SAMMIF (Sensitivity Analysis in Multivariate Methods based on Influence Functions) is a statistical package for sensitivity analysis in multivariate methods in which diagnostics statistics are obtained for detecting influential observations and influential directions on the basis of both influence function approach and Cook's local influence approach. SAMMIF is designed to provide useful graphical user interface and some options for both beginners and specialists. The current version 1.0 performs sensitivity analysis fully in principal component analysis, canonical correlation analysis and exploratory and confirmatory factor analyses with some new diagnostics functions for the analyses. Practical examples illustrate that users can analyze the influence of observations without difficulties.

1. Introduction

The purpose of sensitivity analysis is to detect so-called influential observations or, in other words, to examine whether there exist any subsets of observations such that the obtained results depend heavily upon them. In sensitivity analysis candidates are specified, if any, for not only singly but also jointly influential observations and to confirm whether their influence is really large or not by comparing the results for the sample with and without the specified observations. From the aspect of practical application it is desirable that a useful means of performing sensitivity analysis is provided. Then we are going to develop a new statistical package SAMMIF which is an abbreviation of Sensitivity Analysis in Multivariate Methods based on Influence Functions.

In this paper we give a brief explanation of the backgrounds and theories used in SAMMIF, while the basic concepts have been implemented in the previous versions (e.g., Prototype (version 0.81) in Mori *et al.*, 1998). And we show the features and practical actions of SAMMIF focusing the new diagnostics functions implemented in the latest version (version 1.0).

2. Backgrounds and Motivations

There are two major mathematical tools in sensitivity analysis, Hampel's influence function (Hampel, 1974) and Cook's local influence (Cook, 1986). Methods of sensitivity analysis using either of these tools have been proposed by many authors including Radhakrishnan and Kshirsagar (1981), Critchley (1985), Tanaka (1988), Tanaka and Watadani (1992), and Wang and Lee (1996). The relationship between these two approaches has also been studied by Tanaka (1994), Tanaka and Zhang (1999) among others. Their papers illustrate that two approaches are equivalent where the same perturbation is introduced in both approaches (see the next section). That is, by taking into account the relationship we can detect jointly as well as singly influential observations in most multivariate methods and obtain information on influential directions in the sense of Cook's local influence even if we take influence function approach. In particular, principal component analysis (PCA) of empirical influence function (*EIF*) along with its interpretation from the perspective of Cook's local influence and the varimax method applied to the principal component (PC) scores are very effective for detecting influential subsets of observations. So it is desirable to implement this relationship in some statistical program to perform sensitivity analysis easily.

As for computer programs, Tanaka and his coworkers have proposed sensitivity analysis procedures using influence functions and their analogues and then developed their computer programs based on the proposed procedures, such as SAM (Tarumi and Tanaka, 1986), SAF/B (Odaka *et al.*, 1991), SAM II (Mori and Tarumi, 1993) and SACS (Watadani and Tanaka, 1994). However each of them was developed for a specified family of multivariate methods separately and most of them run on MS-DOS (BASIC) platform. Thus it was desirable to unify these separate programs into one Windows program.

Thus we started to develop a statistical package SAMMIF which utilizes the relationship of the major tools and unifies the separate methods. Then SAMMIF contains the expression "based on influence functions" in its name but it provides information on both the influence function and Cook's local influence using the relationship mentioned above.

3. Main Ideas
3.1 The Relationship between Influence Functions and Cook's Local Influence

Here we show the relationship between influence functions and Cook's local influence briefly (Tanaka, 1994; Mori *et al.*, 1998)

Consider the case where we analyze a sample of n observations $\{x_i; i = 1, \ldots, n\}$ using a multivariate method which contains an m-dimensional parameter vector θ. In the influence function approach a perturbation is introduced to the cumulative distribution function (cdf) from $\hat{F}$ to $(1 - \varepsilon_i)\hat{F} + \varepsilon_i \delta_{x_i}$, where δ is the cdf with a unit point mass at x_i. The first derivative of $\hat{\theta} = \theta(\hat{F})$ with respect to ε at $\varepsilon = 0$, which is simply denoted by $\hat{\theta}_i^{(1)}$, and is called the *EIF* of $\hat{\theta}$ at x_i, is computed to evaluate the influence of observations. We usually summarize the *EIF* vector into some scalar measures to evaluate the influence of a single (i-th) observation (single-case diagnostics, SD). Let $\hat{\theta}$ and $\hat{\theta}_{(A)}$ be the estimates based on the sample with/without a subset A of k observations. Then it is easily verified that the additive relation $\hat{\theta}_{(A)} \cong \hat{\theta} - (n-k)^{-1}\sum_{i \in A}\hat{\theta}_i^{(1)}$ holds. This relation suggests that, as a possible policy to detect influential subsets of observations, we should search for observations which have relatively large EIF vectors with similar directions from the origin (multiple-case diagnostics, MD). To do this by taking into account the correlations among the components of $\hat{\theta}$, we can use PCA with metric $[\widehat{acov}(\hat{\theta})]^{-1}$ and search for the observations as discussed above by inspecting the plots of PCs obtained by solving the eigenvalue problem (EVP)

$$\left(\frac{1}{n}\sum_{i=1}^{n}\hat{\theta}_i^{(1)}\hat{\theta}_i^{(1)^T} - \lambda[\widehat{acov}(\hat{\theta})]\right)u = 0. \tag{1}$$

Instead of the influence function type perturbation we may consider other types of perturbation such as case-weight or variance perturbation and additive perturbation. It is obvious that the additive property as stated above holds and therefore PCA can be applied to the first derivative vectors also in this case.

On the other hand, suppose we introduce a perturbation to the weight vector for n observations from $w_0 = (1,1,\dots,1)^T$ to w. Let $\hat{\theta}$ and $\hat{\theta}_w$ be the estimates for the unperturbed and perturbed cases, respectively. In Cook's local influence approach, the effect of the perturbation from w_0 to w is measured with the likelihood displacement defined as $LD(w) = 2[L(\hat{\theta}|w_0) - L(\hat{\theta}_w|w_0)]$, where $L(.)$ is a log likelihood function, and the effect is represented by a graph called influence graph $(w, LD(w))$. Considering the change of $LD(w)$ along a straight line $w = w_0 + ah$, where $\|h\| = 1$, Cook searches for the direction which has the maximum normal curvature at w_0. The maximum curvature and the most influential direction are obtained as the largest eigenvalue λ_{max} and the associated eigenvector h_{max} of an $n \times n$ EVP as

$$\left(2[\partial\hat{\theta}_w^T/\partial w][\widehat{acov}(\hat{\theta})]^{-1}[\partial\hat{\theta}_w/\partial w^T] - \lambda I\right)h = 0, \tag{2}$$

respectively, where $[\widehat{acov}(\hat{\theta})]^{-1} = -[\partial^2 L/\partial\theta\partial\theta^T]$, which is obtained from the theory of maximum likelihood estimation. Observations with large values of components in h_{max} are regarded as an influential subset of observations. When $\partial^2 L/\partial\theta\partial\theta^T$ is degenerated, the inverse can be replaced by the Moore-Penrose inverse. Usually we may naturally assume that the subspace spanned by the column spaces of $\partial\hat{\theta}_w/\partial w^T$ and $[\widehat{acov}(\hat{\theta})]^{-1}$ are the same. Then, after some algebraic manipulations the above $n \times n$ EVP(2) can be transformed to an $m \times m$ EVP

$$\left(2[\partial\hat{\theta}_w/\partial w^T][\partial\hat{\theta}_w^T/\partial w] - \lambda[\widehat{acov}(\hat{\theta})]\right)u = 0, \tag{3}$$

where u is defined as $h = [\partial\hat{\theta}_w/\partial w^T]u$. This EVP(3) is equivalent to EVP(1) except for the multiplying constant. In particular when we define the weights

$\overset{\bullet}{w_\alpha} = nw_\alpha / \sum_\beta w_\beta$, the relationship $n^{-1}\hat{\theta}_i^{(1)} = \partial\hat{\theta}_w / \partial w_i$ holds and the multiplying constants in EVP(1) and EVP(3) are $1/n$ and $2/n^2$, respectively. This relationship makes it possible to treat the influence function approach and Cook's local influence approach equivalently.

Thus SAMMIF can provide information on influential directions in the sense of Cook's local influence by performing MD in the influence function approach. Moreover, based on this idea such that the problem of searching for influential subsets is similar to the problem of searching for simple structure of the coordinates, the varimax method can be applied to PC scores for detecting influential subsets in SAMMIF.

3.2 General Procedure of Sensitivity Analysis

A general procedure of sensitivity analysis based on influence functions is as follows (Tanaka *et al.*, 1990).

Step 1 Compute the *EIF* vectors.

Step 2 Summarize the *EIF* vectors into scalar influence measures from various aspects such as the influence on the estimate $\hat{\theta}$. Find observations which are individually influential.

Step 3 Search for subsets of observations whose members are individually relatively influential and have similar influence patterns using PCA with/without taking into account the correlation among estimated parameters.

Step 4 Re-estimate the parameters based on the sample without "influential observations" found in *Step 2* and *Step 3*, and evaluate their influences.

4. Statistical Package SAMMIF

We started to develop SAMMIF in 1997. The current version is 1.0. Using SAMMIF, users can perform sensitivity analysis without difficulty according to the general procedure of sensitivity analysis (Tanaka *et al.*, 1990). It provides useful tools to detect influential observations in the Windows manner.

4.1 The Flow of SAMMIF

The flow of SAMMIF is as follows (see, Figure 1):

1) *Data entry (Data)*: SAMMIF reads a data file. Basic statistics and outlier checking based on the covariance matrix can be computed if necessary before the following steps.

2) *Prior analysis (Pre)*: Users choose a multivariate method in which they want to evaluate the influence of observations. SAMMIF applies the method to the data set read in 1) and estimates parameters ordinarily using all observations.

3) *Diagnostics*: User can do 3-1) and 3-2) in parallel. Two types of case-weight perturbation, ordinary case-weight perturbation and influence function type case-weight perturbation, are available to compute *EIF* vectors.

3-1) *Single-case diagnostics (SD)*: The *EIF* or its analogue of each observation for the estimated parameters are computed and then summarized into influence measures such as generalized Cook's distance. Users can display the measures in index plots and/or scatter plots to detect singly influential observations.

3-2) *Multiple-case diagnostics* (*MD*): PCA is applied to the *EIF* for all parameters or for a part of the parameters to detect candidates for influential subsets of observations. Users can display the PC scores in index plots and/or scatter plots. Influential directions in the sense of Cook's local influence can be displayed in index plots and observations which have similar influence and direction to each other can be indicated in scatter plots.

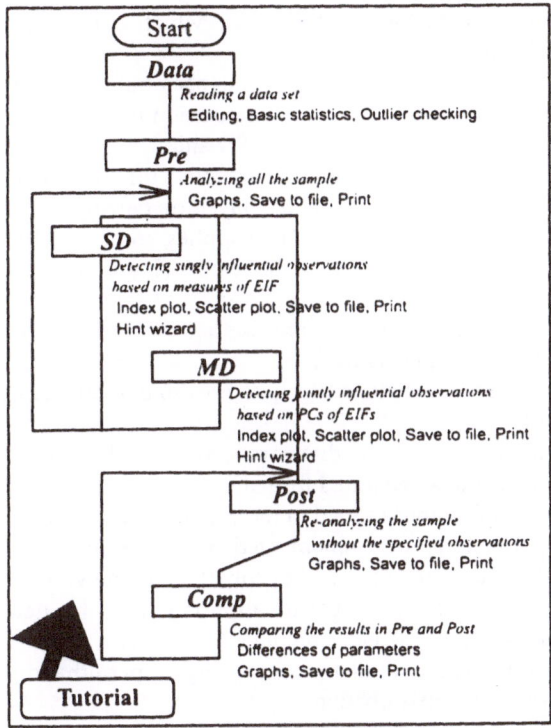

Figure 1: Flow of SAMMIF.

4) *Posterior analysis* (*Post*): Applying the same method as in 2) to the data set without candidate(s) for influential observation(s) found in 3), SAMMIF re-estimates the parameters.

5) *Comparison* (*Comp*): SAMMIF outputs the comparison of the results of 2) and 4) to compare the results with and without the specified set of observations. Users can evaluate whether they are really influential or not, whether omitting them makes the result of the analysis better or worse, and so on.

4.2 The Features of SAMMIF Version 1.0
(1) Both English and Japanese versions are available;
(2) SAMMIF runs on Windows 95/98/Me/NT/2000;
(3) SAMMIF is written in Microsoft Visual Basic 6.0;
(4) The current version handles sensitivity analysis in PCA, canonical correlation

analysis and exploratory and confirmatory factor analysis (FA);

(5) Basic specifications:

a. A flowchart whose shape is the same as Figure 1 is always displayed at the right side of screen to indicate where user is in the flow of sensitivity analysis and what users should do next (see the right side of the SAMMIF window in Figure 2). It is also a clickable-map so as to take the corresponding action when users click one of the buttons on it.

b. SAMMF can read a data file in "de format" which is its own format style and also in the CSV-format.

c. While SAMMIF usually computes the influences based on the default options, users can specify additional options for computation if necessary.

d. When users specify additional options in SD and MD, SAMMIF firstly displays standard outputs which are minimum results for beginners or for initial interpretations. Detailed results can be displayed on demand.

e. To help users' considerations SAMMIF supplies some graphical tools such as index plot and scatter plot to visualize the results (Figure 4 for SD and Figures 5 and 6 for MD).

f. Hint wizards give user how to interpret the results and how to find influential observations in SD and MD (Figure 3).

g. A brief tutorial is also available to illustrate how to operate SAMMIF and how to interpret the results.

h. All results including intermediate reports can be saved to a file in space-separated or comma-separated format.

(6) Supplemental specifications (added in the version 1.0):

The following functions are implemented in the version 1.0. They make the diagnostic of influence more easily and deeply than the previous versions.

i. Since considering more than one perturbation is helpful for detailed diagnostics of influence, the version 1.0 can compute EIF vectors based on two types of case-weight perturbations, ordinary case-weight perturbation and influence function type case-weight perturbation.

j. It is sometime useful to observe the characteristics of the original data set such as the existence of outlier(s) before any multivariate method is applied to it. Then SAMMIF supplies the outlier checking in $Data$.

k. SAMMIF can output not only influence measures summarizing EIF vectors but also raw values of elements in all EIF vectors, if necessary.

l. User can specify some additional options specialized for the corresponding multivariate method. For example, in PCA, axes on which the influence will be evaluated can be specified: all axes, any number of axes from the largest or smallest axis or only one axis (see an example in 5.2).

4.3 Web Pages of SAMMIF

A setup file of SAMMIF can be downloaded freely from and its related information is provided on http://www.f7.ems.okayama-u.ac.jp/sammif/, http://www.soci.ous.ac.jp/~mori/ sammif/ and http://www.kusa.ac.jp/~wat/sammif/.

5. Numerical Examples

5.1 Bodyfat Data for Sensitivity Analysis in FA

To illustrate our procedure we analyze a set of data taken from Bodyfat data (Johnson, 1995) using exploratory FA and its sensitivity analysis procedure. This data set consists of 252 observations on 13 variables.

Figure 2 is the results of *SD* displaying standard outputs, Cook's distance and χ^2 statistics based on all *EIF* vectors. The results window has two buttons, [Spread] button to obtain all the influence measures, and [How to interpret] button to open the hint wizard (a right window in Figure 2) to show how to interpret the results and how to find influential observations. Figure 3 is an index plot of Cook's distances in the first column of the *SD* results in Figure 2, which is drawn by SAMMIF. From this it can be stated that observations C31, C39, C42 and C86 are more influential than others. In *MD* step, we can obtain the results as shown in Figure 4 which are PC scores by applying

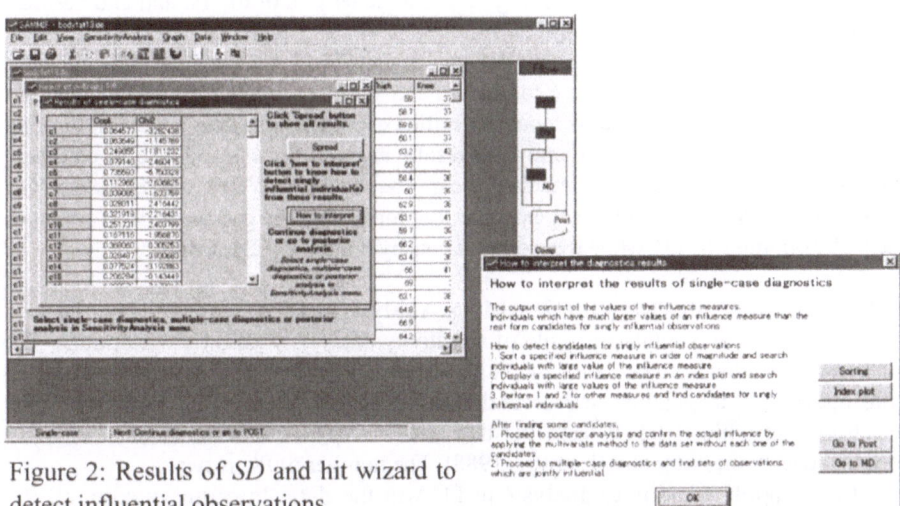

Figure 2: Results of *SD* and hit wizard to detect influential observations.

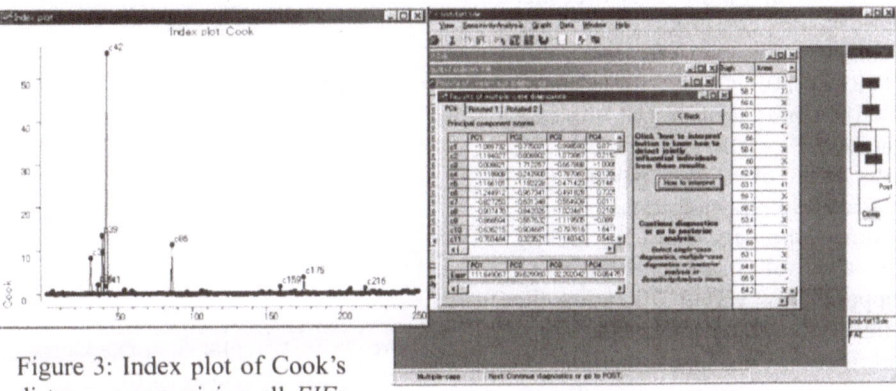

Figure 3: Index plot of Cook's distance summarizing all *EIF*s (*SD*).

Figure 4: Results of *MD*

286

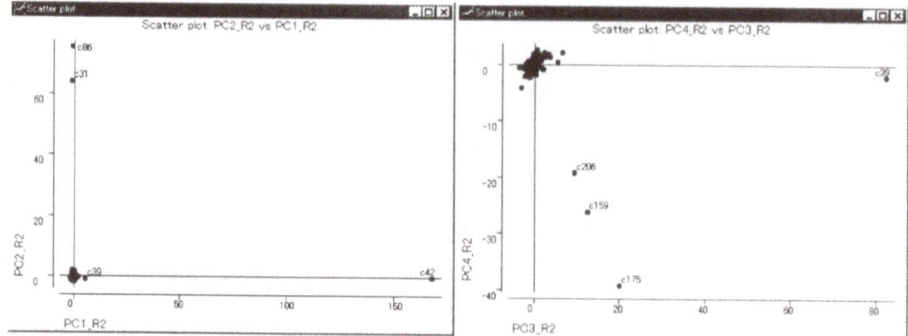

Figure 5: Scatter plot of the 1st and 2nd rotated PCs (*MD*).

Figure 6: Scatter plot of the 3rd and 4th rotated PCs (*MD*).

PCA to the *EIF* for all parameters. Figure 5 is a scatter plot of the 1st and 2nd varimax rotated PCs. This plot illustrates that two subsets {C42} and {C31, C86} can be regarded as candidates for influential subsets. We may search for more candidates by drawing scatter plots of other PCs (Figure 6 for the 3rd and 4th PCs). In the *Post* step exploratory FA is re-applied to the data set without a specified candidate, and then a table illustrating the differences between the *Pre* and *Post* results is given for convenience to compare. When there are more than one candidate, we go back to the *Post* step and repeat the *Post* and *Comp* steps. In our numerical example it is found that the goodness-of-fit becomes better by omitting {C42} while it becomes worse by omitting {C31, C46}.

5.2 Swiss Bank Notes Data for Sensitivity Analysis in PCA

Here we show another example of sensitivity analysis with different options. The data set, Swiss bank notes data, contains 6 variables and 200 observations whose first 100 correspond to genuine and second 100 forged old Swiss 1000 franc bills. The electronic data file is taken from the data sets in XploRe (Härdle *et al.*, 1999) but original data from Flury and Riedwyl (1988). The number of PC is two.

Let us apply sensitivity analysis in PCA to the data. In ordinary case, all the axes specified in the *Pre* step are used for the diagnostics. Here the first two axes are specified. Figure 7 is a scatter plot of the results in *MD*. The figure illustrates that two subsets, {C160, C171, C180, C182, C187} and {C161}, are influential. In case where we wish to consider the influence of observations on particular axes, we can use a new option in the version 1.0 to specify any axes of interest (Figure 8). Suppose we want to know the influences on axes which have small contribution on the results of PCA.

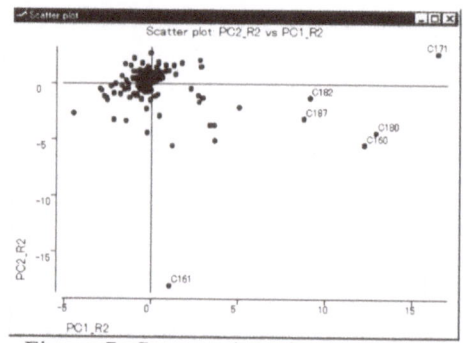

Figure 7: Scatter plot of rotated PCs of *EIF* on the first two axes of PCA.

Here the last four axes are specified and then the scatter plot in *MD* is obtained (Figure 9), in which *EIF*s are computed based on the last four axes. Compared with Figure 7, a subset {C171} is also influential, while {C160, C180, C182, C187} a little influential, but a new influential observation {C40} appears.

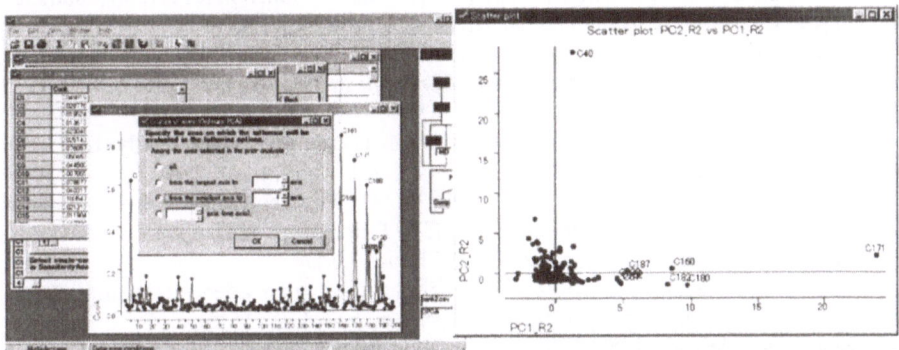

Figure 8: Specifying axes

Figure 9: Scatter plot of rotated PCs of *EIF* on the last four axes of PCA.

6. Concluding Remarks

As illustrated in the numerical examples, the statistical package SAMMIF can provide important information for diagnostic checking of the data in various multivariate methods. This package can be a useful tool for users to analyze the influence of observations easily and a unified platform for programmers to implement new methods for sensitivity analysis.

SAMMIF will be able to deal with sensitivity analysis in other multivariate methods including LISREL type covariance structure analysis, quantification methods and graphical modeling methods. And we are planning to implement the robust version of the general procedure (Tanaka and Watadani, 1994) for the case where the so-called masking effect is very severe.

References:

Cook, R. D. (1986). Assessment of local influence. *J. R. Statist. Soc.*, B48, 133-169.

Critchley, F. (1985). Influence in principal component analysis. *Biometrika*, **72**, 627-636.

Hampel. F. R. (1974). The influence curve and its role in robust estimation. *Journal of the American Statistical Association*, **69**, 383-393.

Härdle, W., Klinke, S. and Müller, M. (1999). Data sets in *XploRe Laerning Guide*. Springer. The original data set is in *Multivariate Statistics, A Practical Approach* (Flury, B. and Riedwyl, H. (1988), Cambridge University Press).

Johnson, W. R. (1995). Body fat data. In *StatLib-Datasets Archive*, http://lib.stat.cmu.edu/datasets/bodyfat.

Mori, Y and Tarumi, T. (1993). Statistical software SAM II: Sensitivity analysis in multivariate methods. *Journal of Japanese Society of Computational Statistics*, 6(2), 21-32.

Mori, Y., Watadani, S., Tarumi, T. and Tanaka, Y. (1998). Development of Statistical Software SAMMIF for Sensitivity Analysis in Multivariate Methods. In: *COMPSTAT98 Proceedings in Computational Statistics* (ed. R.Payne and P.Green), 395-400.

Physica-Verlag.

Odaka, Y., Watadani, S. and Tanaka, Y. (1991). Statistical software SAF/B (sensitivity analysis in factor analysis). *Abstract of the Fifth Conference of Japanese Society of Computational Statistics*, 63-66.

Radhakrishnan, R. and Kshirsagar, A. M. (1981). Influence function for certain parameters in multivariate analysis. *Communication in Statistics*, **A10**, 515-529.

Tanaka, Y. (1988). Sensitivity analysis in principal component analysis: Influence on the subspace spanned by principal components. *Communication in Statistics*, **A17**, 3157-3175. (Corrections, **A18** (1989), 4305).

Tanaka, Y. (1994). Recent advance in sensitivity analysis in multivariate methods. *Journal of Japanese Society of Computational Statistics*, **7**, 1-25.

Tanaka, Y., Castaño-Tostado, E. and Odaka, Y. (1990). Sensitivity analysis in factor analysis: Methods and software. In: *COMPSTAT90 Proceedings in Computational Statistics* (ed. Momirović, K. and Mildner, V.), 205-210. Physica-Verlag.

Tanaka, Y. and Watadani, S. (1992). Sensitivity analysis in covariance structure analysis with equality constraints. *Communication in Statistics*, **A21**, 1501-1515.

Tanaka, Y. and Watadani, S. (1994). Unmasking influential observations in multivariate methods. In: *COMPSTAT94 Proceedings in Computational Statistics* (ed. Dutter, R. and Grossman, W.), 292-297. Heidelberg: Physica-Verlag.

Tanaka, Y and Zhang, F. (1999). R-mode and Q-mode influence analysis in statistical modelling: relationship between influence function approach and local influence approach. *Computational Statistics and Data Analysis*, **32**, 197-218.

Tarumi, T. and Tanaka, Y. (1986). Statistical software SAM: Sensitivity analysis in multivariate methods. In: COMPSTAT86 *Proceedings in Computational Statistics* (ed. De Antoni, F., Lauro, N. and Rizzi, A.), 351-356. Physica-Verlag.

Wang, S.-J. and Lee, S.-Y. (1996). Sensitivity analysis of structural equation models with equality functional constraints. *Computational Statistics and Data Analysis*, **23**, 239-256.

Watadani, S. and Tanaka, Y. (1994). Statistical software SACS: Sensitivity analysis in covariance structure analysis. *Journal of Japanese Society of Computational Statistics*, **7**, 105-118.

Prolog as a Model Search Engine for Data Analysis

Tatsuo Otsu

Hokkaido University
Department of Behavioral Science, N.10 W.7, Kita-ku, Sapporo
Japan 060-0810

Summary: Prolog is a logic programming language for symbolic manipulation. Prolog is able to perform universal pattern matching, which is called *unification*, and backtracking search by indeterministic execution. These features enable simple description of complex statistical model structures and manipulations. I will show some examples of Prolog applications in statistical model handling. They show potential importance of symbolic manipulation in statistical computation.

1. Introduction

Recent advances of statistical methods have increased the necessity of complex statistical modeling. Traditional statistical software has little power for these problems. Although some recent statistical software, for example Splus and R, can handle complex object structures, their ability for symbolic manipulation is rather limited.

Here, I will show the ability of Prolog, which is a logic based programming language for symbolic manipulation, as a tool for statistical modeling. The most attractive features of Prolog as practical programming language are the functions of universal pattern matching and backtracking search by indeterministic control mechanism. These functions are the ubiquitous fundamentals for AI oriented systems. Although there have been some endeavors for building statistical expert systems using Prolog by Japanese researchers (Nakano et al., 1990; Minami et al., 1994), Prolog has not been regarded as a major tool for statistical modeling. There are two reasons for this unpopularity. One is the difficulty of statistical expert systems. The systems mentioned above are primarily intended to implement automatic model diagnosis and statistical consultations. These problems are highly domain specific and need much human knowledge. So the effectiveness of these attempts were limited by the domain specific knowledge that were not usually implemented in the statistical expert systems. Another problem is the large overhead of Prolog for execution. Symbolic manipulation oriented languages have tended to require large memory and computational power. These have imposed severe restrictions on popular use of symbolic languages.

The large improvement of PC power and Prolog interpreter sophistication in last few years solved the latter problem. Although large size numerical problems are still difficult to be handled by Prolog, the overhead is not important in cases of middle size problems. And now many Prolog systems provide convenient utilities for combining the functions written in procedural languages. These utilities provide support to embed numerically intensive procedures in symbolic manipulation programs.

But the first problem still remains. Effective targets seem to be more light automatizations. Here, we restrict our attentions to a model enumeration technique. At first, basic properties of Prolog language are introduced. And some Prolog techniques for data analysis that are not described in Prolog text books are shown. Finally, equivalent DAG (Directed Acyclic Graph) enumeration is described.

2. Basic Properties of Prolog
2.1 Facts and Rules
The design of Prolog is based on the theory of predicate logic. The term *predicate* means a true-false valued function with arguments. Prolog programs are composed of *clauses*. A clause is a fact or a rule. The following three clauses are the examples of fact declarations.

```
holiday.    bird(tweety).    master(charlie, snoopy).
```

A fact describes the relations on the arguments. The first clause shows that the proposition holiday is true. The second shows the predicate bird is true on tweety, i.e. tweety is a bird. The third clause shows the predicate master is true on the pair of charlie and snoopy, i.e. charlie is the master of snoopy.

A rule is composed of *head* and *body*. Head is the part that shows the conclusion. The body shows the premises. The following is an example.

```
fly(X) :- bird(X), \+ penguin(X).
```

The left-hand side of :- shows the *head* part, and the right-hand side shows the *body*. In the above line, the capital letter symbol X shows a *variable*. This clause shows that the predicate fly(X) stands if bird(X) is true and penguin(X) is not true. The operator \+ means the negation of the predicate. The variable in Prolog has very different meaning from other programming languages. Its semantic scope is restricted within a clause, and it shows an abstract symbol that can be unified with any symbols. Facts or rules that have the same predicate name can be declared in multiple clauses.

Prolog program execution is a proof searching process. If the predicate fly(tweety) is enquired, the interpreter unifies the head fly(X) with fly(tweety). Then the variable X has a concrete value tweety. The predicate bird(tweety) is declared as fact, and penguin(tweety) cannot be proved. Therefore, fly(tweety) is proved. If the inquiry has multiple so-lutions, the system can enumerate them. The following shows the response of the Prolog to the inquiry bird(X). Semicolons are user responses to the system for requiring another solution.

```
| ?- bird(X).
X = tweety ? ;
no
```

The system searches a solution, then it responds no. The above examples show unification and predicate evaluations. Input and output are performed by built-in predicates as side-effects.

2.2 List Structure

Prolog does not have usual array structure. When we use large size data, the list plays an important role. The word *list* has a special meaning in computer science. It is a recursive binary tree structure. Usually, the nodes of the tree are represented by *dot operators*. For example, a list notation [a,b,c,d] has the following internal structure.

```
. --- . --- . --- . --- []
|     |     |     |
a     b     c     d
```

The empty list [] has special meaning, and it is treated as an atomic element.

Prolog text books (Sterling & Shapiro,1994; O'Keefe,1990; Bratko,2000) show predicates for basic operations on lists. The following is one of the most popular examples.

```
member(H,[H|_]).
member(I,[H|T]):- member(I,T).
```

This predicate identifies and generates a member of the list. The vertical bar in list notation shows the decomposition of the list into its top element and the rests. The equal symbol = shows the unification of both sides of the expressions. If a unification [a,b,c] = [X|Y] is performed, X is substituted by a, and Y is substituted by [b,c].

Variable unification works for both directions. The above member pred-icate can identify whether the first argument is a member of the second argument or not. And if the inquiry has a variable in the first argument, the predicate enumerates the members.

```
| ?- member(a,[a,b,c]).
yes
| ?- member(a,[p,q,r]).
no
| ?- member(X,[p,q]).
```

```
X = p ? ;
X = q ? ;
no
```

2.3 Importing and Exporting Data

Many Prolog systems have three types of input/output functions. (1) Reading and writing predicates as texts, (2) saving and restoring internal memory states, and (3) character-wise input/output on streams. The first function is convenient for reusing computational results by another Prolog session. Users need not write programs for input/output of complex structured data. Although the second function makes fast input/output possible, the saved file depends on the internal structure of the Prolog system. The third enables usual input-output on text files.

There are some projects for developing XML-parsing systems on Prolog. They seem promising for sharing complex structured data with other application systems (Cabeza & Hermenegildo,2001; Wielemaker,2001).

3. List as Data

3.1 Arithmetic Operations

When we use Prolog for data analysis, list structure is frequently used for representation of vectors and matrices. Usually, we can not directly access the elements of a list. Although this is a severe disadvantage for data analysis, list representation has a large benefit for handling complicated structure.

There are two methods for implementing arithmetic operations on lists. One method is tail recursion, and another is function term copying. The following is an example of element-wise multiplication of two vectors. This uses tail recursion.

```
mlt([X|Xs],[Y|Ys],[Z|Zs]) :- Z is X*Y, mlt(Xs,Ys,Zs),!.
mlt([],[],[]).
```

The infix operator is shows arithmetic evaluation similar to usual programming languages. The cut operator ! inhibits backtracking process. Although the direct interpretation of the above code leads to deep recursion, modern elaborated Prolog compilers transform the tail recursion into an iteration loop. This enables memory conservation and fast computation.

Another method is using built-in function copy_term. The following is a predicate for function mapping. This is similar to Splus/R lapply or mapcar function of LISP language.

```
maps([A|As],X,Pred,Y,[B|Bs]) :-
    copy_term((X,Pred,Y),(X1,Pred1,Y1)),
    X1=A, call(Pred1),!, Y1=B,!, maps(As,X,Pred,Y,Bs),!.
maps([],_,_,_,[]).
```

Table 1: CPU Time for List and Matrix Manipulations

Operation	Data Size	Prolog(sec.)		R-1.2.3(sec.)
		List	Vector	List
Reverse	100000	0.02	0.02	0.11
	200000	0.04	0.04	0.30
Element-wise Multiplication				
(Tail Recursion)	100000	0.13		
	200000	0.29		
(Mapping, R uses lapply)	100000	2.41		2.14
	200000	5.57		4.74
(Vector)*	100000	0.01	0.01	
	200000	0.01	0.02	
Matrix Transposition	100×100	0.01	0.00	
	200×200	0.02	0.00	
Matrix Product	50×50	0.14	0.00	
	100×100	1.16	0.01	

SICStus-Prolog3.8.5 compact mode compiler on RedHat Linux 6.2. CPU Pentium3 500MHz PC with 128MB memory. (*) Prolog used external C-arrays.

The built-in predicate copy_term, which is defined by ISO standard and is equipped by many Prolog systems, generates the second argument that has the same structure as the first argument. Although unification binds the variables in the structure, this predicate does not bind them. The predicate maps is used as follows.

```
| ?- maps([1,2,3], X, Y is X*X, Y, Ys).
Ys = [1,4,9] ?
yes
```

The second argument and its copies are successively unified to the elements of the first argument. The third argument, which should be a predicate, is evaluated for each element, and the values of the forth argument Y are stacked in the list Ys. This predicate provides a very flexible function for data analysis. The cost for flexibility is its performance. This predicate is about 19 times slower than the simple tail recursion for element-wise multiplication (Table 1). Table 1 also shows the comparisons between SICStus-Prolog (The Programming Systems Group, 1995) and R-system (Ihaka & Gentleman, 1996). Although tail recursion is highly optimized on Prolog, that cannot overcome the overhead for matrix computation in list formed data. The performance depends primarily on the adopted data structure rather than the processing software.

A matrix is represented as a nested list. If A is defined as

$$A = \left(\begin{array}{cc} 1 & 2 \\ 3 & 4 \end{array} \right),$$

its list representation is [[1,2],[3,4]]. The following is the predicate for matrix transposition.

```
transpose(Xs,[Y|Ys]) :- tops_and_rests(Xs,Y,Xss),
        transpose(Xss,Ys),!.
transpose([[]|_],[]).

tops_and_rests([[X|Xs]|Xss],[X|Ys],[Xs|Yss]) :-
        tops_and_rests(Xss,Ys,Yss),!.
tops_and_rests([],[],[]).
```

Although there is no standard library for manipulating vectors and matrices, it is not difficult to make libraries. The author wrote (1) fundamental manipulation predicates for vectors and matrices in the list form, and (2) object classes for external vectors and matrices using SICStus Object system.

3.2 Sorting and Tabulation

The ability for data abstraction has a good property for data tabulation. The built-in predicate sort of Prolog sorts a list and deletes duplicated terms.

```
| ?- sort([b,c,a,d,b,a],X).
X = [a,b,c,d] ?
```

We must avoid duplication deletion for data analysis. Using another built-in predicate keysort, we can elude the problem. The function of keysort is to sort a list by key values. The keys are concatenated to the body by minus sign symbols as key-body, where minus symbol does not have any meaning of subtraction. In the following code, the terms in the list are treated as keys. Sequence numbers are concatenated for identification as bodies.

```
sort_data(Ds,Sds) :- length(Ds,N), seq(N,Seq),
    makepairs(Ds,Seq,Dseqs),keysort(Dseqs,SDseqs),
    sort_data_firsts(SDseqs,Sds),!.

sort_data_firsts([V-_|Vis],[V|Vs]):-
    sort_data_firsts(Vis,Vs),!.
sort_data_firsts([],[]).

makepairs([X|Xs],[Y|Ys],[X-Y|Zs]) :- makepairs(Xs,Ys,Zs).
makepairs([],[],[]).
```

This predicate works as follows.

```
| ?- sort_data([b,c,a,c,b],X).
X = [a,b,b,c,c] ?
yes
```

The predicate makepairs makes a list of paired terms from the two lists that have the same length. This simple predicate coupled with keysort is convenient for various operations on list-formed data.

The following short code is the predicate for tabulating the data.

```
frequency(List,ValueFreq) :-
    sort_data(List,Sls),Sls=[V1|_],
    frequency0(Sls,V1,0,ValueFreq),!.
frequency0([V|Sls],V1,N,[V1-F|Vfs]) :-
    ( V = V1 -> N1 is N+1,frequency0(Sls,V1,N1,[V1-F|Vfs])
    ; F is N, frequency0(Sls,V,1,Vfs)
    ),!.
frequency0([],V1,N,[V1-N]):-!.
```

The structure A -> B; C works as if A then B else C control. The above predicate generates a list of value−frequency pairs.

```
| ?- frequency([b,c,a,c,b,c],X).
X = [a-1,b-2,c-3] ?
yes
```

Using makepairs predicate, this also works for cross-classifying the data.

```
| ?- makepairs([b,c,a,c,b,c],[x,x,x,y,y,y],A),
            frequency(A,B).
A = [b-x,c-x,a-x,c-y,b-y,c-y],
B = [a-x-1,b-x-1,b-y-1,c-x-1,c-y-2] ?
yes
```

4. Model Enumeration
4.1 Subset Generation

Prolog has a good control mechanism for model enumerations. The following predicate enumerates the subsets of list elements.

```
subset_gen([X|Xs],Ys) :-
        (Ys=[X|Y1s];Ys=Y1s), subset_gen(Xs,Y1s).
subset_gen([],[]).
```

The semicolon shows *OR* condition.

Using a modified version of the predicate, we can generate subsets of variables in a suitable order for symmetric sweep operations (Beaton,1964; McCullagh & Nelder, 1989, 3.8; Schafer, 1997, 5.2). The built-in predicate findall stacks the possible solutions into a list.

```
subset_gen2([X|Xs],Ys,Zs) :-
      ( Ys=[X|Y1s],Zs=Z1s, subset_gen2(Xs,Y1s,Z1s)
      ;Ys=Y1s,Zs=[X|Z1s], subset_gen2_0(Xs,Y1s,Z1s)).
subset_gen2([],[],[]).
subset_gen2_0([X|Xs],Ys,Zs) :-
      ( Ys=Y1s,Zs=[X|Z1s], subset_gen2(Xs,Y1s,Z1s)
      ;Ys=[X|Y1s],Zs=Z1s, subset_gen2_0(Xs,Y1s,Z1s)).
subset_gen2_0([],[],[]).

| ?- findall(X,subset_gen2([a,b,c],X,_),Xs).
Xs = [[a,b,c],[a,b],[a],[a,c],[c],[],[b],[b,c]] ?
```

The symmetric sweep algorithm includes or deletes one independent variable with one fundamental operation. The above sequence provides the minimum scan procedure for optimal model search.

4.2 Equivalent Directed Acyclic Graphs

A *graph* is a structure composed of nodes and directed edges that connect nodes (Spirtes et al., 1993; Pearl, 2000). If all edges in the graph are directed, i.e. marked by a single arrowhead on an end of every edge, the graph is called *directed*. A DAG (Directed Acyclic Graph) is a directed graph that has no cyclic directed path.

A DAG G specifies a class of probability distributions. In those models, nodes represent random variables. Suppose G has the nodes $\{V_i | i = 1, ..., p\}$, and they correspond to the random variables $\{X_i | i = 1, ..., p\}$. The positive probability distributions that are specified by G are represented as

$$\prod_{i=1}^{p} f_i(X_i | pa_i),$$

where pa_i shows the parents of V_i. A probability distribution on $\{X_i | i = 1, ..., p\}$ that satisfies the above condition is called G-Markov. The correspondence between DAG and the class of distributions may not be unique. There may be some different DAG structures that specify the same distribution class (Spirtes et al., 1993; MacCallum, et al., 1993).

The following theorem is known about the conditional independence of G-Markov distributions (Fryndenberg,1990; Verma & Pearl,1991; Spites et al., 1993; Andersson et al., 1996; Pearl, 1995, 2000).

Table 2: Equivalent connected DAGs on four nodes

Undirected Neighbor Structure	Acyclic Models	Class Size (the Number of Classes)		
1-2,1-3,1-4	8	1(4)	4(1)	
1-2,1-3,2-4	8	2(2)	4(1)	
1-2,1-3,1-4,2-3	12	1(2)	2(1)	8(1)
1-2,1-3,2-4,3-4	14	1(2)	3(4)	
1-2,1-3,1-4,2-3,2-4	18	2(1)	3(2)	10(1)
Saturated	24	24(1)		

Theorem (DAG Markov equivalence): Two DAGs G_1 and G_2 have the same conditional independence on the variables if and only if the following conditions are satisfied.
(1) They have the same nodes.
(2) They have the same undirected connection structure.
(3) They have the same v-structures.
A v-structure is defined as a relationship between three nodes such that $X \rightarrow Y \leftarrow Z$ and there is no edge between X and Z.

If the distribution class is restricted to linear directed path models with normal errors, the equivalence of conditional independence leads to the same zero partial correlations (Stelzl, 1986; Lee & Harshberger, 1990; Luijben, 1991; Mayekawa, 1994).

Using the model enumeration ability of Prolog, we can completely classify the equivalent DAG patterns. This needs two stages of computation. At first, we must classify undirected path topologies. Then the edge directions are considered. Table 2 and Table 3 show the classification of equivalent DAGs. The connected DAGs on four nodes and five nodes are considered. The most computationally cumbersome part is the classification of undirected edge topologies. As for five nodes, the computation takes little time (about 1.5 seconds for six paths on Pentium3 500MHz PC with SICStus-Prolog3.8.5). Although six node DAG classification is possible by PC or small workstation, it took far longer CPU time than the five node classification. In the case of eight paths on six nodes, the classification took about 22 minutes on the same system. Although performance improvement seems to be possible by modifying the algorithm, the exponential increase of time depends on the combinatorial nature of the problem.

5. Conclusion

Prolog, with its high symbolic manipulation ability, seems to be a strong candidate for a computational platform of complex statistical modeling. Model

Table 3: Equivalent connected DAGs on five nodes

Undirected Neighbor Structure	Acyclic Models	Class Size (the Number of Classes)				
1-2,1-3,1-4,1-5	16	1(11)	5(1)			
1-2,1-3,1-4,2-5	16	1(2)	2(3)	3(1)	5(1)	
1-2,1-3,2-4,3-5	16	1(1)	3(2)	4(1)	5(1)	
1-2,1-3,1-4,1-5,2-3	24	1(6)	2(4)	10(1)		
1-2,1-3,1-4,2-3,2-5	24	1(4)	2(2)	3(2)	10(1)	
1-2,1-3,1-4,2-3,4-5	24	2(2)	4(1)	6(1)	10(1)	
1-2,1-3,1-4,2-5,3-5	28	1(4)	2(3)	3(2)	4(3)	
1-2,1-3,2-4,3-5,4-5	30	2(5)	4(5)			
1-2,1-3,1-4,1-5,2-3,2-4	36	1(7)	2(3)	3(2)	4(1)	13(1)
1-2,1-3,1-4,1-5,2-3,4-5	36	1(4)	2(4)	4(1)	20(1)	
1-2,1-3,1-4,2-3,2-4,3-5	36	1(2)	2(2)	4(3)	6(1)	12(1)
1-2,1-3,1-4,2-3,2-5,4-5	42	1(4)	2(4)	3(2)	4(2)	8(2)
1-2,1-3,1-4,2-5,3-5,4-5	46	1(11)	3(9)	4(2)		
1-2,1-3,1-4,1-5,2-3,2-4,2-5	54	1(6)	2(4)	3(6)	4(2)	14(1)
1-2,1-3,1-4,1-5,2-3,2-4,3-4	48	2(6)	6(1)	30(1)		
1-2,1-3,1-4,1-5,2-3,2-4,3-5	54	1(2) 14(1)	2(6)	3(2)	4(1)	8(2)
1-2,1-3,1-4,2-3,2-4,3-5,4-5	60	1(6) 10(1)	2(2)	3(6)	6(1)	8(2)
1-2,1-3,1-4,1-5,2-3,2-4,2-5,3-4	72	1(4) 32(1)	2(2)	3(4)	4(1)	8(2)
1-2,1-3,1-4,1-5,2-3,2-4,3-5,4-5	78	1(2)	3(12)	10(4)		
1-2,1-3,1-4,1-5,2-3,2-4,2-5,3-4,3-5	96	6(4)	10(3)	42(1)		
Saturated	120	120(1)				

enumeration abilities that are shown above provide easy methods for model search programming.

Although these complex manipulations can be written by other procedural languages, they require large elaboration. Prolog gives a simple description of the data and the procedures. These are important features for rapid prototyping. Another important aspect is Prolog's potential for integrating different kinds of knowledge. For example, the description of probability distributions and domain knowledge require different types of data representations. Prolog provides good tools for integrating various information and coping with ill structured data in real world.

Prolog has a rather long history as programming language. There are good text books for programming (Sterling and Shapiro, 1994; Flach, 1994; Bratko, 2001). Several commercial and freely available systems of high qualities are maintained by their developing groups. They have convenient debugging systems. At least some core features of Prolog will remain as important parts of future statistical systems.

References:

Andersson,S.A., Madigan,D. and Perlman,M.D. (1996). A characterisation of Markov equivalence classes for acyclic digraphs. *Annals of Statistics*, **23**, 505–541.

Beaton,A.E. (1964). The use of special matrix operations in statistical calculus. Research Bulletin RB-64-51, Educational Testing Service, Princeton NJ.

Bratko,I. (2001). *PROLOG: Programming for Artificial Intelligence 3rd ed.*. Addison-Wesley.

Cabeza,D. and Hermenegildo,M. (2001). *The PilloW Web Programming Library*. School of Computer Science, Technical University of Madrid.

Flach,P. (1994). *Simply Logical*. Wiley.

Lee,S. and Hershberger,S. (1990). A simple rule for generating equivalent models in covariance structure modeling. *Multivariate Behavioral Research*, **25**, 313–334.

Luijben,T.C.W. (1991). Equivalent models in covariance structure analysis, *Psychometrika*, **56**, 653–665.

MacCallum,R.C., Wegener,D.T., Uchino,B.N., and Fabrigar,L.R. (1993). The problem of equivalent models in applications of covariance structure analysis, *Psychological Bulletin*, **114**, 186–199.

Fryndenberg,M. (1990). The chain graph Markov property, *Scandinavian Journal of Statistics*, **17**, 333–353.

Ihaka,R. and ZGentleman,R. (1996). R: A language for data analysis and graphics, *Journal of Computational and Graphical Statistics*, **5**, 299-314.

Mayekawa,S. (1994). Equivalent path models in linear structural equation models. *Behaviormetrika*, **21**, 79–96.

McCullagh,P. and Nelder,J.A. (1989) *Generalized Linear Models 2nd ed.* Chapman & Hall.

Minami,H., Mizuta,M. and Sato,Y. (1994). Multivariate analysis support system with a hypothesis reasoning mechanism (in Japanese). *Japanese Journal of Applied Statistics*, **23**, 63–79.

Nakano,J., Yamamoto,Y. and Okada,M. (1990). Knowledge base multiple regression analysis system (in Japanese). *Japanese Journal of Applied Statistics*, **20**, 11–23.

O'Keefe (1990). *The Craft of Prolog*, MIT Press.

Pearl,J. (1995). Causal diagrams for empirical research, (with discussions), *Biometrika*. **82**, 669–710.

Pearl,J. (2000). *Causality: Models, Reasoning and Inference*, Cambridge UP.

The Programming Systems Group (1995). *SICStus Prolog User's Manual, Release 3*, Kista,Sweden: Swedish Institute of Computer Science.

Schafer,J.L. (1997). *Analysis of Incomplete Multivariate Data*. Chapman & Hall.

Spirtes,P., Glymour,C. and Scheines,R. (1993). *Causation, Prediction, and Search, Springer Lecture Notes in Statistics 81*, Springer.

300

Stelzl,I. (1986). Changing a causal hypothesis without changing fit: some rules for generating equivalent path models. *Multivariate Behavioral Research*, **21**, 309–331.

Sterling,L. and Shapiro, E. (1994). *The Art of Prolog, 2nd ed.*, MIT Press.

Verma,T. and Pearl,J. (1991). Equivalence and synthesis of causal models. in Bonissone,P.P., Henrion,M., Kanal,L.N., and Lemmer,J.F. (eds.), (1991). *Uncertainty in Artificial Intelligence* **6**, Elsevier Science. 225–268.

Wielemaker,J. (2001). *SWI-Prolog 4.0.8 Reference Manual*. Dept. of Social Science Infomatics, University of Amsterdam.

Gender Differences in Mathematics Achievement

Peter Allerup

The Danish University of Education
101 Emdrupvej , DK 2400 Copenhagen NV, Denmark

Summary: The International TIMSS Study: Third International Mathematics and Science Study revealed large gender differences in Mathematics achievement (Beaton et al.,1996). The present analysis undertaken at the Danish TIMSS data from grade 6,7 demonstrates that six math items, which were used as a common reference set for eight TIMSS booklets distributed among students, are not meeting the intended psychometric requirements set by the Rasch model. Could such item inhomogeneities create TIMSS gender differences in the first place? The paper provides alternative analyses of gender differences based on other statistics and measures than provided by the international math scores.

1. Background and Six Math Items

Both in the middle school, grade 6 and 7 and at the end of secondary schooling, viz. at age 17-18 years Danish students were ranked among the countries showing the largest gap between boys and girls. This conclusion contradicted old national experiences with written examinations at the end of grade 9 and examinations in mathematics by the end of secondary schooling. An immediate explanation to such contradictions seemed to be, that 'true' gender differences were for the first time detected by means of the TIMSS tasks (items), since the items were all, prior to the sampling, thoroughly examined regarding basic psychometric properties defined by the one-parameter (Rasch) and two-parameter (with item discrimination) logistic IRT models (Rasch,1960). A fit ensures consistent item difficulties across any sub grouping of the students, e.g. boys and girls. Observed gender differences would, therefore, in a classical follow-up analysis of TIMSS data, call for interpretations by means of questionnaire background information, this being included as independent variables.

The six math items it1,...,it6 used as rotating reference items for equating student performances across booklets are all in multiple choice format (MC), with only one correct response category. For an item i (i=1,...,6) to fit the one- or two-parameter IRT model, the probability for a response a_{vi} ='correct'(x=1) or 'non-correct'(x=0) from student v to item item i is calculated as follows :

$$P(a_{vi} = x) = \frac{\exp(x(\theta_i + \delta_i \, \sigma_v))}{1 + \exp(\theta_i + \delta_i \, \sigma_v)} \qquad x=0,1 \qquad (1)$$

(Sometimes a minus sign is used in the numerator and the denominator in order for "ability" and "difficulty" to be reflected equally on the θ- and σ axis)

Measures of item difficulties are θ_i (subject to the constraint: $\Sigma\theta_i = 0$), student abilities are the σ_v parameters, and item discriminations are the δ_i 's which on the values $\delta_i \equiv 1$ brings the model to the one parameter Rasch model.

Table 1 summarises results of test statistics for the item fit using grade 6 and 7 data from Denmark. The global test statistics are simultaneous (all items) Conditional Likelihood Ratio test statistics evaluating the hypothesis of consistent θ-values across sub groups (Andersen,1984). Single item test statistics are dealing with the fit of Conditional Item Characteristic Curve (ICC) to the general two-parameter IRT logistic model and fit statistics to the Rasch model in particular (Allerup,1994).

Global test statistics

sub groups:	signf.prob
Gender:	p=0.000
High/Low	p=0.000
Grade 6/7	p=0.007

Single item statistics

	general fit p-value (1)	gender θ_i - diff p-value (2)	Total θ_i (estimate) (3)	Item discrim δ_i (4)	discrim δ_i =1? p-value (5)
ITEM1	0.064	0.670	-0.524	1.349	0.006
ITEM2	0.318	0.222	0.790	0.984	0.753
ITEM3	0.076	0.046	0.262	1.278	0.011
ITEM4	0.135	0.000	0.050	1.094	0.124
ITEM5	0.638	0.001	-0.761	0.792	0.015
ITEM6	0.057	0.000	0.183	0.648	0.003

Table 1: (1) refers to the fit statistic for one- or two-parameter IRT model, (2) lists test results for θ_i differences (exact tests) across gender, (3) lists estimates of item difficulties θ_i , (4) lists estimates of item discriminations δ_i and (5) lists test results from testing: δ_i=1 which on acceptance provides the Rasch Model.

The simultaneous tests are all rejecting the model (p<0.05) and the general fit statistics (1) are sensitive to the choice of the level of significance. From (2) it is seen that, in particular, the last three Items are in particular subject to varying difficulties across boys and girls. From (4) and (5) a strong impression of varying item discriminations is obtained.

For items 4,5 and 6, details of the tests (2) across gender are displayed in table 2. The 'observed' values refer to *the number of girls* in the combined data set of grades 6 and 7, for a given score group (score value across six items), 'expected' values are calculated under the hypothesis of equal item parameters for boys and girls through a hypergeometric distribution (Fisher exact test). It is seen from comparing observed with the expected , that item 4 is systematically too difficult for the girls, while items 5 and 6 are systematically too difficult for the boys.

Item : 4			Item : 5			Item : 6		
Score	Exp	Obs	Score	Exp	Obs	Score	Exp	Obs
0	0.00	0	0	0.00	0	0	0.00	0
1	19.30	20	1	11.92	12	1	31.79	35
2	71.94	66	2	57.21	52	2	127.45	146
3	194.52	173	3	112.62	124	3	218.58	236
4	344.41	314	4	229.26	259	4	349.67	365
5	422.31	410	5	305.50	315	5	385.89	398
6	386.00	386	6	386.00	386	6	386.00	386

Table 2: Exact tests for gender differences. Obs = observed number of girls, Exp = expected number of girls under the hypothesis of equal item parameters for boys and girls.

In table 3 an extract contingency table from the 2^6 =64 cells, containing fit statistics for each cell, is shown; it confirms that response patterns of the type 'xxx01xx' with non-correct answers on item 4 and correct answers on item 5 are observed too often in relation to what is expected by the Rasch model, calculated under a distribution of student abilities identical with estimated.

		Grade6 Frequency(%)			Grade 7 Frequency(%)			Number of students			Rasch model
No.	Response pattern	g p_{11}	b p_{12}	Lodds1	g p_{21}	b p_{22}	loddss2	c10	c20	c120	exp.
1	000010	0.58	0.50	0.15	0.54	0.35	0.43	11	10	21	24.01
2	000011	0.68	0.30	0.81	0.89	0.35	0.95	10	14	24	16.63
3	001010	0.68	0.20	1.22	0.27	0.61	-0.83	9	10	19	17.99
4	001011	0.48	0.40	0.19	0.63	0.79	-0.23	9	16	25	22.79
5	010010	1.06	1.20	-0.13	0.54	0.61	-0.13	23	13	36	30.51
6	010011	1.84	0.90	0.72	1.34	1.05	0.25	28	27	55	38.66
7	011010	0.87	0.80	0.08	1.34	0.87	0.43	17	25	42	41.82
8	011011	2.90	2.01	0.38	3.22	1.40	0.85	50	52	102	70.78
9	100010	0.29	0.20	0.37	0.09	0.35	-1.37	5	5	10	8.19
10	100011	0.39	0.10	1.35	0.09	.	.	5	1	6	10.38
11	101010	0.29	0.20	0.37	0.09	0.09	0.02	5	2	7	11.23
12	101011	1.06	0.30	1.27	0.98	0.35	1.04	14	15	29	19.01
13	110010	0.97	0.10	2.27	0.27	0.17	0.43	11	5	16	19.05

Table 3: Response patterns across six math items, frequencies for boys (b) and girls (g), p11,p12 (grade 6) and p21,p22 (grade 7) deviations in log odds measures lodds1, lodds2.

Log odds values Lodds1 and Lodds2 for the frequencies of girls compared to boys in a specific pattern support the diagnostics concerning item Nos. 4 and 5, e.g. lodds1= log odds1 = log { p11x(1-p12) /p21x(1-p11) }.

It is clear that the six core items fail to fit the one parameter Rasch model and consequently other measures than total scores across the six items must be applied when comparing students.

The statistical impact of the detected inhomogeneity by item 4 on general gender analyses can be judged from the following calculation. Behind the average value for item difficulty θ_4 =0.0496 in table 1 is kept item difficulties θ_4 for girls and boys on –0.24 and 0.19 respectively. For a student with 'average' student ability, i.e. σ_v=0, this leaves girls with a probability of 0.44 for a correct answer to item 4 to compare with 0.55 for the 'average' boy. This can be seen by inserting the values for θ_4 and σ_v in model (1). In table 4 are listed actual gender differences from the TIMSS data.

Grade	All Math	items		6MAT	items	
	Girls	Boys	Diff	Girls	Boys	Diff
6	16.6	17.8	1.2	3.7	3.9	0.2
7	19.6	21.4	1.8	4.2	4.4	0.2
8	23.0	23.5	0.5	4.7	4.8	0.1
9	26.2	27.7	1.5	4.9	5.0	0.1

Table 4: Score differences in TIMSS data using all math items (approx. 40 items) and the six 6MAT core math items.

The actual 6MAT raw score differences are found to be 0.2 for grades 6 and 7 . Although this cannot be compared simply with the expected difference 0.55-0.44 =0.11 from item 4 alone, because the item parameters are constrained: $\Sigma\theta_1$ =0, the magnitude 0.11 is nevertheless contributing significantly to the observed gender TIMSS differences, but solely from reasons of inhomogeneity. Together with similar tendencies from item No. 6 and the fact, that certain response patterns are favoured by girls and other response patterns are favoured by boys, this might be the algebraic reason for a part of the measured TIMSS gender differences.

2. Comparing Students Using other Measures than Scale Scores

One consequence of the lack of scale properties for the six math items, 6MAT is that comparisons between students must be done by other means than simple scores (or Rasch scores). Identifying those students which have all six items correct (6MAT HIGH) and students with only 0 or 1 correct (6 MAT LOW), table 5 summarises frequencies of HIGH/LOW students in grades 6,7,8 and 9; analysing grades 8 and 9 was a national option in Denmark. The table includes information as to whether the students have attempted all six items in the last column.

The first three columns and column 5 are all referring to item 4 and especially the third answer category (M=3) receives attention in columns 2,3 and 5. Item 4 is strongly gender 'biased' , because girls seems to prefer the non-correct distractor 3 (M=3) quite frequently. This category is preferred significantly more often across all grades by the girls. While the boys seem to abandon this choice (less than ten percent) in grades 8 and 9 the girls still prefer this response category .

It is remarkable, that, although the international Rasch scores should be used with caution, column 3 shows, that the girls using category 3 are more able compared with the boys; in fact, this holds true for all grade levels. The average level of ability is, however, generally lower for the M=3 students, which is seen from Rasch scores in columns 3 and 4; this, in turn, is confirmed in column 5, where the frequencies of students using this specific M=3 category are calculated *within* LOW students (to be compared with the marginal frequencies in column 2). A significant difference (marked * in table) between the frequency of HIGH students (28.2% vs. 21.2%) can be observed in grade 7 (column 7), and column 8 shows that in grades 6 and 7 there are significantly fewer girls who attempted all six items of 6MAT.

	Item 4 M=4 correct	Item 4 M=3 wrong	M=3 group Rasch	All stud Rasch	6MAT LOW It4 M=3	6MAT LOW coverage	6MAT HIGH coverage	6MAT attempt all six
grade 6 g	54.8%	28.5%	427.3	460.2	55.6%	9.7%	14.4%	63.9% *
grade 6 b	67.2%	17.9%	417.0	466.4	57.1%	7.4%	15.4%	72.9% *
grade 7 g	64.7%	21.5%	444.2	494.5	65.0%	5.5%	21.2% *	75.2% *
grade 7 b	73.9%	13.0%	435.8	509.7	66.1%	4.9%	28.2% *	79.6% *
grade 8 g	68.7%	18.6%	477.6	508.0	63.2%	3.1%	34.8%	82.2%
grade 8 b	75.9%	8.8%	467.8	547.6	20.0%	2.1%	37.1%	81.8%
grade 9 g	70.4%	18.2%	480.8	552.7	100.0%	2.7%	41.9%	88.6%
grade 9 b	84.5%	7.0%	467.6	574.1	80.0%	2.2%	50.4%	89.7%

Table 5: Columns 1 and 2: Percentage of students responding to categories 3 and 4 of item 4, column 3: Rasch ability scores for students using M=3 in item 4, column 4: Rasch ability scores for all students, column 5: Frequency of students in group 6MAT=LOW who respond M=3 to item 4, columns 6 and 7: Frequency of students in group 6MAT=LOW and HIGH respectively, column 8: Frequency of students who attempt all six items (i.e. no missing). b=boys g=girls.

The HIGH/LOW classification of students can be analysed in relation to variables connected to the class through the measure LODDH, defined as the log odds transformation of the frequency of HIGH-scoring girls pHG to the frequency of HIGH-scoring boys pHB within a class: loddh= log { pHG x (1-pHB) /pHB x (1-pHG) }.The interpretation of loddh is that of a measure of surplus of HIGH girls beyond expectation from the ratio of girls in the class. The within class variation of loddh takes indefinite values when the class consists of zero HIGH girls , which happens in around 21% of the classes, or the class consists of zero HIGH boys, which happens in around 7% of the classes in grade 7. The feature of creating HIGH students by only one sex is therefore unevenly distributed across classes. Apart from these extremes the calculation and tests of loddh values confirm the balance of HIGH girls to HIGH boys in the grade levels 6,8 and 9, while an average loddh value = -0.25 for grade level 7 supports the impression gained from the above table that HIGH girls appear less frequently.

However, an interesting correlation exists between loddh and the ratio of girls in a class: The smaller the ratio of girls the greater the values of loddh, viz. the greater frequency of HIGH girls can be found in the class (viz. Spearman correlation $\rho \approx -0.53$, for points loddh $\in$ [-3.0,3.9]). This is shown for grade 6 in figure 1; a similar pattern is found for grade 7, while grades 8 and 9 have a less destinct structures. The significant difference between the frequency of HIGH girls and boys in grade 7 will consequently be evaluated considering other external variables also.

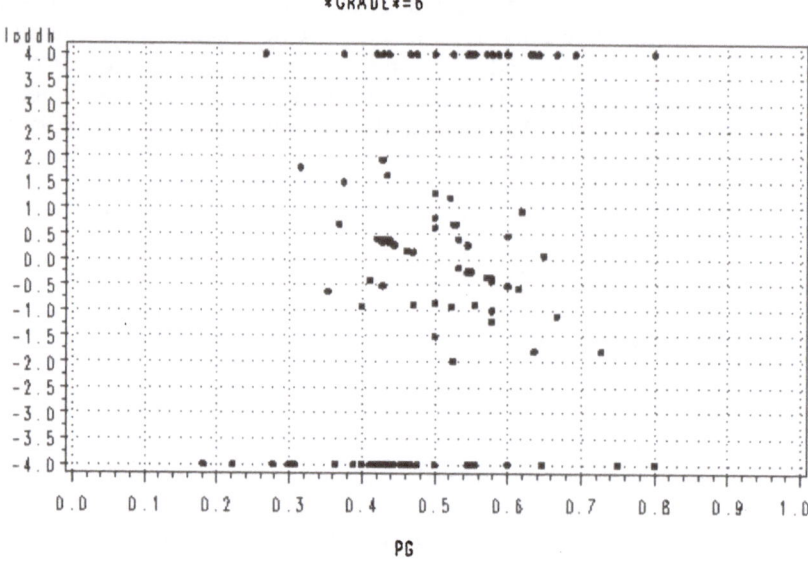

Figure 1: Relation across classes between LODDH= log odds for presence of girls with maximum 6MAT score and percentage girls in class.

Together with the test booklets all students and teachers completed questionnaires comprising attitude questions towards math, personal ambitions in the field and details from daily life experiences in the class room. The various measures to assess high scoring girls and boys, including the surplus measure loddh related to the frequency of high scoring girls in the class have been related to the background variables in the student and teacher questionnaires . Typical results found were that a number of attitudes and motivations are different between girls and boys, irrespective of levels of achievement. One main conclusion is therefore that a large number of student or teacher characteristics can be related properly to the achievement levels.

References:

Allerup, P. (1994). Rasch Measurement, theory of. The International Encyclopedia of Education, second edition, Pergamon Press.

Andersen E.B. (1973). Conditional Inference and Models for Measuring, Copenhagen: Mentalhygiejnisk Forlag.

Beaton , Albert et al (1996): Mathematics Achievement in the Middle School Years. IEA's Third International Mathematics and Science Study. Boston College USA

Rasch G (1960) Probabilistic Models for some intelligence and Attainment tests, Copenhagen, The National Danish Institute for Educational Research.

Analyzing Foreign Financial Statements: A Dual Scaling Approach to the International Ratio Analysis[1]

Jose G. Clavel[1], Isabel Martínez Conesa[2], and
Esther Ortiz Martínez[2]

[1] Department of Quantitative Methods for the Economy
[2] Department of Finance and Accounting
University of Murcia
Faculty of Economy and Business, 30.100 Murcia, Spain.

Summary: Foreign companies listed on the New York Stock Exchange (NYSE) are required by its supervisor organism (Securities Exchange Commission, SEC), to restate their reported net income and shareholders' equity calculated according to domestic accounting standards, into US General Accepted Accounting Principles (US-GAAPs), as a part of a Form 20-f filing. Using these restatements and dual scaling methodology we will be able to distinguish if there are really so important differences among accounting standards around the world, or on the contrary, some other effects, such as the country in which the company is located, also affect financial information in the most important way than the own accounting standards.

1. Introduction

The Form 20-F required by the SEC includes restatement of the net income and shareholders' equity to US-GAAPs. This information provides us with opportunities to examine if those different but generally accepted accounting principles are indeed able to capture value-relevant information. In view of the current debates concerning international disclosure regulation and accounting harmonisation, we will attempt to shed some light on these debates. We have analysed, as Frost and Lang (1996) suggested, the effects of SEC requirements on the functioning of capital markets by applying dual scaling methodology to a multicountry database.

The remainder of this paper is organised into four major sections. The first of these describes the background of primary interest to the study and summarises the relevant research result. The second, the method section, justifies the choice of the database and the methodology. The results of the study are reported in the third section, and the final section contains a discussion of these results and some concluding comments.

---

[1]This work is a part of the research project financed by the DGI: *Influencia de la diversidad contable en el análisis financiero internacional. Un estudio empírico* (SEC2000-0410) in colaboration with Analistas Financieros Internacionales and Morgan Stanley Dean Witter.

308

2. Background

There are two main streams of studies based on the analysis of quantitative differences existing between financial data. The impact of international accounting differences and the extent to which the measurement of company profits or equity are correlated with some other characteristics such as country or industry studied through a *Conservatism Index* by Gray (1980). As his conclusions suggest, the country in which a company is located is a statistically significant factor in quantitatively determining the relative amount of profits disclosed. After his research there are a lot of them which have been based on the same method. This index is the difference between a company's disclosed profits (or in some other most recent papers also equity) and its adjusted profits, divided by the absolute value of the adjusted profits. As in Weetman et al. (1998) it is better to rename the index as index of comparability because it places clearer emphasis on the relative accounting treatment without requiring a judgment as to which is more or less conservative. We will have two different quantities for each financial concept, it means: net income and equity, but these will have been calculated according to domestic standards and then according to US-GAAPs. Hence, using the required 20-F we will be able to examine both pairs of figures without restatement by ourselves, because we understand it is a difficult process. Afterwards this kind of research has been improved adding Wilcoxon test (Norton, 1995)(Weetman et al., 1998) or using instead of domestic standards International Accounting Standards (IASs) issued by the International Accounting Standards Commitee (IASC)(Adams et al., 1993).

The other trend started by Van der Tas (1988) tries to measure harmonisation of financial reporting practice using concentration indices: Herfindahl index (H) which is calculated by weighting the relative frequencies of the alternative opinions against each other; C index that is approximately equal to the H index but has the ability to take account of multiple reporting; I index which is a measure of the degree of international material harmony by multiplying the relative application frequency of a method in country A by the relative application frequency of the same method in country B and subsequently by adding the results of all alternative methods. The most useful index has been I besides the chi-square test, although, after reviewing Van der Tas (1988) and other notable papers on the measurement of international harmonization, Tay and Parker (1990) identified the following as major problems: data sources, statistical methods employed, and distinguishing changes in accounting practice due to compliance with standards from changes due to other reasons. These other reasons are the ones that we are trying to identify through Dual Scaling.

Finally, some other studies have examined aspects of the reconciliations to US GAAP provided by foreign registrant companies because of the debate around the likelihood of IASC's core standards acceptance by the SEC (Adams et al., 1999), or studies such as Choi et al. (1983) in which they

Table 1: Ratios used in the paper. The profitability is measured with ROA —returns on assets— and ROE, returns on equity; leverage is measured with DER and DRC meaning debt ratio and debt return capability.

$$\text{ROA} = \frac{\text{net income}}{\text{assets}} \qquad \text{DER} = \frac{\text{liability}}{\text{equity}}$$

$$\text{ROE} = \frac{\text{net income}}{\text{equity}} \qquad \text{DRC} = \frac{\text{net income} + \text{depreciation}}{\text{liability}}$$

highlight the importance of factors different from accounting standard effect introducing pitfalls in the analyzing process. Even when ratios are based on US-GAAPs they are misinterpreted because the US investor do not understand a particular foreign environment that influences all financial ratios in that environment. Thus, the country effect is remarked by these authors.

3. Database and Method

Whether the information revealed has been obtained from Form 20-F the first condition for being in the sample was to be listed on the NYSE. We have chosen NYSE-listed companies from three different European countries: Spain, United Kingdom and Germany. When making the choice we tried to have a representation of the most different accounting systems in Europe, and we have selected United Kingdom and Germany. The former is an example of an anglosaxon accounting system in Europe while Germany represents the other side: a continental accounting system. The introduction of Spain responds to a desire of having another accounting system represented in the sample.

The second step was to select the companies. Depending on the companies the choice for Spanish and German companies was not difficult because only seven and eight of these companies were listed on the NYSE in 1997. Each company was contacted through their web-sites and a copy of its Form 20-F requested or downloaded. The size of the companies guaranteed the existence of a web-site, from which it was possible to get almost all the available released information (in some cases, the Form 20-F was not available and we asked for it to the investor relations link, but it was impossible to get a copy of it). For the British companies the selection was based on getting as much information as possible, matching at least one British firm to a Spanish and German listed company by the kind of industry because there were a lot of British companies listed on the NYSE at this date (the financial tradition of anglosaxon companies is different from the tradition in continental countries).

Table 2 shows the 17 companies included in the sample. These companies include their 20-F financial statements according to domestic accounting rules and besides this, they have to restate domestic net income and equity into US-

Table 2: Categorization of the information. Numbers in Table indicate if the company ratio is bigger using US (number 2) or domestic (number 1) GAAPs.

Country	Company	ROE	ROA	DRC	DER
Spain	BBV	1	2	2	1
	Banco de Santander	2	2	2	1
	Endesa	2	2	2	2
	Repsol	2	2	2	2
	Telefónica	2	2	2	2
Germany	Veba	1	1	1	2
	SAP	2	1	1	2
	SGL Carbon	1	1	1	1
	Deutsche Telekom	1	1	1	1
	Hoechst	1	1	1	2
United Kingdom	Barclays	2	2	2	2
	British Gas	2	2	2	1
	BOC	1	1	1	2
	British Airways	2	2	2	2
	Cable & Wireless	1	1	2	1
	Vodafone	1	1	1	1
	Cadbury Scweppes	1	1	1	1

GAAPs. With both figures: according to US-GAAPs and non-US or domestic GAAPs, we calculate four relevant ratios: Returns on Assets (ROA), Returns on Equity (ROE), Debt Ratio (DER) and Debt Return Capability (DRC) whose formulas are in Table 1. These are the most common ratios used in empirical studies.

Thus, we have the four ratios expressions for the seventeen companies. The first hypothesis to be tested is if there are real differences on the results according to the way in which the information is elaborated, that is if the normative effect is big enough to be statistically relevant. To test this idea, we use the traditional t-test for the mean where H_0 is *There are no relevant differences*. The p-value for the pairs (ROE according to US GAAPs and domestic GAAPs; ROA according to... and so on) are 0.861, 0.754, 0.892, 0.198 meaning that there is no statistical evidence for the accounting standard effects (differences produced in financial statements because of different accounting standards). In other words, no matter whetter the GAAP is used in the elaboration of the report, the differences are not statistically significant.

But, as can be seen for the example in Figure 1 regarding Returns on Equity (ROE) there are slight differences among countries because not all the points are in the diagonal. That is why we revisited our data looking for any signal of what can be named as the country effect. To do that, we categorize our information according to the principle: if the ratio is bigger using US

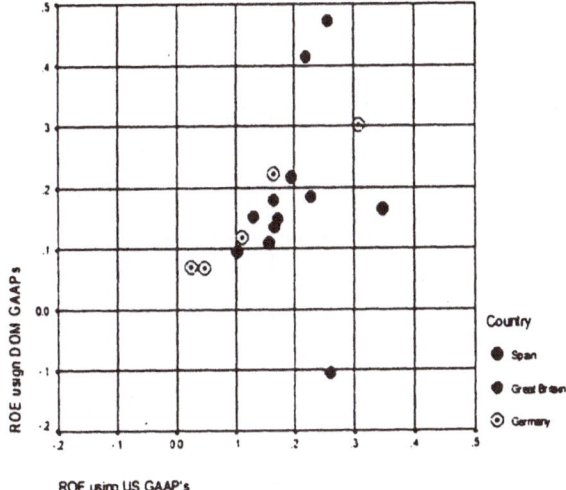

Figure 1: Scatterplot of return on equity, by countries, depending on the gaap used to the diclosure of the financial state: us vs domestic principles.

GAAP than domestic GAAP, it is coded as 2, otherwise 1. The obtained results of this categorization are shown in Table 2. That information can be seen as multiple-choice data. In fact, we also add an item to reflect the nationality of the firms: 1 for Spain, 2 for Germany and 3 for United Kingdom. Dual Scaling is used to analyze the data. As Nishisato (1994) pointed out, the total information of this table is 1.2 and it is equal to the average number of options minus 1. This value is equal to the sum of the squared correlations ratios (η^2) of all the solutions. In our case, all the possible solutions are 6: total number of options (11) minus the number of items (5) but we only use the first two solutions as we are using the criterion of $\eta^2 \geq \frac{1}{n}$ to select the admissible solutions (Nishisato, 1980). Doing that the Cronbach's α is positive and there is internal consistency reliability (86.1% and 25.8%). The total information accounted, —δ criterion on DUAL3 V.4.1—, is 53.6% and 21.0% that adds to 74.6% of the information contained in Table 2.

4. Results

In the Dual output we observe that the biggest sum of squares of weighted responses of each item (SS_j) is in the first solution item 2 (ROA) and item 3 (DRC) and that in second solution, item 4(DER) emerges as the dominant contributor. The representations of the projected weights for options —the principal coordinates—, are in Figure 2. Remember that this map contains 74.6% of the information contained in Table 2.

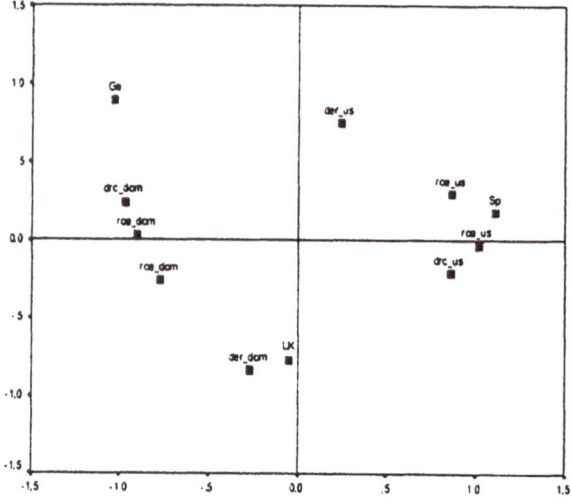

Figure 2: Scatterplot of Solution 1 (horizontal axis) and 2 of projected weights (principal coordinates) for options; 74.6% of the total information.

Looking at dimension 1 (solution 1) we can see the transition of response patterns from better profitability according to US-GAAPs than to domestic. We also note that high item–total correlation yields regularity in the transition of response–patterns. The only low value is related with item 4 (DER) with $r^2 = 0.068$. It is interesting to note that an item which has a high value of r^2 divides the space cleanly by its options. In our case, it is clear that ROA separates the space. On the right part, the Spanish (Sp) companies are better valued using US-GAAPs than domestic. On the other side, one can see United Kingdom (UK) and Germany (Ge). On the other hand, it is easy to see that German GAAPs are more favourable (just to say that they are less conservative in calculating profits) to their companies than, say, Spanish accounting rules. German companies show bigger profitability—talking about the ratio returns on assets or returns on equity— using their domestic GAAPs than US-GAAPs. Hence, they loose their positions if they have to adapt themselves to US-GAAPs as the SEC requires.

Regarding dimension 2, as we have written, the most relevant item that allows us to differentiate the space is the Debt Ratio ($r^2 = 0.63$). The item country is also well represented in the space. Looking at the principal coordinates there are a proportionally bigger number of British companies whose debt ratio is bigger according to domestic standards than to US-GAAPs. While for German and Spanish companies the results are just the opposite, the number of companies whose debt ratio is bigger according US-GAAPs than to domestic standards is higher. Anyway, this axis contains only 21% of the total information included in the table. Although bearing in mind the previous statement it is interesting to remark the similarities between the differences in accounting systems and the results obtained from our dual scaling analysis.

5. Concluding Comments

The aim of this paper is to bring together two areas that could be improved by their mutual influences. Since the last decades, global players companies are claiming for homogeneous accounting principles all over the world. SEC's requirements may be an answer but, as it is shown, they are not neutral at all. There are unavoidable influences of some other factors, such as country and environment of the firms that has to do with the diversity more than the diversity of rules in the world. Although at a first sight it may be said that German standards are more conservative than the others, our evidence shows that as in the famous case of DaimlerBenz (Ordelheide, 1997), domestic standards allow more profitability than US-GAAPs. It is also proved that German and British companies have been influenced by so different variables that they keep being the most different accounting systems in Europe. While Spanish companies present a trend to show more positive profitability according to US-GAAPs than to Spanish standards, it may be due to real ancestral conservative accounting practices.

References:

Adams, C. A., Weetman, P., and Gray, S. J. (1993). Reconciling national with international accounting standards. *European Accounting Review*, **3**, 3, 471–494.

Adams, C. A., Weetman, P., Jones, P. W. and Gray, S. J. (1999). Reducing the burden of US GAAP reconciliations by foreign companies listed in the United States: the key question of materiality. *European Accounting Review*, **8**, 1, 1–22.

Choi, F. D. S., Hino, H., Min, S. K., Nam, S. O., Ujiie, J. and Stonehill, A. I. (1983). Analyzing Foreign Financial Statements: The Use and Misuse of International Ratio Analysis. *Journal of International Business Studies*, Spring, 113–131.

Frost, C. A., and Lang, M. H. (1996). Foreign Companies and U.S. Securities Markets: Financial Reporting Policy Issues and Suggestions for Research. *Accounting Horizons*, **10**, 1, 95–109.

Gray, S. J. (1980). The Impact of International Accounting Differences from a Security-Analysis Perspective: Some European Evidence. *Journal of Accounting Research*,**18**, 1, 64–76.

Nishisato (1984). *Elements of Dual Scaling: an introduction to practical data analysis*. Lawrence Erlbaum Associates. Toronto.

Norton, J.(1995). The Impact of Financial Accounting Practices on the Measurement of Profit and Equity: Australia versus the United States. *ABACUS*, **31**, 2, 178–201.

Ordelheide, D. (1997). Wettbewerb der Rechnungslegungssysteme IAS, US-GAAP und HGB. Plädoyer für eine Reform des deutschen Bilanzrechts. Included in Börsig, C. and Coenenberg, A. G. (Hrsg) (1997). Controlling und Rechnungswesen im Internationalen Wettbewerb. Schmalenbach Gesellschaft für Betriebswirtschaft, Schäffer Poeschel, Stuttgart, 15–53.

Tay, J. S. W. and Parker, R. H. (1990). Measuring International Harmonization and Standarization.*ABACUS*, **26**, 1, 71–88.

Van der Tas, L. G. (1988). Measuring Harmonisation of Financial Reporting Practice.*Accounting and Business Research*, **18**, 70, 157–169.

Weetman, P., Jones, E. A. E., Adams, C. A. and Gray, S. J. (1998). Profit Measurement and UK Accounting Standards: A Case of Increasing Disharmony in Relation to US GAAP and IASs *Accounting and Business Research*, **28**, 3, 189–208.

A Three-Step Approach to Factor Analysis on
Data of Multiple Testlets

Soonmook Lee, Ahyoung Kim

Sungkyunkwan University, Ewha Womans University
Seoul, Korea

Summary: A testlet consists of a text and several items following it. It is often observed that a test consists of multiple testlets. The items within a testlet are more interrelated than they are with other items in the test. In this note, we attempt to show how to estimate common factors in the data of multiple testlets. We start our argument from the rationale of common factor analysis on each testlet. However, we treat the effect of a testlet on the item scores within the testlet as the method effect. If we are able to remove the method effect from subjects' response data, then we can apply ordinary factor analysis on the residualized scores that remain after partialing out the method effect. Application of our approach is also demonstrated.

1. Model of Three-Step Factor Analysis

We will examine the method of factor-analyzing data for a test consisting of multiple testlets. A testlet consists of a text and several items following it. In data of multiple testlets there are two kinds common factors. One kind is the substance factor representing the construct embeded in the test and the other kind is the testlet factor representing the effect of testlets on the items in it. Since the testlet factor is not the factor intended for substantial interpretation, it is a kind of method factor.

Technically a testlet is small compared with a test, but big enough to carry its own context. The accuracy of measurement in using testlet-based tests has been well studied (e.g.: Lee, 2000; Lee & Frisbie, 1999; Sireci, Thissen, & Wainer, 1991). However, there has not been much research regarding factor structure of a testlet-based test. It is especially important to recognize that factor analysis cannot be correctly performed without taking the structure of testlet factors into consideration in the data from testlet-based tests.

We propose a three-step approach to factor-analysing testlet-based data. The rationale starts from the common factor model including substance factors and testlet factors. Testlet factors are types of method factors generated by the format or context in designing the testlet. They are independent of the substance factors measured by the items across the testlets in a test. The key to factor-analysing testlet-based data is to apply factor analysis on the residualized data which remain after the method factors are partialled out.

Suppose a test is made of s testlets and there are t_1, t_2, ..., t_s items for each of the testlets. Then the total number of items is defined as $v=\sum t_i$, $i=1$, ..., s. A testlet consists of a text and several items following it. If some factors other than substance factors are introduced in devising the text, they would work as the major part of method factors in the test and our approach is to do factor analysis after partialling them out from the item response data. Given the method factors due to multiple testlets, the item score (we assume standardized score) from n subjects and v items can be expressed as below, following the common factor model.

$$Z_{nv} = F_{np}P'_{pv} + M_{nq}Q'_{qv} + U_{nv} \qquad (1)$$

where Z_{nv} is the matrix of observed scores, F_{np} is the matrix of p substance factor scores, P_{vp} is the matrix of pattern loadings for substance factors, M_{nq} is the matrix of q method factor scores, Q_{vq} is the matrix of pattern loadings for method factors, and U_{nv} is the matrix of unique factor scores. The subscripts of matrices will be omitted when there is no confusion. If the portion contributed by the testlet effect is removed from the response data, eq (1) is reduced as below.

$$R_{nv} = Z - MQ' = FP' + U \qquad (2)$$

where R_{nv} is the matrix of residualized scores.

The Z_{nv} in eq(1) can be expressed as a set of partitioned matrices as in eq(3).

$$Z_{nv} = [Z_{n,t1} \mid \dots Z_{n,ti} \dots \mid Z_{n,ts}] \qquad (3)$$

$$Z_{n,ti} = W_{n,j} K'_{j,ti} + E_{n,ti} \qquad (4)$$

$$z_{nv} = [E_{n,t1} \mid \dots E_{n,ti} \dots \mid E_{n,ts}] \qquad (5)$$

where $W_{n,j}$ is a matrix of j method factor scores in a testlet, $K_{ti,j}$ is a matrix of pattern loadings for j method factors in the testlet, and E is the matrix of unique factor scores in a testlet. Under "particular conditions", there can be one factor only and it can be treated as the method factor in a testlet. We will present the conditions later. This within-testlet method factor contains the effect that the testlet as a measurement method has over the items in it. When the portions contributed by the within-testlet method factor are combined across all the testlets, then they yield multiple method factors in the test. That is, MQ' in eq (1) is the combination of portions that are reflected in WK' in eq (4) across the s testlets. Then the following equality holds.

$$R_{nv} \approx z_{nv} \qquad (6)$$

Now the structure of substance factors can be obtained by applying ordinary common factor analysis on the data of z_{nv}. The empirical approach we propose here consists of three steps in which determining $E_{n,ti}$ is the first step. Before we explain the first step, we propose to assume just one factor in each testlet and take it for the method factor which is to be partialled out from the response data $Z_{n,ti}$. In order to justify this proposition we need the "particular conditions" mentioned above. They are in the form of assumptions because they need to be verified by the researcher before starting the first step of our approach.

Assumption 1: Levels of design factors are well combined so that no single factor is dominantly reflected in constructing the text of a testlet.

Assumption 2: The items in the testlet are constructed to measure substance factors of major interest in the study which are independent of design factors in developing the text.

Assumption 3: The number of items in a testlet is small enough that items from more than one testlet are needed to define a substance factor.

It would be beneficial to review the plausibility of the assumptions. In designing testlets, it is typical to combine the levels of design factors so that levels of each factor are reflected with equal chance. Suppose there are two design factors A and B with levels (a_1, a_2) and (b_1, b_2) respectively. Then the basic set of testlets are as follows : (a_1, b_1) (a_1, b_2) (a_2, b_1) (a_2, b_2). Then no single factor is dominantly reflected in a testlet, yielding the plausibility of assumption 1.

Given the context created by design factors a problem situation is written in the text of a testlet and some response alternatives are provided as items in the testlet. The design factors comprise method factors of M in eq(1) and response alternatives are developed to measure substance factors of F in eq (1). The independence of substance factors and method factors should be assumed in order to get the model in eq(1). The plausibility of this Assumption 2 in a data set can be examined by testing the significance of correlation between substance factors and method factors later in our three-step approach.

In psychology, it is recommended to have at least three to five items to measure a psychological construct. Even with five items, it is not certain if we can derive one factor because there is always measurement error in the data. Unless the number of items in a testlet is much larger than three to five, the assumption 3 is plausible.

Once the three assumptions are reasonably satisfied in the process of designing the testlet and writing the text and corresponding items, we can follow the three steps of our approach.

In the first step we estimate WK' in eq (4) by factor-analysing the response data to a testlet. As a result we obtain a vector of linear coefficient K' and method factor scores W, and finally $E_{n,ti}$ which are the components in eq (5). At this step our goal is to partial out the proportion contributed by the measurement method from the observed score $Z_{n,ti}$, but not delineating the structure of method factors, which will be done in the third step going back to eq (1).

In the second step, we apply exploratory common factor analysis on the z_{nv} obtained in the second step. As a result, we will get a temporary solution for the substance factors indicated in the items which are independent of the testlet factors. The researcher is interested in interpreting these substance factors. However, it is to be refined in the third step.

In the third and final step, we are going back to eq (1) to delineate method factors in the test and to refine substance factors by the application of confirmatory factor analysis on the original data Z_{nv}. The number of method factors in the test can be defined by considering the factors which are built in the text when testlets are constructed. For example, Wagner and Sternberg (1993) constructed nine testlets in their Tacit Knowledge Inventory for Managers. Each testlet consists of several items, yielding 91 items in total. Each text was written to represent three design factors: content factor consisting of three levels—self, others, and tasks; context factor consisting of the two levels—global, local; orientation factors consisting of two factors—ideal, pragmatic. Of course, these design factors are method factors in the assumption, but not necessarily the factors in reality. The method factors in reality can be empirically determined by applying confirmatory factor analysis on Z_{nv}. Confirmatory factor analysis is one form of covariance structure modeling. In doing this analysis, we designate the design factors of testlets as method factors and the solution obtained in the third step as the substance factors, with method factors and substance factors unrelated. In the process of specification search (Leamer, 1978; MacCallum, 1986), it is expected that the method factors under assumption would reveal their features in reality and the substance factors would develop into a more parsimonious structure. Also the Assumption 2 can be explicitly tested in the confirmatory factor analysis.

2. Application of the Three-Step Approach to Real Data

We apply the three-step approach to extracting common factors in the study of tacit knowledge.

2.1 Introduction to Tacit Knowledge

The tacit knowledge (TK) proposed by philosophers (e.g.: Polanyi, 1958) has been studied experimentally by psychologists for its nature during the last 30 years. Recently there has been increased interest in tacit knowledge as an essential factor for skill development. Proceduralization of knowledge is considered the first step for acquired knowledge to be converted into a skill. Tacit knowledge is a kind of procedural knowledge acquired by oneself without the help of media or environment, and useful for individuals' goal accomplishment (Sternberg, Wagner, Williams, & Horvath, 1995). Thus it is very important to investigate how to identify tacit knowledge in a practical task domain and how to train people so that many others can benefit from the research on tacit knowledge.

2.2 Issues to be Investigated

There is a lot of empirical research on TK for its relationship with learning and job performance (e.g.: Sternberg, Wagner, Okagaki, 1993; Wagner & Sternberg, 1993; Wagner & Sternberg, 1985). Sternberg and his colleagues have concentrated on collecting evidence that TK is different from academic intelligence and is a good predictor of training and performance. However there is very little research on the structure of TK. Although Sternberg and his colleagues devised an inventory of measuring TK based on a framework of content, context, and orientation, the framework is still far from the structure of TK in a particular domain. The framework of content (managing self, others, and tasks), context (global, local), and orientation(ideal, realistic) is very useful in generating TK items. However, they are only the categories, aspects, or scope, but not the factor structures empirically derived. Since TK is defined in a practical domain, it is hard to investigate the accurate relationship between TK and other variables without delineating the different structure of TK in different domains. We tried to estimate the factor structure of TK in the domain of high school students by using the three-step approach.

2.3 Development of Testlets for Measuring TK

We attempted to investigate the factor structure of the tacit knowledge that high school students have in school life. We developed an inventory testing TK, which consists of sixteen testlets and 96 items in total. The texts in the testlets are made of scenarios which might occur and bring about conflict in decision making. We employed a procedure similar to that of Wagner and Sternberg (1993) to develop the scenarios and response alternatives to them. We asked 215 (136 males, 79 females) high school graduates or soon-to-be graduates for problems in their school life and for their behaviors resolving the problems. We also asked 103 freshmen taking introductory psychology courses for the issues and problems they encountered in the high school days and for the ways in which they resolved the difficulties.

Based on the data from the students, we could identify 250 preliminary scenarios for high school life. Our assistants screened them to select 50 scenarios which we examined for the final entry of sixteen. The scenarios were developed on two design factors: Life domain (academic, nonacademic) and context (global, local). The content areas (managing self, others, and tasks) specified in Wagner and Sternberg (1993) were used to develop response alternatives. Finally we developed four scenarios for each of four combinations of life domain (2) by context (2), yielding 16 scenarios. There are, in total, sixteen testlets in this test. We developed six response alternatives (items) in each testlet. We asked 493 high school boys to rate alternatives in all sixteen testlets on 7 point scales for their desirability of resolving the given problem in the text.

2.4 First Step Analysis: Exploratory Factor Analysis within each Testlet

We assume only one method factor in each testlet in this study for many reasons. Firstly, each text is written to reflect one type of life domain and one type of context. The response alternatives are in the content of managing self, others, and tasks, and the item scales are constructed to measure subjects' judgment on the desirability of the alternatives. The number of items in a testlet is six, which is not large enough to define substance factors on its own. Finally, from the examination of the first eigenvalue resulting from principal factor analysis on each of the testlets, we can determine that the variance explained by the first factor was above 75%, supporting the idea of assuming one method factor in each testlet. Thus we performed exploratory factor analysis from which we obtained one factor solution from each set of testlet data. In computing the method factor scores we used regression method (Harman, 1976) which is easily available in statistical programs. By subtracting the product of linear coefficient (K' in eq(4)) and method factor score (W in eq(4)) from the testlet data ($z_{n,ti}$ in eq(4)), we obtained $E_{n,ti}$ and finally z_{nv} in eq(5).

2.5 Second Step Analysis: Exploratory Common Factor Analysis on the Residualized Response Data

From the exploratory common factor analysis (principal axis factoring, VARIMAX rotation), we derived five interpretable factors as in tables 1 and 2.

Table 1: Structure of TK factors

Factor	Items
Advice Seeking For His Own Problems	TK24. Go to school early the next morning and ask a math teacher for the answer.
	TK92. Tell some other friends about my own feelings, and ask them how to cope with that situation.
	TK95. Ask for the advice of my parents or school counselor.
	TK101. Do math homework only during breaks, and then, the teacher checks the homework, explain frankly and ask for his understanding.
	TK114. Tell my dad my feelings man to man, and ask him for advice.
	TK125. Ask other team leaders how this kind of child is treated in their teams.
	TK135. Call up B's mother and request that B be hurried out the door tomorrow morning.
	TK142. Request a math teacher to evaluate my math competency objectively.
	TK155. Ask an English teacher how to improve my English and my study strategies.
Self-Protective Coping	TK11. As it is difficult to force B to accept the advice, try not to get involved, and think that I have done my best as a friend.
	TK22. Decide to solve the math problem later around the exam day, and study history with ease of mind.
	TK43. Pretend not to know about B's situation and keep watching over B until B talks about it on his own.
	TK52. Give up the birthday party, don't take any sides, and keep watching.
	TK122. Conflict may spoil the group as a whole, have patience and ignore it.
	TK134. Use some excuse to suggest that they go to school separately tomorrow.
Advice Seeking For Other's Problems	TK15. Tell the teacher that the reason for B's poor schoolwork is due to a girlfriend problem, and ask for good advice that I could lend him.
	TK41. Tell the teacher the truth but ask that B not be punished, and then request advice on what A

could do for B.

TK44. Meet B's family and leave the problem to be solved by
them.

TK55. Ask the school counselor which way would be the best.

	TK23. Ask a friend who is good at math and try to find out what was the problem.
	TK72. Show aptitude as a cartoonist by participating in a cartoon competition.
Problem Solving Behavior	TK94. Set my own standard and study mathematics hard to reach the goal.
	TK115. Determine never to smoke again.
	TK143. Study mathematics two hours a day regularly to improve grade.
	TK151. Think of a way to study English happily without being stressed by school grade.
	TK14. Think about what could be helpful advice for B.
	TK21. Try once more without becoming nervous.
Reflection for Problem Solving & Future Orientation	TK31. Have confidence that I can do this and follow a step-by-step plan of study.
	TK34. Imagine future self as a college student, then study and endure.
	TK45. Think of ways to persuade B to stop doing the job.
	TK62. Have confidence and plan a week schedule again.
	TK141. Consider again whether it is truly my dream to become a pharmacist by attending a college of pharmacy.

* The last digit indicates the number sequencing items in a testlet and the one or two digits before the last digit is the number of the testlet in our inventory.

Table 2: Intercorrelations of TK Factors

Subfactor	1	2	3	4	5
1. Advice Seeking for His Own Problems	(0.67)				
2. Self-Protective Coping	0.02	(0.59)			
3. Advice Seeking for Other's Problems	0.40	0.12	(0.57)		
4. Problem Solving Behavior	-0.09	-0.14	-0.06	(0.58)	
5. Reflection for Problem Solving and Future Orientation	-0.03	-0.14	-0.02	0.34	(0.64)

* Numbers in the parentheses ae reliability αcoefficients for factor scales.

2.6 Third Step Analysis: Confirmatory Factor Analysis on the Data of Z

At this stage, we will analyze whether we can delineate method factors as assumed in the design stage of the test and TK factors as we obtained in the second step analysis. In order to refine the solution we attempt confirmatory factor analysis coming back to the original data Z_{nv} in eq(1).

In modeling the factor structure, we started from a model of five TK factors and four method factors. The estimation method was GL (generalized least square) and the software was LISREL 8.03 (Jöreskog & Sörbom, 1993). Since we are interested in refining the structure of method factors first, we observed the intercorrelation between the two method factors in each pair: Corr (academic, non-academic)=-.80, corr(local, global)=-.31. This suggests that the academic factor and the non-academic factor can be combined into one factor, and the locality factor and the globality factor should be treated as unique factors although they are correlated to some degree. However, the variances of the locality factor and the globality factor turned out to be too small to be treated as meaningful factors. Then we have only one method factor and four TK factors, which defines the final model in our specification search (Leamer, 1978; MacCallum, 1986). The fit measures were reasonably good and are given in table 3. The variances of four factors were significant and their correlations are given in table 4. As mentioned before, it is possible to examine the plausibility of the independence assumption between the method factor and substance factors. By referring to the value of modification indices provided in LISREL we can determine if the independence assumption is supported or not. In our analysis the modification indices for the possible correlation between the method factor and substance factors were negligible, supporting the Assumption 2.

Table 3: Fit Measures of the Final Solution

χ^2(df=426) = 713.37 (p=.00), RMSEA = .037
ECVI = 1.86,   ECVI for saturated model = 2.15
RMR = .06,   GFI = .91,   AGFI = .89
NFI = .87,   NNFI = .94, GFI = .94

Table 4: Intercorrelation of final factors and their t-values

	1	2	3	4	5	M
F1[a]	1.00					
F2	.00	1.00				
F3	.57	.24	1.00			
	(10.72)[b]	(3.12)				
F4	.29	-.22	-.20	1.00		
	(3.35)	(-2.49)	(-2.19)			
F5	.32	-.21	.00	.76	1.00	
	(4.68)	(-2.73)		(14.35)		
M[c]	.00	.00	.00	.00	.00	1.00

[a] The number of factors F1 through F5 corresponds to the order in table 2.
[b] t-value are given in the parentheses just below the correlations. If the t-value are larger than +2 or smaller than -2, the correlation is considered to be significant.
[c] method factor.

3. Conclusion

We proposed a three-step approach to factor-analysis data from multiple testlets and showed its application on the area of tacit knowledge. The last step is to use confirmatory factor analysis to refine the structure of method factors and substance factors, which is called a specification search in covariance structure modeling (Leamer, 1978; MacCallum, 1986). At the end of the third step, we have arrived at a model including five substance factors and one method factor. It is desirable to cross-validate on new data after specification search. However, since our major purpose was to develop the three-step approach and demonstrate its application to a substantive area, we did not collect new data and cross-validate our procedure.

The critical step in our approach is the first step where the researchers need to examine the three assumptions before they assume one method factor in a testlet. The assumptions should be empirically supported by examining the eigenvalues and the proportion of variance explained by the first eigenvalue in the testlet data. When the common factor model is used, investigators stop factoring process when "75, 80, or 85%" of the common variance is accounted for (Gorsuch, 1983, p.165). One can support one factor structure if the first eigenvalue accounts for "75, 80, or 85%" or more percentage of common variance in the testlet data. And if the number of response alternatives is not large enough to define substance factors on its own, the one factor in the testlet is considered as the method factor caused by the form of the testlet.

By following the steps with sound discretion, we could derive five substance factors from the data of sixteen testlets and ninety-six response alternatives in the data of tacit knowledge for high school boys.

References:

Gorsuch, R. L. (1983). *Factor analysis* (2nd Ed.), Hillsdale, NJ: Lawrence Erlbaum.

Harman, H. H. (1976). *Modern Factor Analysis*. 3rd Ed. Chicago, IL: University of Chicago Press.

Jöreskog, K. G. & Sörbom, D. (1998). *LISREL8*. Chicago, IL: Scientific Software.

Leamer, E. E. (1978). *Specification Searches: Ad hoc inference with nonexperimental data*, New York: Wiley.

Lee, G. (2000). Estimating conditional standard errors of measurement for tests composed of tests. *Applied Measurement in Education*, 13, 161-180.

Lee, G., & Frisbie, D. A. (1999). Estimating reliability under a generalizability theory model for test scores composed of testlets. *Applied Measurement in Education*, 12, 237-255.

MacCallum, R. (1986). Specification searches in covariance structure modeling. *Psychological Bulletin*, 100, 107-120.

Polanyi, M. (1958). *Personal knowledge: Toward a post-critical philosophy*. Chicago: University of Chicago Press

Sireci, S. G., Thissen, D., & Wainer, H. (1991). On the reliability of testlet-based tests. *Journal of Educational Measurement*, 28, 237-247.

Sternberg, R. J., Wagner, R. K. & Okagaki, L. (1993). Practical Intelligence: The nature and role of tacit knowledge in work and at school. In H. Reese & J. Puckett(Eds), *Advances in Lifespan Development*, 205-227. Hillsdale, NJ: Earlbaum.

Sternberg, R. J., Wagner, R. K., Williams, W. M., & Horvath, J. A. (1995). Testing common sense. *American Psychologist*. November, 912-926.

Wagner, R. K. & Sternberg, R. J. (1985). Practical Intelligence in Real-World Pursuits: The role of tacit knowledge. *Journal of Personality and Social Psychology*, 49. 436-458.

Wagner, R. K. & Sternberg, R. J. (1993). *TKIM: The commonsense manager*, user manual. New York: Hartcourt Brace Jovanovich, Inc.

Harman, H.H (1976). *Modern factor analysis*. 3rd Ed. Chicago, ILL: University of Chicago Press.

A Technique for Setting Standards and Maintaining Them over Time

Mark Wilson and Karen Draney

University of California, Berkeley,
Berkeley, CA 94720, USA

Summary: This paper describes a new procedure for standard setting based on item response maps ("Wright maps"). Motivation for the technique is discussed, and variants of the item response formulation are shown. An example based on the Golden State Examinations is used as a context for the discussion, and for some results.

1. Background

The Golden State Examination (GSE) program in the state of California is a set of high school honors examinations in a number of subjects, including mathematics (Algebra, Geometry, High School Mathematics), language (Reading & Literature, Written Composition, Spanish Language), science (Physics, Chemistry, Biology, Coordinated Science), and social science (US History, Government & Civics, Economics). These examinations are composed of multiple choice items and a variety of open-ended, essay, and performance items.

Based on the GSE, examinees are categorized into one of six performance levels—descriptive categories of student performance. The top three levels (4, 5, and 6) are considered "honors" levels (School Recognition, Honors, and High Honors, respectively). If a student achieves one of these honors levels on six exams (including US History, Reading & Literature or Written Composition, a mathematics exam, and a science exam), the student is also eligible for a state honors diploma.

For several years, the California Department of Education (CDE) has been using a so-called "matrix" method of setting standards on the GSE. In this method, a committee of teachers, subject matter experts, and other relevant educators, design a two-dimensional matrix that maps the total scores on the multiple choice items and the scores on the open-ended items into the six performance levels. The committee meets and assigns a performance level to each possible combination of scores on the different item types. This method was seen as a crucial way to bring the judgments of professionals into the arena of test interpretation. The procedure is designed to use professional judgment to balance the technical usefulness of the multiple choice items with the assumed educational validity of the open-ended items.

While this approach has been satisfactory for a number of years, it has several weaknesses that have led to the search for an alternative that provides a more flexible technical solution, and yet preserves the original commitment to relying on teacher professional judgement.

When the traditional technique was developed, most of the examinations consisted of a multiple choice section and a single written response question. Now, however, most of the examinations include at least two written response questions. Thus, the meaning of the score on the written response section is not as clear as it was: for example, if both written response items are scored one to five, a total score of six may mean a score of three on both items; it may also mean a score of five on one item and one on the other item. This is an issue of concern to many committee members, who, for example, might want to make certain that no one who has obtained a score of one on any written response item earns a high performance level on an examination, but would probably deal more leniently with a person who obtained two scores of three.

A deeper problem with the current procedure is that it results in a certain level of arbitrariness. The committee members' ideas of how a score corresponds to a performance level may or may not be based on any conceptual model of what a student at that level knows and can do. In addition, the judgment of particular committees may be influenced by a variety of factors, including a (subjective) evaluation of the difficulty of the items, especially the written response items. It is also possible that the particular membership of the committees setting the standards, along with the characteristics of the committee leader, have a large influence on the composition of the performance levels in a particular year.

Perhaps the most serious problem with the use of the mapping matrix is that it provides a weak case for year-to-year comparability. The committees have had a difficult time maintaining a consistent standard from year-to-year using the matrix method. This is primarily due to marked differences in the difficulties of the written response items from year to year. Efforts to "standardize" the production of these items have not been successful. Reliance on raw scores in the presence of significant changes in the difficulty of the items, along with the aforementioned changes in the test design, has led to a practice of "post-hoc manipulation" of the standards to try to make them less variable from year to year, effectively negating any opportunity to observe changes in average student performance.

We have developed an approach we call "Construct Mapping" to address the problem of standard setting with mixed item types. This process is designed to balance the issues of perceived increases in psychometric precision to be gained by tighter control over task specification and judgment, with the greater instructional validity perceived to be gained by more authentic item types and greater reliance on professional judgment (Wilson, 1994). This method allows the translation of raw scores on various types of items (multiple choice, open-ended, performance, and so on) into two or more hierarchically ordered criterion levels, and allows for the maintenance of the resulting standards or cut scores over time.

The method is based on two sets of tools. The first is a family of item response models, known as the Multidimensional Random Coefficients Multinomial Logit model (MRCML; Adams, Wilson & Wang, 1997). The second is a set of conceptual maps (which we call "Wright" maps as B. D. Wright has pioneered their use) and associated software, based on the results of the modeling process. These maps serve as visual metaphors, allowing the layperson to interpret the results of the modeling process (Draney & Wilson, 1998; Wilson &

Draney, 1998). This technique is different from others developed to address similar problems (Jaeger, 1995; Jaeger, Plake & Hambleton,1993) because of its insistence on the committee members looking across the whole vector of student performance through its concentration on the Wright map.

2. The MRCML Model

The MRCML is a family of LLTM-style models, which allow for the incorporation of a wide variety of effects, including differing rater severities, local item dependencies, and combinations of item modes. The MRCML describes the proficiency of person N as a V-dimensional vector $\theta = (\theta_1,...,\theta_v)$. The relationship between items and dimensions is specified by a scoring function, such that a response in category j to item i is represented by a V-dimensional column vector $b_{ik}=(b_{ik1},...,b_{ikv})'$. These column vectors are then collected into a scoring submatrix B_i for item i, and these submatrices collected into an overall scoring matrix B.

There are p item parameters, which are collected into a parameter vector $\xi=(\xi_1,...,\xi_p)$. The relationship between item responses and parameters is specified by a design matrix A. This is composed of design vectors a_{ij}, each of length p, which when multiplied by the parameter vector ξ form a linear combination of the parameters operative in response j to item i. The probability of a response j to item i is then given by

$$\Pr(X_{ij} = 1;A,B,\xi \mid \theta) = \frac{\exp(b_{ij}\theta + a'_{ij}\xi)}{\sum_{k=1}^{K_i}\exp(b_{ik}\theta + a'_{ik}\xi)}. \tag{1}$$

We implement the model using the Conquest software (Wu, Adams & Wilson, 1998).

Such an approach allows for the incorporation of numerous complications specific to a given testing situation. For example, both dichotomous and polytomous items can easily be incorporated. If item 1 is an item with 4 possible scoring levels, labeled 0 through 3, then Equation 1 becomes:

$$\Pr(X_{10} = 1;A,B,\xi \mid \theta) = 1/D$$
$$\Pr(X_{11} = 1;A,B,\xi \mid \theta) = \exp(\theta + \delta_1)/D$$
$$\Pr(X_{12} = 1;A,B,\xi \mid \theta) = \exp(2\theta + \delta_1 + \delta_2)/D \tag{2}$$
$$\Pr(X_{13} = 1;A,B,\xi \mid \theta) = \exp(3\theta + \delta_1 + \delta_2 + \delta_3)/D$$
$$D = 1 + \exp(\theta + \delta_1) + \exp(2\theta + \delta_1 + \delta_2) + \exp(3\theta + \delta_1 + \delta_2 + \delta_3)$$

where δ_k is the step difficulty for step k in item 1.

In addition , differential rater severities can be incorporated into the model. If the item described above was rated by Rater 1, who has severity ρ_i, then the above equations become:

328

$$\Pr(\mathbf{X}_{10} = 1; \mathbf{A}, \mathbf{B}, \xi \mid \theta) = 1/D$$

$$\Pr(\mathbf{X}_{11} = 1; \mathbf{A}, \mathbf{B}, \xi \mid \theta) = \exp(\theta + \delta_1 + \rho_1)/D$$

$$\Pr(\mathbf{X}_{12} = 1; \mathbf{A}, \mathbf{B}, \xi \mid \theta) = \exp(2\theta + \delta_1 + \delta_2 + \rho_1)/D \qquad (3)$$

$$\Pr(\mathbf{X}_{13} = 1; \mathbf{A}, \mathbf{B}, \xi \mid \theta) = \exp(3\theta + \delta_1 + \delta_2 + \delta_3 + \rho_1)/D$$

and D is defined as the sum of the numerators, as above.

Another, fairly common, complication that can easily be incorporated into the model is local dependence between items. This can occur, for example, if two multiple choice items refer to common stimulus material (for example, two questions about the same reading passage). Say that there are two items which refer to common stimulus material, items 1 and 2. Then there are four possible response patterns to these two items: (0,0), (0,1), (1,0), and (1,1). Label these possible response patterns as a single item bundle, $\mathbf{X}_{ij}$. One possible set of equations to describe this item bundle is

$$\Pr(\mathbf{X}_{00} = 1; \mathbf{A}, \mathbf{B}, \xi \mid \theta) = \exp(\varepsilon_1)/D$$

$$\Pr(\mathbf{X}_{01} = 1; \mathbf{A}, \mathbf{B}, \xi \mid \theta) = \exp(\theta + \delta_1)/D$$

$$\Pr(\mathbf{X}_{10} = 1; \mathbf{A}, \mathbf{B}, \xi \mid \theta) = \exp(\theta + \delta_2)/D \qquad (4)$$

$$\Pr(\mathbf{X}_{11} = 1; \mathbf{A}, \mathbf{B}, \xi \mid \theta) = \exp(2\theta + \delta_1 + \delta_2 + \varepsilon_1)/D$$

where ε_1 is the adjustment to the probability that occurs due to the increased likelihood of getting the same score on both items, because of the local dependence (assuming ε_1 is positive), and D is the sum of the numerators

3. The Construct Mapping Method

The conceptual maps used to display the results of fitting the MRCML model to a given data set are exemplified in Figure 1. This figure illustrates the map for an exam with 50 multiple choice items and two written response items, taken from the software used to set cut scores by the standard-setting committee. The column on the far left contains a logit scale transformed to have a mean of 500, and a range of approximately 0 to 1000. The highlighted row (scale value 540) represents the proficiency currently under investigation by a hypothetical committee member. The two columns immediately to the right of this illustrate the difficulty of the multiple choice items, and the probabilities that a person of proficiency 540 would answer each of these items correctly. Each of the next sets of two columns represents the threshold difficulties of one of the written response items, along with the probability that a person of proficiency 540 would receive each of these scores on this item. These threshold difficulties are cumulative: they indicate the point on the logit scale at which a person becomes 50% likely to receive at least the score in question (i.e. threshold 2 for item 1 indicates the point at which a person is 50% likely to score at least 2 on item 1).

It is also possible to add person proficiency distribution information (including numbers, percents, and cumulative percents of persons at each proficiency level) to such a map. The software also displays, for a person at the selected point on the GSE scale, the expected score total on the multiple choice section, and the expected score on each of the written response items; these are shown at the far right of the Figure.

The standard setting committee undergoes a set of exercises to familiarize them with the exam materials. These include taking all of the multiple choice items, and reading through the written response items, score level descriptions, and corresponding anchor papers. Then the committee receives instruction in the meaning of person proficiencies and item locations for both dichotomous and polytomous items. After this, the software is introduced, along with the set of commands used to operate it. A worksheet with example proficiency levels is passed out, and committee members try out the commands with those proficiency levels and experiment with others.

The display of multiple choice item locations in ascending difficulty, next to the written response thresholds, helps to characterize the scale in terms of what increasing proficiency "looks like" in the pool of test-takers. For example, if a committee was considering 540 (as in Figure 1) as a standard, then they could note that it is a point at which items like 27 are expected to be chosen correctly about 50% of the time, a harder item like 37 is expected to be chosen correctly about 39%, and easier items like 2 are expected to be chosen correctly 86% of the time. The set of multiple choice items, sorted so they are in order of ascending difficulty, is available to the committee so that the members can relate these probabilities to their understanding of the items. From the expected scores, the committee could see that on average, a student at this level would choose a total of approximately 34 multiple choice items correctly. The committee could also note that a student at that point (i.e., 540), would also have a high chance of scoring at least a 3 on the first written response item (74%), and on average would be expected to score 3, and would most likely score either a 2 or a 3 on the second. Examples of student work at these levels would be available to the committee for consideration of the interpretation of these scores.

The committee then uses a consensus-building process to set up cut points on this map, using the item response calibrations to give interpretability in terms of predicted responses to both multiple choice items and open-ended items. Locations of students on the scaled variable are also available for interpretative purposes. This procedure allows both criterion-referenced and norm-referenced interpretations of cut scores. Criterion-referenced interpretations are available via inspection of what students at a given level know and can do, as shown in their probabilities of various types of responses to the items. Norm-referenced interpretations are available by looking at the percentages and cumulative percentages of students at each level of the GSE scale; these are available upon request as part of the maps.

This item response modeling approach used allows relatively easy equating of the GSE scale from year to year, via common item equating. Items used for linking in GSE are multiple choice items, as there are no written response items

330

GSE Scale	Items Multiple Choice Difficulty	Prob	Written Response 1 Threshold	Prob	Written Response 2 Threshold	Prob
750						
740					2.5	.00
730						
720			1.5	.01		
710						
700						
690						
680						
670					2.4	.00
660						
650						
640						
630						
620						
610			1.4	.14		
600						
590						
580	37	.39				
570	15	.44				
560						
550	28 39	.48				
540	27	.51			2.3	.46
530	38	.53				
520	19	.55				
510						
500	34 43 45 48	.60	1.3	.59		
490	17 18 20 40 50	.62				
480	31	.65				
470	4 11 33 44	.66				
460	5 9 12 32 46 47	.69				
450	3 7 16	.71				
440	6 10 29	.72				
430	22 35 36	.75				
420	8 14 23 26	.76			2.2	.53
410	13 24 25	.78				
400	41 42	.79				
390	1 21 49	.81	1.2	.25		
380	30	.82				
370						
360						
350	2	.86				

Expected score for:
MC 33.9
WR 1 2.9
WR 2 2.5

Figure 1: Example Wright map

repeated in more than one year. The common logit-based scale, so equated, allows detection of trends in student proficiency over time.

4. An Example of the Construct Mapping Method

This study involved the Government and Civics GSE test, which consisted of 50 multiple choice and two written response items, each scored on a scale of one to five. The content of this test includes concepts and principles of government, particularly US government.

The study included two committees of 7 to 10 persons each, the usual number for a GSE standard setting session. The same person led both committees; training in the particulars of the method was provided by members of the development staff, also the same persons for each committee. The particular version of the software developed for this study allowed differential weights to be applied to the two written response items, and to the written response versus the multiple choice items.

Table 1: Analysis of classification matches

Category	Category
1st group 2 lower	0.00
1st group 1 lower	0.02
both groups same	0.86
1st group 1 higher	0.12
1st group 2 higher	0.02

This study allowed examination of the replicability of the new process in its current form. In addition, since a standard setting session using the traditional method had previously taken place, it allowed a comparison of the results of this process to the standards that had previously resulted from the traditional process.

Table 1 contains an analysis of the "matches" between the performance level to which students were classified by the different groups within a study. A match indicates that a student was assigned to the same performance level by the two committees. As, ultimately, the only information reported to the student is the performance level, a 100% match would indicate perfect agreement between the two committees. Notice that the two groups only 14% of students classified differently.

5. Conclusions

Based on the results of the study described above, the Construct Mapping approach seems to work quite well as a method for consistently setting standards. Feedback from committee members was quite positive. Members stated that they felt the method to be easily understandable, the software to be straightforward to use, and the resulting cut scores to be defensible.

It seems that there are a variety of factors affecting the decisions of standard setting committees. These include the experience of committee members, both in teaching and in working with the GSE program. In addition, characteristics of the committee leader may have an effect on the committee's decisions.

However, conceptual models between student knowledge and performance levels are still under-developed. Particularly with multiple choice items, committee members tended to focus on overall expected scores ("students should get at least 40 of 50 multiple choice items correct"), rather than the specific knowledge represented by particular items. Although they had the items available to them in order of difficulty, they tended not to pay much attention to which items a person at a given level was likely to answer correctly. We are hoping to improve this by making it easier to focus on the items as part of using the software. For example, clicking on the multiple choice items could reveal the item text. Similarly, clicking on the threshold for a particular score on one of the written response items could show an exemplary paper for that response level.

References:

Adams, R. J., Wilson, M., & Wang, W. C. (1997). The Multidimensional Random Coefficients Multinomial Logit Model. *Applied Psychological Measurement, 21*, 1-23.

Draney, K., & Wilson, M. (1998, June). *Creating composite scores and setting performance levels: Use of a scaling procedure.* Council of Chief State School Officers National Conference on Large Scale Assessment, Colorado Springs, CO.

Jaeger, R. M. (1995). Setting standards for complex performances: An iterative, judgmental policy-capturing strategy. *Educational Measurement, Issues, and Practice, 14*(4), 16-20.

Jaeger, R. M., Plake B. S., & Hambleton, R. K. (1993). *Integrating multidimensional performance and setting performance standards.* Paper presented at the annual meeting of the National Council on Measurement in Education, Atlanta, GA.

Wilson, M. (1994). Assessment nets: An alternative approach to assessment in mathematics education.. In T. Romberg (Ed.), *Assessment in School Mathematics* New York: SUNY Press.

Wilson, M., & Draney, K. (1998, June). *Creating composite scores and setting performance levels: Comparison of raw score and scaling procedures.* Council of Chief State School Officers National Conference on Large Scale Assessment, Colorado Springs, CO.

Wu, M., Adams, R. J., & Wilson M. (1998). *ACER-Conquest* [computer program and manual]. Hawthorn, Australia: ACER Press